Studies in Systems, Decision and Control

Volume 441

The series "Studies in Systems, Decision and Control" (SSDC) covers both new developments and advances, as well as the state of the art, in the various areas of broadly perceived systems, decision making and control–quickly, up to date and with a high quality. The intent is to cover the theory, applications, and perspectives on the state of the art and future developments relevant to systems, decision making, control, complex processes and related areas, as embedded in the fields of engineering, computer science, physics, economics, social and life sciences, as well as the paradigms and methodologies behind them. The series contains monographs, textbooks, lecture notes and edited volumes in systems, decision making and control spanning the areas of Cyber-Physical Systems, Autonomous Systems, Sensor Networks, Control Systems, Energy Systems, Automotive Systems, Biological Systems, Vehicular Networking and Connected Vehicles, Aerospace Systems, Automation, Manufacturing, Smart Grids, Nonlinear Systems, Power Systems, Robotics, Social Systems, Economic Systems and other. Of particular value to both the contributors and the readership are the short publication timeframe and the world-wide distribution and exposure which enable both a wide and rapid dissemination of research output.

Indexed by SCOPUS, DBLP, WTI Frankfurt eG, zbMATH, SCImago.

All books published in the series are submitted for consideration in Web of Science.

More information about this series at https://link.springer.com/bookseries/13304

George A. Anastassiou

Fractional Inequalities In Banach Algebras

Springer

George A. Anastassiou
Department of Mathematical Sciences
University of Memphis
Memphis, TN, USA

ISSN 2198-4182 ISSN 2198-4190 (electronic)
Studies in Systems, Decision and Control
ISBN 978-3-031-05147-0 ISBN 978-3-031-05148-7 (eBook)
https://doi.org/10.1007/978-3-031-05148-7

This Springer imprint is published by the registered company Springer Nature Switzerland AG
The registered company address is: Gewerbestrasse 11, 6330 Cham, Switzerland

Dedicated to Covid-19 vaccine creators

Preface

Fractional Calculus and its long history started in 1695 by L'Hospital and Leibnitz. Over the years, many different kinds of it emerged according to the needs of science given by many famous mathematicians. In the last 50 years, fractional calculus due to its wide applications to many applied sciences has gained extreme popularity.

By employing regular and iterated (sequential) generalized Caputo and Canavati fractional left and right vectorial Taylor formulae, we establish a great variety of mixed regular and iterated generalized fractional inequalities. In particular, we derive generalized Caputo Fractional Ostrowski and Grüss type inequalities involving several Banach algebra valued functions, the regular and iterated (sequential) versions. Also, we establish generalized Canavati fractional Ostrowski, Opial, Grüss and Hilbert-Pachpatte type inequalities for multiple Banach algebra valued functions. The same kind of the above types of inequalities are studied at univariate and multivariate settings for ordinary vectorial derivatives and vectorial partial derivatives, respectively. The counterpart studies continue, extensively, by using the p-Schatten norms over the von Neumann-Schatten classes (∗-ideals) $\mathcal{B}_p(H)$, $p >\_1$, and we produce all the analogous inequalities. We provide a great variety of applications per chapter.

So in this monograph, all that is presented is the original work by the author given at a very general level to cover a maximum number of different fractional calculus applications.

As a result, this monograph is the natural and expected evolution of the author's recent research work put in a book form for the first time. The presented approaches are original, and chapters are self-contained and can be read independently. This monograph is suitable to be used in related graduate classes and research projects.

The motivation to write this monograph came by the following: various issues related to the modeling and analysis of ordinary and fractional order systems have gained increased popularity, as witnesses by many books and volumes in Springer's program:

http://www.springer.com/gp/search?query=fractional&submit=Prze%C5%9Blij

and the purpose of our book is to capture at a very general level a deeper formal analysis on some issues that are relevant to many areas, for instance: decision making, complex processes, systems modeling and control, and related areas.

The above are deeply embedded in the fields of mathematics, engineering, computer science, physics, economics, social and life sciences.

Next is listed the author's recent monographs in Fractional Analysis and applications:

- George Anastassiou, "Intelligent Computations: Abstract Fractional Calculus, Inequalities, Approximations", Springer, Studies in Computational Intelligence 734, Heidelberg, New York, 2018.
- George Anastassiou, "Nonlinearity: Ordinary and Fractional Approximations by Sublinear and Max-product Operators", Springer, Studies in Systems, Decision and Control 147, Heidelberg, New York, 2018.
- George Anastassiou, "Ordinary and Fractional approximation by non-additive integrals: Choquet, Shilkret and Sugeno integral approximators", Springer, Studies in Systems, Decision and Control 190, Heidelberg, New York, 2019.
- George Anastassiou, "Intelligent Analysis: Fractional Inequalities and Approximations Expanded", Springer, Studies in Computational Intelligence 886, Heidelberg, New York, 2020.
- George Anastassiou, "Generalized Fractional Calculus: New Advancements and Applications", Springer, Studies in Systems, Decision and Control 305, Heidelberg, New York, 2021.
- George Anastassiou, "Constructive Fractional Analysis with Applications", Springer, Studies in Systems, Decision and Control 362, Heidelberg, New York, 2021.
- George Anastassiou, "Unification of Fractional Calculi with Applications", Springer, Studies in Systems, Decision and Control 398, Heidelberg, New York, 2022.

The complete list of presented topics here follows:

- Generalized Fractional Ostrowski and Grüss type inequalities involving several Banach algebra valued functions,
- Sequential Generalized Fractional Ostrowski and Grüss type inequalities for several Banach algebra valued functions,
- Generalized Canavati Fractional Ostrowski, Opial and Grüss type inequalities for Banach algebra valued functions,
- Generalized Canavati Fractional Hilbert-Pachpatte type inequalities for Banach algebra valued functions,
- Generalized Ostrowski, Opial and Hilbert-Pachpatte type inequalities for Banach algebra valued functions involving integer vectorial derivatives,
- Multivariate Ostrowski type inequalities for several Banach algebra valued functions,
- p-Schatten norm generalized fractional Ostrowski and Grüss type inequalities involving several functions,

- p-Schatten norm sequential generalized fractional Ostrowski and Grüss type inequalities for several functions,
- p-Schatten norm generalized Canavati fractional Ostrowski, Opial and Grüss type inequalities involving several functions,
- p-Schatten norm Generalized Canavati Fractional Hilbert-Pachpatte type inequalities for von Neumann-Schatten class $\mathcal{B}_p(H)$ valued functions,
- p-Schatten norm Generalized Ostrowski, Opial and Hilbert Pachpatte type inequalities for von Neumann-Schatten class $\mathcal{B}_p(H)$ valued functions with integer vectorial derivatives,
- p-Schatten norm Multivariate Ostrowski type inequalities for several Neumann-Schatten class $\mathcal{B}_p(H)$ valued functions.

The book's results are expected to find applications in many areas of pure and applied mathematics, especially in fractional inequalities and fractional differential equations. Other possible applications can be in applied sciences like geophysics, physics, chemistry, economics, engineering, etc. All in all, what is presented here is a valuable tool for a large range of applications. Therefore, this monograph is suitable for researchers, graduate students, practitioners and seminars of the above disciplines, and also to be in all science and engineering libraries.

The preparation of the book took place during 2021–22 at the University of Memphis, during the author's home stay during the Covid-19 and Omicron outbreak keeping him alive, happy and in control!

The author likes to thank Prof. Alina Lupas of University of Oradea, Romania, for checking and reading the manuscript.

Memphis, TN, USA
March 2022

George A. Anastassiou

Contents

Chapter 1
Generalized Fractional Ostrowski and Grüss Inequalities with Multiple Banach Algebra Valued Functions

Using generalized Caputo fractional left and right vectorial Taylor formulae we establish mixed fractional Ostrowski and Grüss type inequalities involving several Banach algebra valued functions. The estimates are with respect to all norms $\|\cdot\|_p$, $1 \leq p \leq \infty$. It follows [6].

1.1 Introduction

The following results motivate our chapter.

Theorem 1.1 (1938, Ostrowski [11]) *Let $f : [a, b] \to \mathbb{R}$ be continuous on $[a, b]$ and differentiable on (a, b) whose derivative $f' : (a, b) \to \mathbb{R}$ is bounded on (a, b), i.e., $\|f'\|_{\infty}^{\sup} := \sup_{t \in (a,b)} |f'(t)| < +\infty$. Then*

$$\left| \frac{1}{b-a} \int_a^b f(t)\, dt - f(x) \right| \leq \left[\frac{1}{4} + \frac{\left(x - \frac{a+b}{2}\right)^2}{(b-a)^2} \right] (b-a) \|f'\|_{\infty}^{\sup}, \tag{1.1}$$

for any $x \in [a, b]$. The constant $\frac{1}{4}$ is the best possible.

Ostrowski type inequalities have great applications to integral approximations in Numerical Analysis.

Theorem 1.2 (1882, Čebyšev [7]) *Let $f, g : [a, b] \to \mathbb{R}$ be absolutely continuous functions with $f', g' \in L_{\infty}([a, b])$. Then*

G. A. Anastassiou, *Fractional Inequalities In Banach Algebras*,
Studies in Systems, Decision and Control 441,
https://doi.org/10.1007/978-3-031-05148-7_1

$$\left|\frac{1}{b-a}\int_a^b f(x)\,g(x)\,dx - \left(\frac{1}{b-a}\int_a^b f(x)\,dx\right)\left(\frac{1}{b-a}\int_a^b g(x)\,dx\right)\right|$$

$$\leq \frac{1}{12}(b-a)^2 \left\|f'\right\|_\infty \left\|g'\right\|_\infty . \tag{1.2}$$

The above integrals are assumed to exist.

The related Grüss type inequalities have many applications to Probability Theory. We presented also ([4], Chaps. 8, 9) mixed fractional Ostrowski and Grüss-Cebysev type inequalities for several functions, acting to all possible directions. The estimates involve the left and right Caputo fractional derivatives. See also the monographs written by the author [2], Chaps. 24–26 and [3], Chaps. 2–6.

In this chapter we generalize [4], Chaps. 8, 9 for several Banach algebra valued functions. Now our left and right Caputo fractional derivatives are for Banach space valued functions and our integrals are of Bochner type. Several applications finish this chapter. Inspiration came also from [8, 9].

1.2 Vectorial Background Fractional Calculus

Here all come from [5].

We need

Definition 1.3 ([5], *p. 106*) Let $\alpha > 0$, $\lceil \alpha \rceil = n$, $\lceil \cdot \rceil$ the ceiling of the number. Let $f \in C^n([a,b], X)$, where $[a,b] \subset \mathbb{R}$, and $(X, \|\cdot\|)$ is a Banach space. Let $g \in C^1([a,b])$, strictly increasing, such that $g^{-1} \in C^n([g(a), g(b)])$.

We define the left generalized g-fractional derivative X-valued of f of order α as follows:

$$\left(D^\alpha_{a+;g} f\right)(x) := \frac{1}{\Gamma(n-\alpha)} \int_a^x (g(x) - g(t))^{n-\alpha-1} g'(t) \left(f \circ g^{-1}\right)^{(n)} (g(t))\,dt, \tag{1.3}$$

$\forall\, x \in [a,b]$, where Γ is the gamma function. The last integral is of Bochner type ([1], pp. 422–428; [10]).

If $\alpha \notin \mathbb{N}$, by Theorem 4.10 ([5], p. 98), we have that $\left(D^\alpha_{a+;g} f\right) \in C([a,b], X)$.

We set

$$D^n_{a+;g} f(x) := \left(\left(f \circ g^{-1}\right)^n \circ g\right)(x) \in C([a,b], X),\ n \in \mathbb{N}, \tag{1.4}$$

$$D^0_{a+;g} f(x) = f(x),\ \forall\, x \in [a,b].$$

When $g = id$, then

$$D^\alpha_{a+;g} f = D^\alpha_{a+;id} f = D^\alpha_{*a} f, \tag{1.5}$$

the usual left X-valued Caputo fractional derivative, see [5], Chap. 1.

We also need

Definition 1.4 ([5], *p. 107*) Let $\alpha > 0$, $\lceil \alpha \rceil = n$, $\lceil \cdot \rceil$ the ceiling of the number. Let $f \in C^n([a, b], X)$, where $[a, b] \subset \mathbb{R}$, and $(X, \|\cdot\|)$ is a Banach space. Let $g \in C^1([a, b])$, strictly increasing, such that $g^{-1} \in C^n([g(a), g(b)])$.

We define the right generalized g-fractional derivative X-valued of f of order α as follows:

$$\left(D_{b-;g}^{\alpha} f\right)(x) := \frac{(-1)^n}{\Gamma(n-\alpha)} \int_x^b (g(t) - g(x))^{n-\alpha-1} g'(t) \left(f \circ g^{-1}\right)^{(n)} (g(t))\, dt, \tag{1.6}$$

$\forall\, x \in [a, b]$. The last integral is of Bochner type.

If $\alpha \notin \mathbb{N}$, by Theorem 4.11 ([5], p. 101), we have that $\left(D_{b-;g}^{\alpha} f\right) \in C([a, b], X)$.

We set

$$D_{b-;g}^{n} f(x) := (-1)^n \left(\left(f \circ g^{-1}\right)^n \circ g\right)(x) \in C([a, b], X), n \in \mathbb{N}, \tag{1.7}$$

$$D_{b-;g}^{0} f(x) := f(x), \ \forall\, x \in [a, b].$$

When $g = id$, then

$$D_{b-;g}^{\alpha} f(x) = D_{b-;id}^{\alpha} f(x) = D_{b-}^{\alpha} f, \tag{1.8}$$

the usual right X-valued Caputo fractional derivative, see [5], Chap. 2.

We mention the following generalized fractional Taylor formulae with integral remainders over Banach spaces.

Theorem 1.5 ([5], p. 107) *Let $\alpha > 0$, $n = \lceil \alpha \rceil$, and $f \in C^n([a, b], X)$, where $[a, b] \subset \mathbb{R}$ and $(X, \|\cdot\|)$ is a Banach space. Let $g \in C^1([a, b])$, strictly increasing, such that $g^{-1} \in C^n([g(a), g(b)])$, $a \le x \le b$. Then*

$$f(x) = f(a) + \sum_{i=1}^{n-1} \frac{(g(x) - g(a))^i}{i!} \left(f \circ g^{-1}\right)^{(i)} (g(a)) +$$

$$\frac{1}{\Gamma(\alpha)} \int_a^x (g(x) - g(t))^{\alpha-1} g'(t) \left(D_{a+;g}^{\alpha} f\right)(t)\, dt =$$

$$f(a) + \sum_{i=1}^{n-1} \frac{(g(x) - g(a))^i}{i!} \left(f \circ g^{-1}\right)^{(i)} (g(a)) + \tag{1.9}$$

$$\frac{1}{\Gamma(\alpha)} \int_{g(a)}^{g(x)} (g(x) - z)^{\alpha-1} \left(\left(D_{a+;g}^{\alpha} f\right) \circ g^{-1}\right)(z)\, dz.$$

We also mention

Theorem 1.6 ([5], p. 108) *Let* $\alpha > 0$, $n = \lceil \alpha \rceil$, *and* $f \in C^n([a,b], X)$, *where* $[a,b] \subset \mathbb{R}$ *and* $(X, \|\cdot\|)$ *is a Banach space. Let* $g \in C^1([a,b])$, *strictly increasing, such that* $g^{-1} \in C^n([g(a), g(b)])$, $a \leq x \leq b$. *Then*

$$f(x) = f(b) + \sum_{i=1}^{n-1} \frac{(g(x) - g(b))^i}{i!} \left(f \circ g^{-1}\right)^{(i)} (g(b)) +$$

$$\frac{1}{\Gamma(\alpha)} \int_x^b (g(t) - g(x))^{\alpha-1} g'(t) \left(D_{b-;g}^{\alpha} f\right)(t)\, dt =$$

$$f(b) + \sum_{i=1}^{n-1} \frac{(g(x) - g(b))^i}{i!} \left(f \circ g^{-1}\right)^{(i)} (g(b)) + \tag{1.10}$$

$$\frac{1}{\Gamma(\alpha)} \int_{g(x)}^{g(b)} (z - g(x))^{\alpha-1} \left(\left(D_{b-;g}^{\alpha} f\right) \circ g^{-1}\right)(z)\, dz.$$

If $0 < \alpha \leq 1$, then the sums in (1.9), (1.10) disappear.

Also in (1.9), (1.10), we have that $\left\|\left\|\left(D_{a+;g}^{\alpha} f\right) \circ g^{-1}\right\|\right\|_{\infty,[g(a),g(b)]}$, $\left\|\left\|\left(D_{b-;g}^{\alpha} f\right) \circ g^{-1}\right\|\right\|_{\infty,[g(a),g(b)]} < \infty$.

1.3 Banach Algebras Background

All here come from [12].

We need

Definition 1.7 ([12], *p. 245*) A complex algebra is a vector space A over the complex filed $\mathbb{C}$ in which a multiplication is defined that satisfies

$$x(yz) = (xy)z, \tag{1.11}$$

$$(x + y)z = xz + yz, \; x(y + z) = xy + xz, \tag{1.12}$$

and

$$\alpha(xy) = (\alpha x)y = x(\alpha y), \tag{1.13}$$

for all x, y and z in A and for all scalars α.

Additionally if A is a Banach space with respect to a norm that satisfies the multiplicative inequality

$$\|xy\| \leq \|x\| \, \|y\| \;\; (x \in A, y \in A) \tag{1.14}$$

and if A contains a unit element e such that

$$xe = ex = x \; (x \in A) \tag{1.15}$$

and

$$\|e\| = 1, \tag{1.16}$$

then A is called a Banach algebra.

A is commutative iff $xy = yx$ for all $x, y \in A$.

We make

Remark 1.8 Commutativity of A will be explicited stated when needed.

There exists at most one $e \in A$ that satisfies (1.15).

Inequality (1.14) makes multiplication to be continuous, more precisely left and right continuous, see [12], p. 246.

Multiplication in A is not necessarily the numerical multiplication, it is something more general and it is defined abstractly, that is for $x, y \in A$ we have $xy \in A$, e.g. composition or convolution, etc.

For nice examples about Banach algebras see [12], p. 247–248, Sect. 10.3.

We also make

Remark 1.9 Next we mention about integration of A-valued functions, see [12], p. 259, Sect. 10.22:

If A is a Banach algebra and f is a continuous A-valued function on some compact Hausdorff space Q on which a complex Borel measure μ is defined, then $\int f d\mu$ exists and has all the properties that were discussed in Chap. 3 of [12], simply because A is a Banach space. However, an additional property can be added to these, namely: If $x \in A$, then

$$x \int_Q f \, d\mu = \int_Q xf(p) \, d\mu(p) \tag{1.17}$$

and

$$\left(\int_Q f \, d\mu \right) x = \int_Q f(p) \, x \, d\mu(p) . \tag{1.18}$$

The Bochner integrals we will involve in our chapter follow (1.17) and (1.18). Also, let $f \in C([a, b], X)$, where $[a, b] \subset \mathbb{R}$, $(X, \|\cdot\|)$ is a Banach space. By [5], p. 3, f is Bochner integrable.

1.4 Main Results

We start with mixed generalized fractional Ostrowski type inequalities for several functions over a Banach algebra. A uniform estimate follows.

Theorem 1.10 *Let $(A, \|\cdot\|)$ be a Banach algebra, $x_0 \in [a,b] \subset \mathbb{R}$, $\alpha > 0$, $n = \lceil \alpha \rceil$, $f_i \in C^n([a,b], A)$, $i = 1, ..., r \in \mathbb{N} - \{1\}$; $g \in C^1([a,b])$, strictly increasing, such that $g^{-1} \in C^n([g(a), g(b)])$, with $\left(f_i \circ g^{-1}\right)^{(k)}(g(x_0)) = 0$, $k = 1, ..., n-1$; $i = 1, ..., r$. Denote by*

$$\theta(f_1, ..., f_r)(x_0) :=$$

$$\sum_{i=1}^{r} \left[\int_a^b \left(\prod_{\substack{j=1 \\ j \neq i}}^{r} f_j(x) \right) f_i(x)\, dx - \left(\int_a^b \left(\prod_{\substack{j=1 \\ j \neq i}}^{r} f_j(x) \right) dx \right) f_i(x_0) \right]. \quad (1.19)$$

Then

$$\|\theta(f_1, ..., f_r)(x_0)\| \leq \frac{1}{\Gamma(\alpha+1)} \sum_{i=1}^{r} \left[\left[\left\| \left\| \left(D_{x_0 -; g}^{\alpha} f_i \right) \circ g^{-1} \right\| \right\|_{\infty, [g(a), g(x_0)]} \right.\right.$$

$$\left. (g(x_0) - g(a))^{\alpha} \left(\int_a^{x_0} \left(\prod_{\substack{j=1 \\ j \neq i}}^{r} \|f_j(x)\| \right) dx \right) \right] + \quad (1.20)$$

$$\left. \left[\left\| \left\| \left(D_{x_0 +; g}^{\alpha} f_i \right) \circ g^{-1} \right\| \right\|_{\infty, [g(x_0), g(b)]} (g(b) - g(x_0))^{\alpha} \left(\int_{x_0}^{b} \left(\prod_{\substack{j=1 \\ j \neq i}}^{r} \|f_j(x)\| \right) dx \right) \right] \right].$$

Proof Since $\left(f_i \circ g^{-1}\right)^{(k)}(g(x_0)) = 0$, $k = 1, ..., n-1$; $i = 1, ..., r$, we have by Theorem 1.5 that

$$f_i(x) - f_i(x_0) = \frac{1}{\Gamma(\alpha)} \int_{g(x_0)}^{g(x)} (g(x) - z)^{\alpha - 1} \left(\left(D_{x_0 +; g}^{\alpha} f_i \right) \circ g^{-1} \right)(z)\, dz, \quad (1.21)$$

$\forall\, x \in [x_0, b]$,
and by Theorem 1.6 that

$$f_i(x) - f_i(x_0) = \frac{1}{\Gamma(\alpha)} \int_{g(x)}^{g(x_0)} (z - g(x))^{\alpha - 1} \left(\left(D_{x_0 -; g}^{\alpha} f_i \right) \circ g^{-1} \right)(z)\, dz, \quad (1.22)$$

$\forall\, x \in [a, x_0]$;

for all $i = 1, ..., r$.

Left multiplying (1.21) and (1.22) with $\left(\prod_{\substack{j=1 \\ j\neq i}}^{r} f_j(x)\right)$ we get, respectively,

$$\left(\prod_{\substack{j=1 \\ j\neq i}}^{r} f_j(x)\right) f_i(x) - \left(\prod_{\substack{j=1 \\ j\neq i}}^{r} f_j(x)\right) f_i(x_0) = \tag{1.23}$$

$$\frac{\left(\prod_{\substack{j=1 \\ j\neq i}}^{r} f_j(x)\right)}{\Gamma(\alpha)} \int_{g(x_0)}^{g(x)} (g(x) - z)^{\alpha-1} \left(\left(D_{x_0+;g}^{\alpha} f_i\right) \circ g^{-1}\right)(z)\, dz,$$

$\forall\, x \in [x_0, b]$,

and

$$\left(\prod_{\substack{j=1 \\ j\neq i}}^{r} f_j(x)\right) f_i(x) - \left(\prod_{\substack{j=1 \\ j\neq i}}^{r} f_j(x)\right) f_i(x_0) = \tag{1.24}$$

$$\frac{\left(\prod_{\substack{j=1 \\ j\neq i}}^{r} f_j(x)\right)}{\Gamma(\alpha)} \int_{g(x)}^{g(x_0)} (z - g(x))^{\alpha-1} \left(\left(D_{x_0-;g}^{\alpha} f_i\right) \circ g^{-1}\right)(z)\, dz,$$

$\forall\, x \in [a, x_0]$;

for all $i = 1, ..., r$.

Adding (1.23) and (1.24) as separate groups, we obtain

$$\sum_{i=1}^{r} \left(\prod_{\substack{j=1 \\ j\neq i}}^{r} f_j(x)\right) f_i(x) - \sum_{i=1}^{r} \left(\prod_{\substack{j=1 \\ j\neq i}}^{r} f_j(x)\right) f_i(x_0) =$$

$$\frac{1}{\Gamma(\alpha)} \sum_{i=1}^{r} \left(\prod_{\substack{j=1 \\ j\neq i}}^{r} f_j(x)\right) \int_{g(x_0)}^{g(x)} (g(x) - z)^{\alpha-1} \left(\left(D_{x_0+;g}^{\alpha} f_i\right) \circ g^{-1}\right)(z)\, dz, \tag{1.25}$$

$\forall\, x \in [x_0, b]$,

and

$$\sum_{i=1}^{r} \left(\prod_{\substack{j=1 \\ j\neq i}}^{r} f_j(x)\right) f_i(x) - \sum_{i=1}^{r} \left(\prod_{\substack{j=1 \\ j\neq i}}^{r} f_j(x)\right) f_i(x_0) =$$

$$\frac{1}{\Gamma(\alpha)} \sum_{i=1}^{r} \left(\prod_{\substack{j=1 \\ j \neq i}}^{r} f_j(x) \right) \int_{g(x)}^{g(x_0)} (z - g(x))^{\alpha-1} \left(\left(D_{x_0-;g}^{\alpha} f_i \right) \circ g^{-1} \right)(z)\, dz, \quad (1.26)$$

$\forall\, x \in [a, x_0]$.

Next we integrate (1.25) and (1.26) with respect to $x \in [a, b]$. We have

$$\sum_{i=1}^{r} \int_{x_0}^{b} \left(\prod_{\substack{j=1 \\ j \neq i}}^{r} f_j(x) \right) f_i(x)\, dx - \sum_{i=1}^{r} \left(\int_{x_0}^{b} \left(\prod_{\substack{j=1 \\ j \neq i}}^{r} f_j(x) \right) dx \right) f_i(x_0) =$$

$$\frac{1}{\Gamma(\alpha)} \sum_{i=1}^{r} \left[\int_{x_0}^{b} \left(\prod_{\substack{j=1 \\ j \neq i}}^{r} f_j(x) \right) \left(\int_{g(x_0)}^{g(x)} (g(x) - z)^{\alpha-1} \left(\left(D_{x_0+;g}^{\alpha} f_i \right) \circ g^{-1} \right)(z)\, dz \right) dx \right], \quad (1.27)$$

and

$$\sum_{i=1}^{r} \int_{a}^{x_0} \left(\prod_{\substack{j=1 \\ j \neq i}}^{r} f_j(x) \right) f_i(x)\, dx - \sum_{i=1}^{r} \left(\int_{a}^{x_0} \left(\prod_{\substack{j=1 \\ j \neq i}}^{r} f_j(x) \right) dx \right) f_i(x_0) =$$

$$\frac{1}{\Gamma(\alpha)} \sum_{i=1}^{r} \left[\int_{a}^{x_0} \left(\prod_{\substack{j=1 \\ j \neq i}}^{r} f_j(x) \right) \left(\int_{g(x)}^{g(x_0)} (z - g(x))^{\alpha-1} \left(\left(D_{x_0-;g}^{\alpha} f_i \right) \circ g^{-1} \right)(z)\, dz \right) dx \right]. \quad (1.28)$$

Finally, adding (1.27) and (1.28) we obtain the useful identity

$$\theta(f_1, \ldots, f_r)(x_0) :=$$

$$\sum_{i=1}^{r} \left[\int_{a}^{b} \left(\prod_{\substack{j=1 \\ j \neq i}}^{r} f_j(x) \right) f_i(x)\, dx - \left(\int_{a}^{b} \left(\prod_{\substack{j=1 \\ j \neq i}}^{r} f_j(x) \right) dx \right) f_i(x_0) \right] =$$

$$\frac{1}{\Gamma(\alpha)} \sum_{i=1}^{r} \left[\left[\int_{a}^{x_0} \left(\prod_{\substack{j=1 \\ j \neq i}}^{r} f_j(x) \right) \left(\int_{g(x)}^{g(x_0)} (z - g(x))^{\alpha-1} \left(\left(D_{x_0-;g}^{\alpha} f_i \right) \circ g^{-1} \right)(z)\, dz \right) dx \right] \right.$$

$$+\left[\int_{x_0}^{b}\left(\prod_{\substack{j=1\\j\neq i}}^{r} f_j(x)\right)\left(\int_{g(x_0)}^{g(x)}(g(x)-z)^{\alpha-1}\left(\left(D_{x_0+;g}^{\alpha}f_i\right)\circ g^{-1}\right)(z)\,dz\right)dx\right]\Bigg]. \tag{1.29}$$

Therefore we get that

$$\|\theta(f_1,\ldots,f_r)(x_0)\| =$$

$$\left\|\sum_{i=1}^{r}\left[\int_a^b\left(\prod_{\substack{j=1\\j\neq i}}^{r} f_j(x)\right)f_i(x)\,dx-\left(\int_a^b\left(\prod_{\substack{j=1\\j\neq i}}^{r} f_j(x)\right)dx\right)f_i(x_0)\right]\right\| \leq \frac{1}{\Gamma(\alpha)} \tag{1.30}$$

$$\sum_{i=1}^{r}\left[\left\|\left[\int_a^{x_0}\left(\prod_{\substack{j=1\\j\neq i}}^{r} f_j(x)\right)\left(\int_{g(x)}^{g(x_0)}(z-g(x))^{\alpha-1}\left(\left(D_{x_0-;g}^{\alpha}f_i\right)\circ g^{-1}\right)(z)\,dz\right)dx\right]\right\|\right.$$

$$\left.+\left\|\left[\int_{x_0}^{b}\left(\prod_{\substack{j=1\\j\neq i}}^{r} f_j(x)\right)\left(\int_{g(x_0)}^{g(x)}(g(x)-z)^{\alpha-1}\left(\left(D_{x_0+;g}^{\alpha}f_i\right)\circ g^{-1}\right)(z)\,dz\right)dx\right]\right\|\right]\leq$$

$$\frac{1}{\Gamma(\alpha)}\sum_{i=1}^{r}\left[\left[\int_a^{x_0}\left\|\left(\prod_{\substack{j=1\\j\neq i}}^{r} f_j(x)\right)\left(\int_{g(x)}^{g(x_0)}(z-g(x))^{\alpha-1}\left(\left(D_{x_0-;g}^{\alpha}f_i\right)\circ g^{-1}\right)(z)\,dz\right)\right\|dx\right]\right. \tag{1.31}$$

$$\left.+\left[\int_{x_0}^{b}\left\|\left(\prod_{\substack{j=1\\j\neq i}}^{r} f_j(x)\right)\left(\int_{g(x_0)}^{g(x)}(g(x)-z)^{\alpha-1}\left(\left(D_{x_0+;g}^{\alpha}f_i\right)\circ g^{-1}\right)(z)\,dz\right)\right\|dx\right]\right]\leq$$

$$\frac{1}{\Gamma(\alpha)}\sum_{i=1}^{r}\left[\left[\int_a^{x_0}\left(\prod_{\substack{j=1\\j\neq i}}^{r}\|f_j(x)\|\right)\left(\int_{g(x)}^{g(x_0)}(z-g(x))^{\alpha-1}\left\|\left(\left(D_{x_0-;g}^{\alpha}f_i\right)\circ g^{-1}\right)(z)\right\|dz\right)dx\right]\right.$$

$$\left.+\left[\int_{x_0}^{b}\left(\prod_{\substack{j=1\\j\neq i}}^{r}\|f_j(x)\|\right)\left(\int_{g(x_0)}^{g(x)}(g(x)-z)^{\alpha-1}\left\|\left(\left(D_{x_0+;g}^{\alpha}f_i\right)\circ g^{-1}\right)(z)\right\|dz\right)dx\right]\right]=:(*). \tag{1.32}$$

Hence it holds

$$\|\theta(f_1,\ldots,f_r)(x_0)\| \leq (*). \tag{1.33}$$

We have that

$$(*) \leq \frac{1}{\Gamma(\alpha+1)}$$

$$\sum_{i=1}^{r} \left[\left[\left\| \left(\left(D_{x_0-;g}^{\alpha} f_i \right) \circ g^{-1} \right) \right\|_{\infty,[g(a),g(x_0)]} \int_a^{x_0} \left(\prod_{\substack{j=1 \\ j \neq i}}^{r} \| f_j(x) \| \right) (g(x_0) - g(x))^{\alpha} dx \right] \right.$$

$$\left. + \left[\left\| \left(\left(D_{x_0+;g}^{\alpha} f_i \right) \circ g^{-1} \right) \right\|_{\infty,[g(x_0),g(b)]} \int_{x_0}^{b} \left(\prod_{\substack{j=1 \\ j \neq i}}^{r} \| f_j(x) \| \right) (g(x) - g(x_0))^{\alpha} dx \right] \right] \leq \tag{1.34}$$

$$\frac{1}{\Gamma(\alpha+1)} \sum_{i=1}^{r} \left[\left[\left\| \left(\left(D_{x_0-;g}^{\alpha} f_i \right) \circ g^{-1} \right) \right\|_{\infty,[g(a),g(x_0)]} \right. \right.$$

$$\left. (g(x_0) - g(a))^{\alpha} \left(\int_a^{x_0} \left(\prod_{\substack{j=1 \\ j \neq i}}^{r} \| f_j(x) \| \right) dx \right) \right] + \tag{1.35}$$

$$\left. \left[\left\| \left(\left(D_{x_0+;g}^{\alpha} f_i \right) \circ g^{-1} \right) \right\|_{\infty,[g(x_0),g(b)]} (g(b) - g(x_0))^{\alpha} \left(\int_{x_0}^{b} \left(\prod_{\substack{j=1 \\ j \neq i}}^{r} \| f_j(x) \| \right) dx \right) \right] \right],$$

proving (1.20). ■

Next comes an L_1 estimate.

Theorem 1.11 *All as in Theorem 1.10, with $\alpha \geq 1$. Then*

$$\| \theta (f_1, ..., f_r)(x_0) \| \leq \frac{1}{\Gamma(\alpha)}$$

$$\sum_{i=1}^{r} \left[\left[\left\| \left(D_{x_0-;g}^{\alpha} f_i \right) \circ g^{-1} \right\|_{L_1([g(a),g(x_0)])} \int_a^{x_0} \left(\prod_{\substack{j=1 \\ j \neq i}}^{r} \| f_j(x) \| \right) (g(x_0) - g(x))^{\alpha-1} dx \right] \right. \tag{1.36}$$

$$\left. + \left[\left\| \left(D_{x_0+;g}^{\alpha} f_i \right) \circ g^{-1} \right\|_{L_1([g(x_0),g(b)])} \int_{x_0}^{b} \left(\prod_{\substack{j=1 \\ j \neq i}}^{r} \| f_j(x) \| \right) (g(x) - g(x_0))^{\alpha-1} dx \right] \right].$$

Proof Next we use that $\alpha \geq 1$. We observe that

$$(*) \overset{(1.33)}{\leq} \frac{1}{\Gamma(\alpha)}$$

$$\sum_{i=1}^{r}\left[\left[\left\|\left(D_{x_0-;g}^{\alpha}f_i\right)\circ g^{-1}\right\|_{L_1([g(a),g(x_0)])}\int_a^{x_0}\left(\prod_{\substack{j=1\\j\neq i}}^{r}\|f_j(x)\|\right)(g(x_0)-g(x))^{\alpha-1}dx\right]\right. \tag{1.37}$$

$$\left.+\left[\left\|\left(D_{x_0+;g}^{\alpha}f_i\right)\circ g^{-1}\right\|_{L_1([g(x_0),g(b)])}\int_{x_0}^{b}\left(\prod_{\substack{j=1\\j\neq i}}^{r}\|f_j(x)\|\right)(g(x)-g(x_0))^{\alpha-1}dx\right]\right],$$

proving Theorem 1.11. ■

An L_p estimate follows.

Theorem 1.12 *All as in Theorem 1.10. Let now $p, q > 1 : \frac{1}{p} + \frac{1}{q} = 1$, with $\alpha > \frac{1}{q}$. Then*

$$\|\theta(f_1, ..., f_r)(x_0)\| \leq \frac{1}{(p(\alpha-1)+1)^{\frac{1}{p}}\Gamma(\alpha)}$$

$$\sum_{i=1}^{r}\left[\left\|\left(D_{x_0-;g}^{\alpha}f_i\right)\circ g^{-1}\right\|_{q,[g(a),g(x_0)]}\left(\int_a^{x_0}(g(x_0)-g(x))^{\alpha-\frac{1}{q}}\left(\prod_{\substack{j=1\\j\neq i}}^{r}\|f_j(x)\|\right)dx\right)\right. \tag{1.38}$$

$$+\left\|\left(D_{x_0+;g}^{\alpha}f_i\right)\circ g^{-1}\right\|_{q,[g(x_0),g(b)]}$$

$$\left.\left(\int_{x_0}^{b}(g(x)-g(x_0))^{\alpha-\frac{1}{q}}\left(\prod_{\substack{j=1\\j\neq i}}^{r}\|f_j(x)\|\right)dx\right)\right] =: \Phi(x_0).$$

Proof We have that

$$(*) \overset{(1.33)}{\leq} \frac{1}{\Gamma(\alpha)}\sum_{i=1}^{r}\left[\left[\int_a^{x_0}\left(\prod_{\substack{j=1\\j\neq i}}^{r}\|f_j(x)\|\right)\left(\int_{g(x)}^{g(x_0)}(z-g(x))^{p(\alpha-1)}dz\right)^{\frac{1}{p}}\right.\right.$$

$$\left.\left(\int_{g(x)}^{g(x_0)}\left\|\left(\left(D_{x_0-;g}^{\alpha}f_i\right)\circ g^{-1}\right)(z)\right\|^q dz\right)^{\frac{1}{q}}dx\right]+$$

$$\left[\int_{x_0}^{b}\left(\prod_{\substack{j=1\\j\neq i}}^{r}\|f_j(x)\|\right)\left(\int_{g(x_0)}^{g(x)}(g(x)-z)^{p(\alpha-1)}dz\right)^{\frac{1}{p}}\right.$$

$$\left.\left.\left(\int_{g(x_0)}^{g(x)}\left\|\left(\left(D_{x_0+;g}^{\alpha}f_i\right)\circ g^{-1}\right)(z)\right\|^q dz\right)^{\frac{1}{q}}dx\right]\right]\leq \tag{1.39}$$

$$\frac{1}{\Gamma(\alpha)}\sum_{i=1}^{r}\left[\left[\int_a^{x_0}\left(\prod_{\substack{j=1\\j\neq i}}^{r}\|f_j(x)\|\right)\frac{(g(x_0)-g(x))^{\alpha-1+\frac{1}{p}}}{(p(\alpha-1)+1)^{\frac{1}{p}}}\left\|\left(\left(D_{x_0-;g}^{\alpha}f_i\right)\circ g^{-1}\right)\right\|_{q,[g(a),g(x_0)]}dx\right]\right.$$

$$\left.+\left[\int_{x_0}^{b}\left(\prod_{\substack{j=1\\j\neq i}}^{r}\|f_j(x)\|\right)\frac{(g(x)-g(x_0))^{\alpha-1+\frac{1}{p}}}{(p(\alpha-1)+1)^{\frac{1}{p}}}\left\|\left(\left(D_{x_0+;g}^{\alpha}f_i\right)\circ g^{-1}\right)(z)\right\|_{q,[g(x_0),g(b)]}dx\right]\right]$$

$$=\frac{1}{(p(\alpha-1)+1)^{\frac{1}{p}}\Gamma(\alpha)}$$

$$\sum_{i=1}^{r}\left[\left\|\left(D_{x_0-;g}^{\alpha}f_i\right)\circ g^{-1}\right\|_{q,[g(a),g(x_0)]}\left(\int_a^{x_0}(g(x_0)-g(x))^{\alpha-\frac{1}{q}}\left(\prod_{\substack{j=1\\j\neq i}}^{r}\|f_j(x)\|\right)dx\right)\right. \tag{1.40}$$

$$\left.+\left\|\left(D_{x_0+;g}^{\alpha}f_i\right)\circ g^{-1}\right\|_{q,[g(x_0),g(b)]}\left(\int_{x_0}^{b}(g(x)-g(x_0))^{\alpha-\frac{1}{q}}\left(\prod_{\substack{j=1\\j\neq i}}^{r}\|f_j(x)\|\right)dx\right)\right],$$

proving Theorem 1.12. ∎

We continue with generalized fractional Grüss-Cebysev type inequalities for several functions over a Banach algebra. A uniform estimate follows.

Theorem 1.13 *Let* $(A, \|\cdot\|)$ *be a Banach algebra,* $0<\alpha\leq 1$, $f_i\in C^1([a,b],A)$, $i=1,...,r\in\mathbb{N}-\{1\}$; $g\in C^1([a,b])$, *strictly increasing, such that* $g^{-1}\in C^1([g(a),g(b)])$; $x_0\in[a,b]\subset\mathbb{R}$ *and* $\theta(f_1,...,f_r)(x_0)$ *as in (1.19). Assume that*

$$M(f_1,...,f_r):=\max_{i=1,...,r}\left\{\sup_{x_0\in[a,b]}\left\|\left(D_{x_0-;g}^{\alpha}f_i\right)\circ g^{-1}\right\|_{\infty,[g(a),g(x_0)]},\right. \tag{1.41}$$

$$\left.\sup_{x_0\in[a,b]}\left\|\left(D_{x_0+;g}^{\alpha}f_i\right)\circ g^{-1}\right\|_{\infty,[g(x_0),g(b)]}\right\}<\infty.$$

Denote by

$$\Delta (f_1, ..., f_r) := \int_a^b \theta (f_1, ..., f_r) (x) \, dx =$$

$$\sum_{i=1}^{r} \left[(b-a) \left(\int_a^b \left(\prod_{\substack{j=1 \\ j \neq i}}^{r} f_j (x) \right) f_i (x) \, dx \right) - \right.$$

$$\left. \left(\int_a^b \left(\prod_{\substack{j=1 \\ j \neq i}}^{r} f_j (x) \right) dx \right) \left(\int_a^b f_i (x) \, dx \right) \right]. \quad (1.42)$$

Then it holds

$$\| \Delta (f_1, ..., f_r) \| =$$

$$\left\| \sum_{i=1}^{r} \left[(b-a) \left(\int_a^b \left(\prod_{\substack{j=1 \\ j \neq i}}^{r} f_j (x) \right) f_i (x) \, dx \right) - \right. \right.$$

$$\left. \left. \left(\int_a^b \left(\prod_{\substack{j=1 \\ j \neq i}}^{r} f_j (x) \right) dx \right) \left(\int_a^b f_i (x) \, dx \right) \right] \right\| \leq$$

$$\frac{M (f_1, ..., f_r) (b-a)^2 (g(b) - g(a))^{\alpha}}{\Gamma (\alpha + 1)} \left(\sum_{i=1}^{r} \left\| \prod_{\substack{j=1 \\ j \neq i}}^{r} \| f_j \| \right\|_{\infty, [a,b]} \right). \quad (1.43)$$

Proof We have that

$$\| \Delta (f_1, ..., f_r) \| \leq \int_a^b \| \theta (f_1, ..., f_r) (x_0) \| \, dx_0 \overset{(1.33)}{\leq}$$

$$\frac{1}{\Gamma (\alpha + 1)} \sum_{i=1}^{r} \left[\left[\sup_{x_0 \in [a,b]} \left\| \left\| \left(D_{x_0 -; g}^{\alpha} f_i \right) \circ g^{-1} \right\| \right\|_{\infty, [g(a), g(x_0)]} \right. \right.$$

$$\left. \left. \left\{ \int_a^b \left[(g(x_0) - g(a))^{\alpha} \left(\int_a^{x_0} \left(\prod_{\substack{j=1 \\ j \neq i}}^{r} \| f_j (x) \| \right) dx \right) \right] dx_0 \right\} \right] + \quad (1.44)$$

$$\left[\sup_{x_0\in[a,b]}\left\|\left\|\left(D^{\alpha}_{x_0+;g}f_i\right)\circ g^{-1}\right\|\right\|_{\infty,[g(x_0),g(b)]}\right.$$

$$\left.\left.\left\{\int_a^b\left[(g(b)-g(x_0))^{\alpha}\left(\int_{x_0}^b\left(\prod_{\substack{j=1\\j\neq i}}^r\|f_j(x)\|\right)dx\right)\right]dx_0\right\}\right]\right]\leq$$

$$\frac{(g(b)-g(a))^{\alpha}}{\Gamma(\alpha+1)}\sum_{i=1}^r\left[\left[\sup_{x_0\in[a,b]}\left\|\left\|\left(D^{\alpha}_{x_0-;g}f_i\right)\circ g^{-1}\right\|\right\|_{\infty,[g(a),g(x_0)]}\right.\right.$$

$$\left.\left\{\int_a^b\left(\int_a^{x_0}\left(\prod_{\substack{j=1\\j\neq i}}^r\|f_j(x)\|\right)dx\right)dx_0\right\}\right]+ \tag{1.45}$$

$$\left[\sup_{x_0\in[a,b]}\left\|\left\|\left(D^{\alpha}_{x_0+;g}f_i\right)\circ g^{-1}\right\|\right\|_{\infty,[g(x_0),g(b)]}\right.$$

$$\left.\left.\left\{\int_a^b\left(\int_{x_0}^b\left(\prod_{\substack{j=1\\j\neq i}}^r\|f_j(x)\|\right)dx\right)dx_0\right\}\right]\right]=:(**).$$

Hence it holds

$$(**)\leq\frac{M(f_1,...,f_r)(g(b)-g(a))^{\alpha}(b-a)^2}{2\Gamma(\alpha+1)}$$

$$\sum_{i=1}^r\left[\left[\sup_{x_0\in[a,b]}\left\|\prod_{\substack{j=1\\j\neq i}}^r\|f_j(x)\|\right\|_{\infty,[a,x_0]}\right]+\right.$$

$$\left.\left[\sup_{x_0\in[a,b]}\left\|\prod_{\substack{j=1\\j\neq i}}^r\|f_j(x)\|\right\|_{\infty,[x_0,b]}\right]\right]$$

$$\leq \frac{M(f_1, ..., f_r)(b-a)^2 (g(b)-g(a))^{\alpha}}{\Gamma(\alpha+1)} \left(\sum_{i=1}^{r} \left\| \prod_{\substack{j=1 \\ j \neq i}}^{r} \|f_j\| \right\|_{\infty,[a,b]} \right), \tag{1.46}$$

proving (1.43). ∎

Next comes an L_p estimate

Theorem 1.14 *Let* $(A, \|\cdot\|)$ *be a Banach algebra,* $0 < \alpha \leq 1$, $f_i \in C^1([a,b], A)$, $i = 1, ..., r \in \mathbb{N} - \{1\}$; $g \in C^1([a,b])$, *strictly increasing, such that* $g^{-1} \in C^1([g(a), g(b)])$. *Let* $p, q > 1 : \frac{1}{p} + \frac{1}{q} = 1$, $\frac{1}{q} < \alpha \leq 1$, $x_0 \in [a,b] \subset \mathbb{R}$. *Assume that*

$$N(f_1, ..., f_r) := \max_{i=1,...,r} \left\{ \sup_{x_0 \in [a,b]} \left\| \left\| \left(D_{x_0-;g}^{\alpha} f_i \right) \circ g^{-1} \right\| \right\|_{L_q([g(a), g(x_0)])}, \right.$$

$$\left. \sup_{x_0 \in [a,b]} \left\| \left\| \left(D_{x_0+;g}^{\alpha} f_i \right) \circ g^{-1} \right\| \right\|_{L_q([g(x_0), g(b)])} \right\} < \infty, \tag{1.47}$$

and set

$$Z(f_1, ..., f_r) := \max_{i=1,...,r} \left\{ \left\| \left(\prod_{\substack{j=1 \\ j \neq i}}^{r} \|f_j\| \right) \right\|_{q,[a,b]} \right\}. \tag{1.48}$$

Then

$$\left\| \sum_{i=1}^{r} \left[(b-a) \left(\int_a^b \left(\prod_{\substack{j=1 \\ j \neq i}}^{r} f_j(x) \right) f_i(x) \, dx \right) - \right. \right.$$

$$\left. \left. \left(\int_a^b \left(\prod_{\substack{j=1 \\ j \neq i}}^{r} f_j(x) \right) dx \right) \left(\int_a^b f_i(x) \, dx \right) \right] \right\| \leq$$

$$\frac{2rpN(f_1, ..., f_r) Z(f_1, ..., f_r)}{(p+1)(p(\alpha-1)+1)^{\frac{1}{p}} \Gamma(\alpha)} (b-a)^{\frac{1}{p}+1} (g(b)-g(a))^{\alpha-\frac{1}{q}}. \tag{1.49}$$

Proof Here $\Phi(x_0)$ is as in (1.38).

Clearly then it holds

$$\Phi(x_0) \leq \frac{N(f_1, ..., f_r)}{(p(\alpha-1)+1)^{\frac{1}{p}} \Gamma(\alpha)}$$

$$\sum_{i=1}^{r}\left[\left(\int_a^{x_0}\left(g\left(x_0\right)-g\left(x\right)\right)^{\alpha-\frac{1}{q}}\left(\prod_{\substack{j=1\\j\neq i}}^{r}\left\|f_j\left(x\right)\right\|\right)dx\right)+\right. \tag{1.50}$$

$$\left.\left(\int_{x_0}^{b}\left(g\left(x\right)-g\left(x_0\right)\right)^{\alpha-\frac{1}{q}}\left(\prod_{\substack{j=1\\j\neq i}}^{r}\left\|f_j\left(x\right)\right\|\right)dx\right)\right].$$

Therefore we obtain

$$\left\|\Delta\left(f_1,...,f_r\right)\right\|\leq\int_a^b\left\|\theta\left(f_1,...,f_r\right)\left(x_0\right)\right\|dx_0\leq$$

$$\int_a^b\Phi\left(x_0\right)dx_0\leq\frac{N\left(f_1,...,f_r\right)}{\left(p\left(\alpha-1\right)+1\right)^{\frac{1}{p}}\Gamma\left(\alpha\right)}$$

$$\sum_{i=1}^{r}\left[\left[\int_a^b\left(\int_a^{x_0}\left(g\left(x_0\right)-g\left(x\right)\right)^{\left(\frac{p(\alpha-1)+1}{p}\right)}\left(\prod_{\substack{j=1\\j\neq i}}^{r}\left\|f_j\left(x\right)\right\|\right)dx\right)dx_0\right]\right.$$

$$\left.+\left[\int_a^b\left(\int_{x_0}^{b}\left(g\left(x\right)-g\left(x_0\right)\right)^{\left(\frac{p(\alpha-1)+1}{p}\right)}\left(\prod_{\substack{j=1\\j\neq i}}^{r}\left\|f_j\left(x\right)\right\|\right)dx\right)dx_0\right]\right]\leq$$

$$\frac{N\left(f_1,...,f_r\right)}{\left(p\left(\alpha-1\right)+1\right)^{\frac{1}{p}}\Gamma\left(\alpha\right)}\sum_{i=1}^{r}\left[\left\|\left(\prod_{\substack{j=1\\j\neq i}}^{r}\left\|f_j\right\|\right)\right\|_{q,[a,b]}\right.$$

$$\left[\int_a^b\left(\int_a^{x_0}\left(g\left(x_0\right)-g\left(x\right)\right)^{(p(\alpha-1)+1)}dx\right)^{\frac{1}{p}}dx_0\right]+ \tag{1.51}$$

$$\left.\left[\int_a^b\left(\int_{x_0}^{b}\left(g\left(x\right)-g\left(x_0\right)\right)^{(p(\alpha-1)+1)}dx\right)^{\frac{1}{p}}dx_0\right]\right]=:\left(\eta\right).$$

Hence we get

$$\left(\eta\right)\leq\frac{rN\left(f_1,...,f_r\right)Z\left(f_1,...,f_r\right)}{\left(p\left(\alpha-1\right)+1\right)^{\frac{1}{p}}\Gamma\left(\alpha\right)}$$

$$\left[\left[\int_a^b \left(\int_a^{x_0} (g(x_0) - g(x))^{(p(\alpha-1)+1)} dx\right)^{\frac{1}{p}} dx_0\right] +\right.$$

$$\left.\left[\int_a^b \left(\int_{x_0}^b (g(x) - g(x_0))^{(p(\alpha-1)+1)} dx\right)^{\frac{1}{p}} dx_0\right]\right] \leq \tag{1.52}$$

$$\frac{2rpN(f_1, ..., f_r) Z(f_1, ..., f_r)}{(p+1)(p(\alpha-1)+1)^{\frac{1}{p}} \Gamma(\alpha)} (g(b) - g(a))^{\alpha - \frac{1}{q}} (b-a)^{\frac{1}{p}+1},$$

proving (1.49). ■

1.5 Applications

We make

Remark 1.15 Assume from now on that $(A, \|\cdot\|)$ is a commutative Banach algebra. Then, we get that

$$\theta(f_1, ..., f_r)(x_0) \overset{(1.19)}{=} r \int_a^b \left(\prod_{j=1}^r f_j(x)\right) dx - \sum_{i=1}^r \left(\int_a^b \left(\prod_{\substack{j=1 \\ j \neq i}}^r f_j(x)\right) dx\right) f_i(x_0), \tag{1.53}$$

$x_0 \in [a, b]$.

Furthermore, it holds ($0 < \alpha \leq 1$ case)

$$\Delta(f_1, ..., f_r) \overset{(1.42)}{=} r(b-a) \int_a^b \left(\prod_{j=1}^r f_j(x)\right) dx -$$

$$\sum_{i=1}^r \left[\left(\int_a^b \left(\prod_{\substack{j=1 \\ j \neq i}}^r f_j(x)\right) dx\right) \left(\int_a^b f_i(x) dx\right)\right]. \tag{1.54}$$

When $r = 2$, we get that

$$\theta(f_1, f_2)(x_0) = 2 \int_a^b f_1(x) f_2(x) dx - f_1(x_0) \int_a^b f_2(x) dx - f_2(x_0) \int_a^b f_1(x) dx, \tag{1.55}$$

$x_0 \in [a, b]$,

and

$$\Delta(f_1, f_2) = 2\left[(b-a)\int_a^b f_1(x) f_2(x)\,dx - \left(\int_a^b f_1(x)\,dx\right)\left(\int_a^b f_2(x)\,dx\right)\right], \tag{1.56}$$

$0 < \alpha \leq 1$.

We give

Corollary 1.16 *All as in Theorem 1.10, A is a commutative Banach algebra, $r = 2$. Then*

$$\|\theta(f_1, f_2)(x_0)\| \leq \frac{1}{\Gamma(\alpha+1)} \sum_{i=1}^{2} \Bigg[\Bigg[\left\| \left\| \left(\left(D_{x_0-;g}^{\alpha} f_i \right) \circ g^{-1} \right) \right\| \right\|_{\infty,[g(a),g(x_0)]}$$

$$(g(x_0) - g(a))^{\alpha} \left(\int_a^{x_0} \left(\prod_{\substack{j=1\\ j\neq i}}^{2} \|f_j(x)\| \right) dx \right) \Bigg] +$$

$$\left[\left\| \left\| \left(\left(D_{x_0+;g}^{\alpha} f_i \right) \circ g^{-1} \right) \right\| \right\|_{\infty,[g(x_0),g(b)]} (g(b) - g(x_0))^{\alpha} \left(\int_{x_0}^{b} \left(\prod_{\substack{j=1\\ j\neq i}}^{2} \|f_j(x)\| \right) dx \right) \right] \Bigg]. \tag{1.57}$$

Proof By Theorem 1.10. ■

We continue with

Corollary 1.17 *All as in Theorem 1.13, A is a commutative Banach algebra, $r = 2$, $0 < \alpha \leq 1$, $M(f_1, f_2)$ as in (1.41). Then*

$$\|\Delta(f_1, f_2)\| \leq \frac{M(f_1, f_2)(b-a)^2 (g(b) - g(a))^{\alpha}}{\Gamma(\alpha+1)} \left[\| \|f_1\| \|_{\infty,[a,b]} + \| \|f_2\| \|_{\infty,[a,b]} \right]. \tag{1.58}$$

Proof By Theorem 1.13. ■

Finally we derive:

Corollary 1.18 *All as in Corollary 1.16, for $g(t) = e^t$. Then*

$$\|\theta(f_1, f_2)(x_0)\| \leq \frac{1}{\Gamma(\alpha+1)} \sum_{i=1}^{2} \Bigg[\Bigg[\left\| \left\| \left(\left(D_{x_0-;e^t}^{\alpha} f_i \right) \circ \log \right) \right\| \right\|_{\infty,[e^a,e^{x_0}]}$$

$$(e^{x_0} - e^{a})^{\alpha} \left(\int_a^{x_0} \left(\prod_{\substack{j=1\\ j\neq i}}^{2} \|f_j(x)\| \right) dx \right) \Bigg] +$$

$$\left[\left\| \left\| \left(\left(D^{\alpha}_{x_0+;e'} f_i \right) \circ \log \right) \right\| \right\|_{\infty,[e^{x_0},e^b]} \left(e^b - e^{x_0} \right)^{\alpha} \left(\int_{x_0}^{b} \left(\prod_{\substack{j=1 \\ j \neq i}}^{2} \| f_j (x) \| \right) dx \right) \right] \right]. \tag{1.59}$$

Proof By Corollary 1.16. ■

References

1. Aliprantis, C., Border, K.: Infinite Dimensional Analysis, 3rd edn. Springer, New York (2006)
2. Anastassiou, G.A.: Fractional Differentiation Inequalities. Research Monograph. Springer, New York (2009)
3. Anastassiou, G.A.: Advances on Fractional Inequalities. Research Monograph. Springer, New York (2011)
4. Anastassiou, G.A.: Intelligent Comparisons: Analytic Inequalities. Springer, Heidelberg, New York (2016)
5. Anastassiou, G.A.: Intelligent Computations: Abstract Fractional Calculus, Inequalities, Approximations. Springer, Heidelberg, New York (2018)
6. Anastassiou, G.A.: Generalized Fractional Ostrowski and Grüss type inequalities involving several Banach algebra valued functions. submitted for publication (2021)
7. Čebyšev: Sur les expressions approximatives des intégrales définies par les aures prises entre les mêmes limites
8. Dragomir, S.S.: Noncommutative Ostrowski type inequalities for functions in Banach algebras. RGMIA Res. Rep. Coll. **24**, Art. 10, 24 pp (2021)
9. Dragomir, S.S.: New Noncommuttive Grüss type inequalities for functions in Banach algebras. RGMIA Res. Rep. Coll. **24**, Art. 13, 19 pp (2021)
10. Mikusinski, J.: The Bochner Integral. Academic Press, New York (1978)
11. Ostrowski, A.: Über die Absolutabweichung einer differentiabaren Funcktion von ihrem Integralmittelwert. Comment. Math. Helv. **10**, 226–227 (1938)
12. Rudin, W.: Functional Analysis, 2nd edn. McGraw-Hill Inc, New York (1991)

Chapter 2
Iterated Generalized Fractional Ostrowski and Grüss Inequalities with Multiple Banach Algebra Valued Functions

Employing iterated generalized Caputo fractional left and right vectorial Taylor formulae we establish mixed sequential generalized fractional Ostrowski and Grüss type inequalities for several Banach algebra valued functions. The estimates are with respect to all norms $\|\cdot\|_p$, $1 \le p \le \infty$. We finish with applications. It follows [5].

2.1 Introduction

The following results motivate our chapter.

Theorem 2.1 (1938, Ostrowski [10]) *Let $f : [a, b] \to \mathbb{R}$ be continuous on $[a, b]$ and differentiable on (a, b) whose derivative $f' : (a, b) \to \mathbb{R}$ is bounded on (a, b), i.e., $\|f'\|_\infty^{\sup} := \sup_{t \in (a,b)} |f'(t)| < +\infty$. Then*

$$\left| \frac{1}{b-a} \int_a^b f(t)\, dt - f(x) \right| \le \left[\frac{1}{4} + \frac{\left(x - \frac{a+b}{2}\right)^2}{(b-a)^2} \right] (b-a) \|f'\|_\infty^{\sup}, \tag{2.1}$$

for any $x \in [a, b]$. The constant $\frac{1}{4}$ is the best possible.

Ostrowski type inequalities have great applications to integral approximations in Numerical Analysis.

Theorem 2.2 (1882, Čebyšev [6]) *Let $f, g : [a, b] \to \mathbb{R}$ be absolutely continuous functions with $f', g' \in L_\infty([a, b])$. Then*

G. A. Anastassiou, *Fractional Inequalities In Banach Algebras*,
Studies in Systems, Decision and Control 441,
https://doi.org/10.1007/978-3-031-05148-7_2

$$\left|\frac{1}{b-a}\int_a^b f(x)\,g(x)\,dx-\left(\frac{1}{b-a}\int_a^b f(x)\,dx\right)\left(\frac{1}{b-a}\int_a^b g(x)\,dx\right)\right|$$
$$\leq\frac{1}{12}(b-a)^2\left\|f'\right\|_\infty\left\|g'\right\|_\infty. \tag{2.2}$$

The above integrals are assumed to exist.

The related Grüss type inequalities have many applications to Probability Theory.

In this article we present sequential generalized fractional Ostrowski and Cebysev-Grüss type inequalities for several Banach algebra valued functions. Now our sequential generalized left and right Caputo fractional derivatives are for Banach space valued functions and our integrals are of Bochner type. The main motivation here is Theorem 2.13 at the end of Sect. 2.2. Several applications finish this article. Inspiration came also from [7, 8]. See also [1–3].

2.2 Vectorial Sequential Generalized Fractional Calculus Background

We need

Definition 2.3 ([4], p. 106) Let $0<\alpha\leq 1$, $f\in C^1([a,b],X)$, where $[a,b]\subset\mathbb{R}$, and $(X,\|\cdot\|)$ is a Banach space. Let $g\in C^1([a,b])$, strictly increasing, such that $g^{-1}\in C^1([g(a),g(b)])$.

We define the left generalized g-fractional derivative X-valued of f of order α as follows:

$$\left(D^\alpha_{a+;g}f\right)(x):=\frac{1}{\Gamma(1-\alpha)}\int_a^x (g(x)-g(t))^{-\alpha}\,g'(t)\left(f\circ g^{-1}\right)'(g(t))\,dt, \tag{2.3}$$

$\forall\, x\in[a,b]$, where Γ is the gamma function. The last integral is of Bochner type ([9]).

If $0<\alpha<1$, by Theorem 4.10, p. 98, [4], we have that $\left(D^\alpha_{a+;g}f\right)\in C([a,b],X)$.

We set

$$D^1_{a+;g}f(x):=\left(\left(f\circ g^{-1}\right)'\circ g\right)(x)\in C([a,b],X), \tag{2.4}$$

$$D^0_{a+;g}f(x)=f(x),\ \forall\, x\in[a,b].$$

When $g=id$, then

$$D^\alpha_{a+;g}f=D^\alpha_{a+;id}f=D^\alpha_{*a}f, \tag{2.5}$$

the usual left X-valued Caputo fractional derivative, see [4], Chap. 1.

We make

Remark 2.4 By (2.3) we have

$$\left\|\left(D^{\alpha}_{a+;g}f\right)(x)\right\| \leq \frac{1}{\Gamma(1-\alpha)}\int_a^x (g(x)-g(t))^{-\alpha} g'(t)\left\|\left(f\circ g^{-1}\right)'(g(t))\right\| dt \leq$$

$$\frac{\left\|\left\|\left(f\circ g^{-1}\right)'\circ g\right\|\right\|_{\infty,[a,b]}}{\Gamma(1-\alpha)}\int_a^x (g(x)-g(t))^{-\alpha} g'(t)\,dt =$$

$$\frac{\left\|\left\|\left(f\circ g^{-1}\right)'\circ g\right\|\right\|_{\infty,[a,b]}}{\Gamma(1-\alpha)}\frac{(g(x)-g(t))^{1-\alpha}}{1-\alpha} = \tag{2.6}$$

$$\frac{\left\|\left\|\left(f\circ g^{-1}\right)'\circ g\right\|\right\|_{\infty,[a,b]}}{\Gamma(2-\alpha)}(g(x)-g(a))^{1-\alpha},\ \forall\, x\in[a,b].$$

Hence

$$\left(D^{\alpha}_{a+;g}f\right)(a)=0. \tag{2.7}$$

We need

Definition 2.5 ([4], p. 107) Let $0<\alpha\leq 1$, $f\in C^1([a,b],X)$, where $[a,b]\subset\mathbb{R}$, and $(X,\|\cdot\|)$ is a Banach space. Let $g\in C^1([a,b])$, strictly increasing, such that $g^{-1}\in C^1([g(a),g(b)])$.

We define the right generalized g-fractional derivative X-valued of f of order α as follows:

$$\left(D^{\alpha}_{b-;g}f\right)(x) := \frac{-1}{\Gamma(1-\alpha)}\int_x^b (g(t)-g(x))^{-\alpha} g'(t)\left(f\circ g^{-1}\right)'(g(t))\,dt, \tag{2.8}$$

$\forall\, x\in[a,b]$. The last integral is of Bochner type.

If $0<\alpha<1$, by Theorem 4.11, p. 101 ([4]), we have that $\left(D^{\alpha}_{b-;g}f\right)\in C([a,b],X)$.

We set

$$D^{1}_{b-;g}f(x) := -\left(\left(f\circ g^{-1}\right)'\circ g\right)(x)\in C([a,b],X), \tag{2.9}$$

$$D^{0}_{b-;g}f(x) := f(x),\ \forall\, x\in[a,b].$$

When $g=id$, then

$$D^{\alpha}_{b-;g}f(x) = D^{\alpha}_{b-;id}f(x) = D^{\alpha}_{b-}f, \tag{2.10}$$

the usual right X-valued Caputo fractional derivative, see [4], Chap. 2.

We make

Remark 2.6 By (2.8) we have

$$\left\| \left(D_{b-;g}^{\alpha} f \right)(x) \right\| \leq \frac{1}{\Gamma(1-\alpha)} \int_x^b (g(t) - g(x))^{-\alpha} g'(t) \left\| \left(f \circ g^{-1} \right)'(g(t)) \right\| dt \leq$$

$$\frac{\left\| \left\| \left(f \circ g^{-1} \right)' \circ g \right\| \right\|_{\infty,[a,b]}}{\Gamma(1-\alpha)} \int_x^b (g(t) - g(x))^{-\alpha} g'(t) \, dt = \tag{2.11}$$

$$\frac{\left\| \left\| \left(f \circ g^{-1} \right)' \circ g \right\| \right\|_{\infty,[a,b]}}{\Gamma(1-\alpha)} \frac{(g(b) - g(x))^{1-\alpha}}{1-\alpha} =$$

$$\frac{\left\| \left\| \left(f \circ g^{-1} \right)' \circ g \right\| \right\|_{\infty,[a,b]}}{\Gamma(2-\alpha)} (g(b) - g(x))^{1-\alpha}, \ \forall x \in [a, b].$$

Hence

$$\left(D_{b-;g}^{\alpha} f \right)(b) = 0. \tag{2.12}$$

We need

Definition 2.7 ([4], p. 115) Denote by $(0 < \alpha \leq 1)$

$$D_{a+;g}^{n\alpha} := D_{a+;g}^{\alpha} D_{a+;g}^{\alpha} ... D_{a+;g}^{\alpha} \ (n \text{ times}), \ n \in \mathbb{N} \tag{2.13}$$

and $D_{a+;g}^{0} = I$ (identity operator).

We also need

Definition 2.8 ([4], p. 118)

$$D_{b-;g}^{n\alpha} := D_{b-;g}^{\alpha} D_{b-;g}^{\alpha} ... D_{b-;g}^{\alpha} \ (n \text{ times}), \ n \in \mathbb{N} \tag{2.14}$$

and $D_{b-;g}^{0} = I$ (identity operator).

Based on (2.7) and Theorem 4.30, p. 117, ([4]), we have the following g-left generalized modified X-valued Taylor's formula:

Theorem 2.9 *Let* $0 < \alpha \leq 1$, $n \in \mathbb{N}$, $f \in C^1([a, b], X)$, $(X, \|\cdot\|)$ *a Banach space,* $g \in C^1([a, b])$, *strictly increasing, such that* $g^{-1} \in C^1([g(a), g(b)])$. *Let* $F_k := D_{a+;g}^{k\alpha} f$, $k = 1, ..., n$, *that fulfill* $F_k \in C^1([a, b], X)$, *and* $F_{n+1} \in C([a, b], X)$. *Then*

$$f(x) - f(a) = \sum_{i=2}^{n} \frac{(g(x) - g(a))^{i\alpha}}{\Gamma(i\alpha + 1)} \left(D_{a+;g}^{i\alpha} f \right)(a) +$$

$$\frac{1}{\Gamma((n+1)\alpha)} \int_a^x (g(x) - g(t))^{(n+1)\alpha - 1} g'(t) \left(D_{a+;g}^{(n+1)\alpha} f\right)(t) \, dt, \tag{2.15}$$

$\forall \, x \in [a, b]$.

When $n = 1$ we obtain

Corollary 2.10 *Let* $0 < \alpha \leq 1$, $f \in C^1([a, b], X)$, $(X, \|\cdot\|)$ *is a Banach space,* $g \in C^1([a, b])$, *strictly increasing, such that* $g^{-1} \in C^1([g(a), g(b)])$. *Assume that* $D_{a+;g}^{\alpha} f \in C^1([a, b], X)$, *and* $D_{a+;g}^{2\alpha} f \in C([a, b], X)$. *Then*

$$f(x) - f(a) = \frac{1}{\Gamma(2\alpha)} \int_a^x (g(x) - g(t))^{2\alpha - 1} g'(t) \left(D_{a+;g}^{2\alpha} f\right)(t) \, dt, \tag{2.16}$$

$\forall \, x \in [a, b]$.

Based on (2.12) and Theorem 4.33, p. 120, ([4]), we have the following g-right generalized modified X-valued Taylor's formula:

Theorem 2.11 *Let* $f \in C^1([a, b], X)$, $(X, \|\cdot\|)$ *a Banach space,* $g \in C^1([a, b])$, *strictly increasing, such that* $g^{-1} \in C^1([g(a), g(b)])$. *Suppose* $F_k := D_{b-;g}^{k\alpha} f$, $k = 1, ..., n$, *fulfill* $F_k \in C^1([a, b], X)$, *and* $F_{n+1} \in C([a, b], X)$, *where* $0 < \alpha \leq 1$, $n \in \mathbb{N}$. *Then*

$$f(x) - f(b) = \sum_{i=2}^{n} \frac{(g(b) - g(x))^{i\alpha}}{\Gamma(i\alpha + 1)} \left(D_{b-;g}^{i\alpha} f\right)(b) +$$

$$\frac{1}{\Gamma((n+1)\alpha)} \int_x^b (g(t) - g(x))^{(n+1)\alpha - 1} g'(t) \left(D_{b-;g}^{(n+1)\alpha} f\right)(t) \, dt, \tag{2.17}$$

$\forall \, x \in [a, b]$.

When $n = 1$ we obtain

Corollary 2.12 *Let* $0 < \alpha \leq 1$, $f \in C^1([a, b], X)$, $(X, \|\cdot\|)$ *is a Banach space,* $g \in C^1([a, b])$, *strictly increasing, such that* $g^{-1} \in C^1([g(a), g(b)])$. *Assume that* $D_{b-;g}^{\alpha} f \in C^1([a, b], X)$, *and* $D_{b-;g}^{2\alpha} f \in C([a, b], X)$. *Then*

$$f(x) - f(b) = \frac{1}{\Gamma(2\alpha)} \int_x^b (g(t) - g(x))^{2\alpha - 1} g'(t) \left(D_{b-;g}^{2\alpha} f\right)(t) \, dt, \tag{2.18}$$

$\forall \, x \in [a, b]$.

We are greatly motivated by the following sequential generalized fractional Ostrowski type inequality:

Theorem 2.13 (p. 140, [4]) *Let* $g \in C^1([a,b])$ *and strictly increasing, such that* $g^{-1} \in C^1([g(a), g(b)])$, *and* $0 < \alpha < 1$, $n \in \mathbb{N}$, $f \in C^1([a,b], X)$, *where* $(X, \|\cdot\|)$ *is a Banach space. Let* $x_0 \in [a,b]$ *be fixed. Assume that* $F_k^{x_0} := D_{x_0-;g}^{k\alpha} f$, *for* $k = 1, ..., n$, *fulfill* $F_k^{x_0} \in C^1([a,b], X)$ *and* $F_{n+1}^{x_0} \in C([a, x_0], X)$ *and* $\left(D_{x_0-;g}^{i\alpha} f\right)(x_0) = 0$, $i = 1, ..., n$.

Similarly, we assume that $G_k^{x_0} := D_{x_0+;g}^{k\alpha} f$, *for* $k = 1, ..., n$, *fulfill* $G_k^{x_0} \in C^1([x_0, b], X)$ *and* $G_{n+1}^{x_0} \in C([x_0, b], X)$ *and* $\left(D_{x_0+;g}^{i\alpha} f\right)(x_0) = 0$, $i = 1, ..., n$.

Then

$$\left\| \frac{1}{b-a} \int_a^b f(x)\, dx - f(x_0) \right\| \leq \frac{1}{(b-a)\,\Gamma((n+1)\alpha+1)} \cdot$$

$$\left\{ (g(b) - g(x_0))^{(n+1)\alpha} (b - x_0) \left\| D_{x_0+;g}^{(n+1)\alpha} f \right\|_{\infty,[x_0,b]} + \right. \tag{2.19}$$

$$\left. (g(x_0) - g(a))^{(n+1)\alpha} (x_0 - a) \left\| D_{x_0-;g}^{(n+1)\alpha} f \right\|_{\infty,[a,x_0]} \right\}.$$

2.3 Banach Algebras Background

All here come from [11].

We need

Definition 2.14 ([11], p. 245) A complex algebra is a vector space A over the complex filed $\mathbb{C}$ in which a multiplication is defined that satisfies

$$x(yz) = (xy)z, \tag{2.20}$$

$$(x+y)z = xz + yz,\; x(y+z) = xy + xz, \tag{2.21}$$

and

$$\alpha(xy) = (\alpha x)y = x(\alpha y), \tag{2.22}$$

for all x, y and z in A and for all scalars α.

Additionally if A is a Banach space with respect to a norm that satisfies the multiplicative inequality

$$\|xy\| \leq \|x\|\,\|y\| \;\; (x \in A,\; y \in A) \tag{2.23}$$

and if A contains a unit element e such that

$$xe = ex = x \;\; (x \in A) \tag{2.24}$$

and

$$\|e\| = 1, \tag{2.25}$$

then A is called a Banach algebra.

A is commutative iff $xy = yx$ for all $x, y \in A$.

We make

Remark 2.15 Commutativity of A will be explicited stated when needed.

There exists at most one $e \in A$ that satisfies (2.24).

Inequality (2.23) makes multiplication to be continuous, more precisely left and right continuous, see [11], p. 246.

Multiplication in A is not necessarily the numerical multiplication, it is something more general and it is defined abstractly, that is for $x, y \in A$ we have $xy \in A$, e.g. composition or convolution, etc.

For nice examples about Banach algebras see [11], pp. 247–248, Sect. 10.3.

We also make

Remark 2.16 Next we mention about integration of A-valued functions, see [11], p. 259, Sect. 10.22:

If A is a Banach algebra and f is a continuous A-valued function on some compact Hausdorff space Q on which a complex Borel measure μ is defined, then $\int f d\mu$ exists and has all the properties that were discussed in Chap. 3 of [11], simply because A is a Banach space. However, an additional property can be added to these, namely: If $x \in A$, then

$$x \int_Q f \, d\mu = \int_Q xf(p) \, d\mu(p) \tag{2.26}$$

and

$$\left(\int_Q f \, d\mu \right) x = \int_Q f(p) \, x \, d\mu(p) . \tag{2.27}$$

The Bochner integrals we will involve in our chapter follow (2.26) and (2.27). Also, let $f \in C([a, b], X)$, where $[a, b] \subset \mathbb{R}$, $(X, \|\cdot\|)$ is a Banach space. By [4], p. 3, f is Bochner integrable.

2.4 Main Results

We start with mixed sequential generalized fractional Ostrowski type inequalities for several functions over a Banach algebra. A uniform estimate follows.

Theorem 2.17 *Let $(A, \|\cdot\|)$ be a Banach algebra, $x_0 \in [a, b] \subset \mathbb{R}$, $0 < \alpha < 1$, $n \in \mathbb{N}$, $f_i \in C^1([a, b], A)$, $i = 1, ..., r$. Let $g \in C^1([a, b])$ and strictly increasing, such that $g^{-1} \in C^1([g(a), g(b)])$. Assume that $F_{ki}^{x_0} := D_{x_0-;g}^{k\alpha} f_i$, for $k = 1, ..., n$, fulfill*

$F_{ki}^{x_0} \in C^1([a,x_0],A)$ and $F_{(n+1)i}^{x_0} \in C([a,x_0],A)$ and $\left(D_{x_0-;g}^{j\alpha} f_i\right)(x_0) = 0$, $j = 2,...,n$; $i = 1,...,r$. Similarly, we assume that $G_{ki}^{x_0} := D_{x_0+;g}^{k\alpha} f_i$, $k = 1,...,n$, fulfill $G_{ki}^{x_0} \in C^1([x_0,b],A)$ and $G_{(n+1)i}^{x_0} \in C([x_0,b],A)$ and $\left(D_{x_0+;g}^{j\alpha} f_i\right)(x_0) = 0$, $j = 2,...,n$; $i = 1,...,r$.

Denote by

$$\Omega(f_1,...,f_r)(x_0) :=$$

$$\sum_{i=1}^{r}\left[\int_a^b\left(\prod_{\substack{j=1\\j\neq i}}^{r} f_j(x)\right) f_i(x)\,dx - \left(\int_a^b\left(\prod_{\substack{j=1\\j\neq i}}^{r} f_j(x)\right)dx\right) f_i(x_0)\right]. \tag{2.28}$$

Then

$$\|\Omega(f_1,...,f_r)(x_0)\| \leq \frac{1}{\Gamma((n+1)\alpha+1)}\sum_{i=1}^{r}\left[\left[\left\|\left(D_{x_0-;g}^{(n+1)\alpha} f_i\right)\right\|_{\infty,[a,x_0]}\right.\right.$$

$$\left.(g(x_0)-g(a))^{(n+1)\alpha}\left(\int_a^{x_0}\left(\prod_{\substack{j=1\\j\neq i}}^{r}\|f_j(x)\|\right)dx\right)\right] + \tag{2.29}$$

$$\left.\left[\left\|\left(D_{x_0+;g}^{(n+1)\alpha} f_i\right)\right\|_{\infty,[x_0,b]}(g(b)-g(x_0))^{(n+1)\alpha}\left(\int_{x_0}^{b}\left(\prod_{\substack{j=1\\j\neq i}}^{r}\|f_j(x)\|\right)dx\right)\right]\right].$$

Proof By Theorem 2.11 we obtain

$$f_i(x) - f_i(x_0) = \frac{1}{\Gamma((n+1)\alpha)}\int_x^{x_0}(g(t)-g(x))^{(n+1)\alpha-1} g'(t)\left(D_{x_0-;g}^{(n+1)\alpha} f_i\right)(t)\,dt, \tag{2.30}$$

$\forall\, x \in [a,x_0]$, $i = 1,...,r$.

Also, by Theorem 2.9, we get

$$f_i(x) - f_i(x_0) = \frac{1}{\Gamma((n+1)\alpha)}\int_{x_0}^{x}(g(x)-g(t))^{(n+1)\alpha-1} g'(t)\left(D_{x_0+;g}^{(n+1)\alpha} f_i\right)(t)\,dt, \tag{2.31}$$

$\forall\, x \in [x_0,b]$, $i = 1,...,r$.

Left multiplying (2.30) and (2.31) with $\left(\prod_{\substack{j=1\\j\neq i}}^{r} f_j(x)\right)$ we get, respectively,

$$\left(\prod_{\substack{j=1\\j\neq i}}^{r} f_j(x)\right) f_i(x) - \left(\prod_{\substack{j=1\\j\neq i}}^{r} f_j(x)\right) f_i(x_0) = \tag{2.32}$$

$$\frac{\left(\prod_{\substack{j=1\\j\neq i}}^{r} f_j(x)\right)}{\Gamma((n+1)\alpha)} \int_x^{x_0} (g(t) - g(x))^{(n+1)\alpha-1} g'(t) \left(D_{x_0-;g}^{(n+1)\alpha} f_i\right)(t)\, dt,$$

$\forall\, x \in [a, x_0]$, $i = 1, ..., r$,

and

$$\left(\prod_{\substack{j=1\\j\neq i}}^{r} f_j(x)\right) f_i(x) - \left(\prod_{\substack{j=1\\j\neq i}}^{r} f_j(x)\right) f_i(x_0) =$$

$$\frac{\left(\prod_{\substack{j=1\\j\neq i}}^{r} f_j(x)\right)}{\Gamma((n+1)\alpha)} \int_{x_0}^{x} (g(x) - g(t))^{(n+1)\alpha-1} g'(t) \left(D_{x_0+;g}^{(n+1)\alpha} f_i\right)(t)\, dt, \tag{2.33}$$

$\forall\, x \in [x_0, b]$, $i = 1, ..., r$.

Adding (2.32) and (2.33) as separate groups, we obtain

$$\sum_{i=1}^{r}\left(\prod_{\substack{j=1\\j\neq i}}^{r} f_j(x)\right) f_i(x) - \sum_{i=1}^{r}\left(\prod_{\substack{j=1\\j\neq i}}^{r} f_j(x)\right) f_i(x_0) =$$

$$\frac{\sum_{i=1}^{r}\left(\prod_{\substack{j=1\\j\neq i}}^{r} f_j(x)\right)}{\Gamma((n+1)\alpha)} \int_x^{x_0} (g(t) - g(x))^{(n+1)\alpha-1} g'(t) \left(D_{x_0-;g}^{(n+1)\alpha} f_i\right)(t)\, dt, \tag{2.34}$$

$\forall\, x \in [a, x_0]$,

and

$$\sum_{i=1}^{r}\left(\prod_{\substack{j=1\\j\neq i}}^{r} f_j(x)\right) f_i(x) - \sum_{i=1}^{r}\left(\prod_{\substack{j=1\\j\neq i}}^{r} f_j(x)\right) f_i(x_0) =$$

$$\frac{\sum_{i=1}^{r}\left(\prod_{\substack{j=1\\j\neq i}}^{r} f_j(x)\right)}{\Gamma((n+1)\alpha)} \int_{x_0}^{x} (g(x) - g(t))^{(n+1)\alpha-1} g'(t) \left(D_{x_0+;g}^{(n+1)\alpha} f_i\right)(t)\, dt, \tag{2.35}$$

$\forall\, x \in [x_0, b]$.

Next, we integrate (2.34) and (2.35) with respect to $x \in [a, b]$. We have

$$\sum_{i=1}^{r} \int_{a}^{x_0} \left(\prod_{\substack{j=1 \\ j \neq i}}^{r} f_j(x) \right) f_i(x)\, dx - \sum_{i=1}^{r} \left(\int_{a}^{x_0} \left(\prod_{\substack{j=1 \\ j \neq i}}^{r} f_j(x) \right) dx \right) f_i(x_0) =$$

$$\frac{1}{\Gamma((n+1)\alpha)} \sum_{i=1}^{r} \left[\int_{a}^{x_0} \left(\prod_{\substack{j=1 \\ j \neq i}}^{r} f_j(x) \right) \right.$$

$$\left. \left(\int_{x}^{x_0} (g(t) - g(x))^{(n+1)\alpha - 1} g'(t) \left(D_{x_0-;g}^{(n+1)\alpha} f_i \right)(t)\, dt \right) dx \right], \tag{2.36}$$

and

$$\sum_{i=1}^{r} \int_{x_0}^{b} \left(\prod_{\substack{j=1 \\ j \neq i}}^{r} f_j(x) \right) f_i(x)\, dx - \sum_{i=1}^{r} \left(\int_{x_0}^{b} \left(\prod_{\substack{j=1 \\ j \neq i}}^{r} f_j(x) \right) dx \right) f_i(x_0) =$$

$$\frac{1}{\Gamma((n+1)\alpha)} \sum_{i=1}^{r} \left[\int_{x_0}^{b} \left(\prod_{\substack{j=1 \\ j \neq i}}^{r} f_j(x) \right) \right.$$

$$\left. \left(\int_{x_0}^{x} (g(x) - g(t))^{(n+1)\alpha - 1} g'(t) \left(D_{x_0+;g}^{(n+1)\alpha} f_i \right)(t)\, dt \right) dx \right]. \tag{2.37}$$

Finally, adding (2.36) and (2.37) we obtain the useful identity

$$\Omega(f_1, ..., f_r)(x_0) :=$$

$$\sum_{i=1}^{r} \left[\int_{a}^{b} \left(\prod_{\substack{j=1 \\ j \neq i}}^{r} f_j(x) \right) f_i(x)\, dx - \left(\int_{a}^{b} \left(\prod_{\substack{j=1 \\ j \neq i}}^{r} f_j(x) \right) dx \right) f_i(x_0) \right] =$$

$$\frac{1}{\Gamma((n+1)\alpha)} \sum_{i=1}^{r} \left[\left[\int_{a}^{x_0} \left(\prod_{\substack{j=1 \\ j \neq i}}^{r} f_j(x) \right) \right. \right.$$

$$\left(\int_x^{x_0} (g(t) - g(x))^{(n+1)\alpha-1} g'(t) \left(D_{x_0-;g}^{(n+1)\alpha} f_i\right)(t)\, dt\right) dx\Bigg]$$

$$+ \left[\int_{x_0}^b \left(\prod_{\substack{j=1 \\ j\neq i}}^r f_j(x)\right) \left(\int_{x_0}^x (g(x) - g(t))^{(n+1)\alpha-1} g'(t) \left(D_{x_0+;g}^{(n+1)\alpha} f_i\right)(t)\, dt\right) dx\right]\Bigg]. \tag{2.38}$$

Therefore, we get that

$$\|\Omega(f_1, ..., f_r)(x_0)\| \leq \frac{1}{\Gamma((n+1)\alpha)} \sum_{i=1}^r \Bigg[\Bigg\|\Bigg[\int_a^{x_0} \left(\prod_{\substack{j=1 \\ j\neq i}}^r f_j(x)\right)$$

$$\left(\int_x^{x_0} (g(t) - g(x))^{(n+1)\alpha-1} g'(t) \left(D_{x_0-;g}^{(n+1)\alpha} f_i\right)(t)\, dt\right) dx\Bigg]\Bigg\| +$$

$$\left\|\left[\int_{x_0}^b \left(\prod_{\substack{j=1 \\ j\neq i}}^r f_j(x)\right) \left(\int_{x_0}^x (g(x) - g(t))^{(n+1)\alpha-1} g'(t) \left(D_{x_0+;g}^{(n+1)\alpha} f_i\right)(t)\, dt\right) dx\right]\right\|\Bigg] \tag{2.39}$$

$$\leq \frac{1}{\Gamma((n+1)\alpha)} \sum_{i=1}^r \Bigg[\Bigg[\int_a^{x_0} \Bigg\|\left(\prod_{\substack{j=1 \\ j\neq i}}^r f_j(x)\right)$$

$$\left(\int_x^{x_0} (g(t) - g(x))^{(n+1)\alpha-1} g'(t) \left(D_{x_0-;g}^{(n+1)\alpha} f_i\right)(t)\, dt\right)\Bigg\| dx\Bigg] +$$

$$\left[\int_{x_0}^b \left\|\left(\prod_{\substack{j=1 \\ j\neq i}}^r f_j(x)\right) \left(\int_{x_0}^x (g(x) - g(t))^{(n+1)\alpha-1} g'(t) \left(D_{x_0+;g}^{(n+1)\alpha} f_i\right)(t)\, dt\right)\right\| dx\right]\Bigg]$$

$$\leq \frac{1}{\Gamma((n+1)\alpha)} \sum_{i=1}^r \Bigg[\Bigg[\int_a^{x_0} \left(\prod_{\substack{j=1 \\ j\neq i}}^r \|f_j(x)\|\right)$$

$$\left(\int_x^{x_0} (g(t) - g(x))^{(n+1)\alpha-1} g'(t) \left\|\left(D_{x_0-;g}^{(n+1)\alpha} f_i\right)(t)\right\| dt\right) dx\Bigg] +$$

$$\left[\int_{x_0}^{b}\left(\prod_{\substack{j=1\\ j\neq i}}^{r}\|f_j(x)\|\right)\left(\int_{x_0}^{x}(g(x)-g(t))^{(n+1)\alpha-1}g'(t)\left\|\left(D_{x_0+;g}^{(n+1)\alpha}f_i\right)(t)\right\|dt\right)dx\right]\Bigg]$$

$$=: (*) . \tag{2.40}$$

Hence it holds

$$\|\Omega(f_1, ..., f_r)(x_0)\| \leq (*) . \tag{2.41}$$

We have that

$$(*) \leq \frac{1}{\Gamma((n+1)\alpha+1)}$$

$$\sum_{i=1}^{r}\left[\left[\left\|\left(D_{x_0-;g}^{(n+1)\alpha}f_i\right)\right\|_{\infty,[a,x_0]}\int_a^{x_0}\left(\prod_{\substack{j=1\\ j\neq i}}^{r}\|f_j(x)\|\right)(g(x_0)-g(x))^{(n+1)\alpha}dx\right]\right.$$

$$\left.+\left[\left\|\left(D_{x_0+;g}^{(n+1)\alpha}f_i\right)\right\|_{\infty,[x_0,b]}\int_{x_0}^{b}\left(\prod_{\substack{j=1\\ j\neq i}}^{r}\|f_j(x)\|\right)(g(x)-g(x_0))^{(n+1)\alpha}dx\right]\right] \leq$$

$$\frac{1}{\Gamma((n+1)\alpha+1)}\sum_{i=1}^{r}\left[\left[\left\|\left(D_{x_0-;g}^{(n+1)\alpha}f_i\right)\right\|_{\infty,[a,x_0]}\right.\right.$$

$$\left.(g(x_0)-g(a))^{(n+1)\alpha}\left(\int_a^{x_0}\left(\prod_{\substack{j=1\\ j\neq i}}^{r}\|f_j(x)\|\right)dx\right)\right]+ \tag{2.42}$$

$$\left.\left[\left\|\left(D_{x_0+;g}^{(n+1)\alpha}f_i\right)\right\|_{\infty,[x_0,b]}(g(b)-g(x_0))^{(n+1)\alpha}\left(\int_{x_0}^{b}\left(\prod_{\substack{j=1\\ j\neq i}}^{r}\|f_j(x)\|\right)dx\right)\right]\right],$$

proving (2.29). ∎

Next comes an L_1 estimate.

Theorem 2.18 *All as in Theorem 2.17, plus* $\frac{1}{(n+1)} \leq \alpha < 1$, $n \in \mathbb{N}$. *Then*

$$\|\Omega(f_1, ..., f_r)(x_0)\| \leq \frac{\|g'\|_{\infty,[a,b]}}{\Gamma((n+1)\alpha)}$$

$$\sum_{i=1}^{r}\left[\left[\left\|\left(D_{x_0-;g}^{(n+1)\alpha}f_i\right)\right\|_{L_1([a,x_0])}(g(x_0)-g(a))^{(n+1)\alpha-1}\left(\int_a^{x_0}\left(\prod_{\substack{j=1\\j\neq i}}^{r}\|f_j(x)\|\right)dx\right)\right]\right.$$

$$\left.+\left[\left\|\left(D_{x_0+;g}^{(n+1)\alpha}f_i\right)\right\|_{L_1([x_0,b])}(g(b)-g(x_0))^{(n+1)\alpha-1}\int_{x_0}^{b}\left(\prod_{\substack{j=1\\j\neq i}}^{r}\|f_j(x)\|\right)dx\right]\right]. \tag{2.43}$$

Proof We have that (by (2.40), (2.41))

$$(*)\leq\frac{\|g'\|_{\infty,[a,b]}}{\Gamma((n+1)\alpha)}$$

$$\sum_{i=1}^{r}\left[\left[\left\|\left(D_{x_0-;g}^{(n+1)\alpha}f_i\right)\right\|_{L_1([a,x_0])}\int_a^{x_0}\left(\prod_{\substack{j=1\\j\neq i}}^{r}\|f_j(x)\|\right)(g(x_0)-g(x))^{(n+1)\alpha-1}dx\right]\right.$$

$$\left.+\left[\left\|\left(D_{x_0+;g}^{(n+1)\alpha}f_i\right)\right\|_{L_1([x_0,b])}\int_{x_0}^{b}\left(\prod_{\substack{j=1\\j\neq i}}^{r}\|f_j(x)\|\right)(g(x)-g(x_0))^{(n+1)\alpha-1}dx\right]\right]$$

$$\leq\frac{\|g'\|_{\infty,[a,b]}}{\Gamma((n+1)\alpha)}$$

$$\sum_{i=1}^{r}\left[\left[\left\|\left(D_{x_0-;g}^{(n+1)\alpha}f_i\right)\right\|_{L_1([a,x_0])}(g(x_0)-g(a))^{(n+1)\alpha-1}\left(\int_a^{x_0}\left(\prod_{\substack{j=1\\j\neq i}}^{r}\|f_j(x)\|\right)dx\right)\right]\right. \tag{2.44}$$

$$\left.+\left[\left\|\left(D_{x_0+;g}^{(n+1)\alpha}f_i\right)\right\|_{L_1([x_0,b])}(g(b)-g(x_0))^{(n+1)\alpha-1}\left(\int_{x_0}^{b}\left(\prod_{\substack{j=1\\j\neq i}}^{r}\|f_j(x)\|\right)dx\right)\right]\right],$$

proving the claim. ■

An L_p estimate follows.

Theorem 2.19 *All as in Theorem 2.17, plus* $p,q>1:\frac{1}{p}+\frac{1}{q}=1$, *and* $\frac{1}{q(n+1)}<\alpha<1$. *Then*

$$\left\|\Omega\left(f_1,\ldots,f_r\right)(x_0)\right\| \leq \frac{1}{\Gamma\left((n+1)\alpha\right)\left(p\left((n+1)\alpha-1\right)+1\right)^{\frac{1}{p}}}$$

$$\sum_{i=1}^{r}\left[\left[\left\|\left(D_{x_0-;g}^{(n+1)\alpha}f_i\right)\circ g^{-1}\right\|_{q,[g(a),g(x_0)]}\right.\right.$$

$$\left.(g(x_0)-g(a))^{(n+1)\alpha-\frac{1}{q}}\left(\int_a^{x_0}\left(\prod_{\substack{j=1\\j\neq i}}^{r}\left\|f_j(x)\right\|\right)dx\right)\right]+$$

$$\left[\left\|\left(D_{x_0+;g}^{(n+1)\alpha}f_i\right)\circ g^{-1}\right\|_{q,[g(x_0),g(b)]}(g(b)-g(x_0))^{(n+1)\alpha-\frac{1}{q}}\right.$$

$$\left.\left.\left(\int_{x_0}^{b}\left(\prod_{\substack{j=1\\j\neq i}}^{r}\left\|f_j(x)\right\|\right)dx\right)\right]\right]. \tag{2.45}$$

Proof By (2.40) and (2.41) we obtain

$$(*)=\frac{1}{\Gamma\left((n+1)\alpha\right)}\sum_{i=1}^{r}\left[\left[\int_a^{x_0}\left(\prod_{\substack{j=1\\j\neq i}}^{r}\left\|f_j(x)\right\|\right)\right.\right.$$

$$\left.\left(\int_x^{x_0}(g(t)-g(x))^{(n+1)\alpha-1}\left\|\left(\left(D_{x_0-;g}^{(n+1)\alpha}f_i\right)\circ g^{-1}\right)(g(t))\right\|dg(t)\right)dx\right]+$$

$$\left[\int_{x_0}^{b}\left(\prod_{\substack{j=1\\j\neq i}}^{r}\left\|f_j(x)\right\|\right)\right.$$

$$\left.\left.\left(\int_{x_0}^{x}(g(x)-g(t))^{(n+1)\alpha-1}\left\|\left(\left(D_{x_0+;g}^{(n+1)\alpha}f_i\right)\circ g^{-1}\right)(g(t))\right\|dg(t)\right)dx\right]\right] \tag{2.46}$$

$$=\frac{1}{\Gamma\left((n+1)\alpha\right)}\sum_{i=1}^{r}\left[\left[\int_a^{x_0}\left(\prod_{\substack{j=1\\j\neq i}}^{r}\left\|f_j(x)\right\|\right)\right.\right.$$

$$\left(\int_{g(x)}^{g(x_0)} (z - g(x))^{(n+1)\alpha-1} \left\| \left(\left(D_{x_0-;g}^{(n+1)\alpha} f_i \right) \circ g^{-1} \right)(z) \right\| dz \right) dx \Bigg] +$$

$$\left[\int_{x_0}^{b} \left(\prod_{\substack{j=1 \\ j \neq i}}^{r} \| f_j(x) \| \right) \left(\int_{g(x_0)}^{g(x)} (g(x) - z)^{(n+1)\alpha-1} \left\| \left(\left(D_{x_0+;g}^{(n+1)\alpha} f_i \right) \circ g^{-1} \right)(z) \right\| dz \right) dx \right]\Bigg]$$

$$\leq \frac{1}{\Gamma((n+1)\alpha)} \sum_{i=1}^{r} \left[\left[\int_{a}^{x_0} \left(\prod_{\substack{j=1 \\ j \neq i}}^{r} \| f_j(x) \| \right) \left(\int_{g(x)}^{g(x_0)} (z - g(x))^{p((n+1)\alpha-1)} dz \right)^{\frac{1}{p}} \right.\right.$$

$$\left(\int_{g(x)}^{g(x_0)} \left\| \left(\left(D_{x_0-;g}^{(n+1)\alpha} f_i \right) \circ g^{-1} \right)(z) \right\|^q dz \right)^{\frac{1}{q}} dx \Bigg] +$$

$$\left[\int_{x_0}^{b} \left(\prod_{\substack{j=1 \\ j \neq i}}^{r} \| f_j(x) \| \right) \left(\int_{g(x_0)}^{g(x)} (g(x) - z)^{p((n+1)\alpha-1)} dz \right)^{\frac{1}{p}} \right.$$

$$\left.\left. \left(\int_{g(x_0)}^{g(x)} \left\| \left(\left(D_{x_0+;g}^{(n+1)\alpha} f_i \right) \circ g^{-1} \right)(z) \right\|^q dz \right)^{\frac{1}{q}} dx \right]\right] \leq \tag{2.47}$$

$$\frac{1}{\Gamma((n+1)\alpha)(p((n+1)\alpha - 1) + 1)^{\frac{1}{p}}}$$

$$\sum_{i=1}^{r} \left[\left[\left\| \left(\left(D_{x_0-;g}^{(n+1)\alpha} f_i \right) \circ g^{-1} \right) \right\|_{q,[g(a),g(x_0)]} \right.\right.$$

$$\left(\int_{a}^{x_0} \left(\prod_{\substack{j=1 \\ j \neq i}}^{r} \| f_j(x) \| \right) (g(x_0) - g(x))^{\frac{p((n+1)\alpha-1)+1}{p}} dx \right) \Bigg] +$$

$$\left[\left\| \left(\left(D_{x_0+;g}^{(n+1)\alpha} f_i \right) \circ g^{-1} \right) \right\|_{q,[g(x_0),g(b)]} \right.$$

$$\left.\left. \left(\int_{x_0}^{b} \left(\prod_{\substack{j=1 \\ j \neq i}}^{r} \| f_j(x) \| \right) (g(x) - g(x_0))^{\frac{p((n+1)\alpha-1)+1}{p}} dx \right) \right]\right]$$

$$\leq \frac{1}{\Gamma\left(\left(n+1\right)\alpha\right)\left(p\left(\left(n+1\right)\alpha-1\right)+1\right)^{\frac{1}{p}}}$$

$$\sum_{i=1}^{r}\left[\left\|\left(D_{x_0-;g}^{(n+1)\alpha}f_i\right)\circ g^{-1}\right\|_{q,[g(a),g(x_0)]}\right.$$

$$\left(g\left(x_0\right)-g\left(a\right)\right)^{(n+1)\alpha-\frac{1}{q}}\left(\int_a^{x_0}\left(\prod_{\substack{j=1\\j\neq i}}^{r}\left\|f_j\left(x\right)\right\|\right)dx\right)$$

$$+\left\|\left(D_{x_0+;g}^{(n+1)\alpha}f_i\right)\circ g^{-1}\right\|_{q,[g(x_0),g(b)]}\left(g\left(b\right)-g\left(x_0\right)\right)^{(n+1)\alpha-\frac{1}{q}}$$

$$\left.\left(\int_{x_0}^{b}\left(\prod_{\substack{j=1\\j\neq i}}^{r}\left\|f_j\left(x\right)\right\|\right)dx\right)\right], \tag{2.48}$$

proving the claim. ■

When $n=1$ we get the following results:

Corollary 2.20 (to Theorem 2.17) *Let $(A,\|\cdot\|)$ be a Banach algebra, $x_0\in[a,b]\subset\mathbb{R}$, $0<\alpha<1$, $f_i\in C^1([a,b],A)$, $i=1,...,r$. Let $g\in C^1([a,b])$ and strictly increasing, such that $g^{-1}\in C^1([g(a),g(b)])$. Assume that $D_{x_0-;g}^{\alpha}f_i\in C^1([a,x_0],A)$ and $D_{x_0-;g}^{2\alpha}f_i\in C([a,x_0],A)$; $i=1,...,r$. Similarly, assume that $D_{x_0+;g}^{\alpha}f_i\in C^1([x_0,b],A)$ and $D_{x_0+;g}^{2\alpha}f_i\in C([x_0,b],A)$; $i=1,...,r$.*

Denote by

$$\Omega\left(f_1,...,f_r\right)\left(x_0\right):=$$

$$\sum_{i=1}^{r}\left[\int_a^b\left(\prod_{\substack{j=1\\j\neq i}}^{r}f_j\left(x\right)\right)f_i\left(x\right)dx-\left(\int_a^b\left(\prod_{\substack{j=1\\j\neq i}}^{r}f_j\left(x\right)\right)dx\right)f_i\left(x_0\right)\right]. \tag{2.49}$$

Then

$$\left\|\Omega\left(f_1,...,f_r\right)\left(x_0\right)\right\|\leq\frac{1}{\Gamma\left(2\alpha+1\right)}\sum_{i=1}^{r}\left[\left[\left\|\left(D_{x_0-;g}^{2\alpha}f_i\right)\right\|_{\infty,[a,x_0]}\right.\right.$$

$$\left.\left(g\left(x_0\right)-g\left(a\right)\right)^{2\alpha}\left(\int_a^{x_0}\left(\prod_{\substack{j=1\\j\neq i}}^{r}\left\|f_j\left(x\right)\right\|\right)dx\right)\right]+ \tag{2.50}$$

$$\left[\left\|\left\|\left(D^{2\alpha}_{x_0+;g}f_i\right)\right\|\right\|_{\infty,[x_0,b]}(g(b)-g(x_0))^{2\alpha}\left(\int_{x_0}^{b}\left(\prod_{\substack{j=1\\j\neq i}}^{r}\|f_j(x)\|\right)dx\right)\right]\Bigg] =: \Lambda_1(x_0).$$

Proof By Theorem 2.17 and Corollaries 2.10, 2.12. ∎

We continue with

Corollary 2.21 (to Theorem 2.18) *All as in Corollary 2.20, plus* $\frac{1}{2} \leq \alpha < 1$. *Then*

$$\|\Omega(f_1,...,f_r)(x_0)\| \leq \frac{\|g'\|_{\infty,[a,b]}}{\Gamma(2\alpha)}\sum_{i=1}^{r}\Bigg[\left[\left\|\left\|\left(D^{2\alpha}_{x_0-;g}f_i\right)\right\|\right\|_{L_1([a,x_0])}\right.$$

$$\left.(g(x_0)-g(a))^{2\alpha-1}\left(\int_a^{x_0}\left(\prod_{\substack{j=1\\j\neq i}}^{r}\|f_j(x)\|\right)dx\right)\right]+$$

$$\left[\left\|\left\|\left(D^{2\alpha}_{x_0+;g}f_i\right)\right\|\right\|_{L_1([x_0,b])}(g(b)-g(x_0))^{2\alpha-1}\left(\int_{x_0}^{b}\left(\prod_{\substack{j=1\\j\neq i}}^{r}\|f_j(x)\|\right)dx\right)\right]\Bigg] = \Lambda_2(x_0). \tag{2.51}$$

Proof By Theorem 2.18. ∎

We also give

Corollary 2.22 (to Theorem 2.19) *All as in Corollary 2.20, plus* $p, q > 1 : \frac{1}{p} + \frac{1}{q} = 1$, *and* $\frac{1}{2q} < \alpha < 1$. *Then*

$$\|\Omega(f_1,...,f_r)(x_0)\| \leq \frac{1}{\Gamma(2\alpha)(p(2\alpha-1)+1)^{\frac{1}{p}}}$$

$$\sum_{i=1}^{r}\Bigg[\left[\left\|\left\|\left(D^{2\alpha}_{x_0-;g}f_i\right)\circ g^{-1}\right\|\right\|_{q,[g(a),g(x_0)]}\right.$$

$$\left.(g(x_0)-g(a))^{2\alpha-\frac{1}{q}}\left(\int_a^{x_0}\left(\prod_{\substack{j=1\\j\neq i}}^{r}\|f_j(x)\|\right)dx\right)\right]+$$

$$\left[\left\|\left\|\left(D^{2\alpha}_{x_0+;g}f_i\right)\circ g^{-1}\right\|\right\|_{q,[g(x_0),g(b)]}(g(b)-g(x_0))^{2\alpha-\frac{1}{q}}\right.$$

$$\left(\int_{x_0}^{b}\left(\prod_{\substack{j=1\\ j\neq i}}^{r}\left\|f_j(x)\right\|\right)dx\right)\Bigg]\Bigg] = \Lambda_3(x_0). \tag{2.52}$$

Proof By Theorem 2.19. ■

We make

Remark 2.23 (1) Call and assume

$$N_1(f_1, ..., f_r) := \max_{i=1,...,r}\left\{\sup_{x_0\in[a,b]}\left\|\left\|\left(D_{x_0-;g}^{2\alpha}f_i\right)\right\|\right\|_{\infty,[a,x_0]},\right.$$

$$\left.\sup_{x_0\in[a,b]}\left\|\left\|\left(D_{x_0+;g}^{2\alpha}f_i\right)\right\|\right\|_{\infty,[x_0,b]}\right\} < \infty. \tag{2.53}$$

Then (by (2.50))

$$\Lambda_1(x_0) \leq \frac{(g(b)-g(a))^{2\alpha}N_1(f_1,...,f_r)}{\Gamma(2\alpha+1)}\sum_{i=1}^{r}\left(\int_a^b\left(\prod_{\substack{j=1\\ j\neq i}}^{r}\left\|f_j(x)\right\|\right)dx\right). \tag{2.54}$$

(2) Call and assume

$$N_2(f_1, ..., f_r) := \max_{i=1,...,r}\left\{\sup_{x_0\in[a,b]}\left\|\left\|\left(D_{x_0-;g}^{2\alpha}f_i\right)\right\|\right\|_{L_1([a,x_0])},\right.$$

$$\left.\sup_{x_0\in[a,b]}\left\|\left\|\left(D_{x_0+;g}^{2\alpha}f_i\right)\right\|\right\|_{L_1([x_0,b])}\right\} < \infty. \tag{2.55}$$

Then (by (2.51))

$$\Lambda_2(x_0) \leq \frac{\left\|g'\right\|_{\infty,[a,b]}(g(b)-g(a))^{2\alpha-1}N_2(f_1,...,f_r)}{\Gamma(2\alpha)}$$

$$\sum_{i=1}^{r}\left(\int_a^b\left(\prod_{\substack{j=1\\ j\neq i}}^{r}\left\|f_j(x)\right\|\right)dx\right). \tag{2.56}$$

(3) Call and assume

$$N_3(f_1, ..., f_r) := \max_{i=1,...,r} \left\{ \sup_{x_0 \in [a,b]} \left\| \left\| \left(D^{2\alpha}_{x_0-;g} f_i \right) \circ g^{-1} \right\| \right\|_{q,[g(a),g(x_0)]}, \right.$$

$$\left. \sup_{x_0 \in [a,b]} \left\| \left\| \left(D^{2\alpha}_{x_0+;g} f_i \right) \circ g^{-1} \right\| \right\|_{q,[g(x_0),g(b)]} \right\} < \infty. \tag{2.57}$$

Then (by (2.52))

$$\Lambda_3(x_0) \leq \frac{(g(b) - g(a))^{2\alpha - \frac{1}{q}} N_3(f_1, ..., f_r)}{\Gamma(2\alpha)(p(2\alpha - 1) + 1)^{\frac{1}{p}}} \sum_{i=1}^{r} \left(\int_a^b \left(\prod_{\substack{j=1 \\ j \neq i}}^{r} \| f_j(x) \| \right) dx \right). \tag{2.58}$$

Next denote by

$$\Phi(f_1, ..., f_r) := \int_a^b \Omega(f_1, ..., f_r)(x_0) \, dx_0 =$$

$$\sum_{i=1}^{r} \left[(b - a) \left(\int_a^b \left(\prod_{\substack{j=1 \\ j \neq i}}^{r} f_j(x) \right) f_i(x) \, dx \right) - \right.$$

$$\left. \left(\int_a^b \left(\prod_{\substack{j=1 \\ j \neq i}}^{r} f_j(x) \right) dx \right) \left(\int_a^b f_i(x) \, dx \right) \right]. \tag{2.59}$$

Furthermore, it holds (see (2.50)–(2.52))

$$\|\Phi(f_1, ..., f_r)\| \leq \int_a^b \|\Omega(f_1, ..., f_r)(x_0)\| \, dx_0 \leq \int_a^b \Lambda_i(x_0) \, dx_0, \ i = 1, 2, 3, \tag{2.60}$$

under proper assumptions, respectively, and

$$\|\Phi(f_1, ..., f_r)\| =$$

$$\left\| \sum_{i=1}^{r} \left[(b - a) \left(\int_a^b \left(\prod_{\substack{j=1 \\ j \neq i}}^{r} f_j(x) \right) f_i(x) \, dx \right) - \right. \right.$$

$$\left(\int_a^b \left(\prod_{\substack{j=1 \\ j\neq i}}^r f_j(x)\right) dx\right)\left(\int_a^b f_i(x)\,dx\right)\Bigg]\Bigg\|. \tag{2.61}$$

Therefore, by (2.54) we get that

$$\|\Phi(f_1,...,f_r)\| \leq \frac{(b-a)(g(b)-g(a))^{2\alpha} N_1(f_1,...,f_r)}{\Gamma(2\alpha+1)}$$

$$\sum_{i=1}^r \left(\int_a^b \left(\prod_{\substack{j=1 \\ j\neq i}}^r \|f_j(x)\|\right) dx\right), \tag{2.62}$$

and by (2.56) we obtain

$$\|\Phi(f_1,...,f_r)\| \leq \frac{\|g'\|_{\infty,[a,b]}(b-a)(g(b)-g(a))^{2\alpha-1} N_2(f_1,...,f_r)}{\Gamma(2\alpha)}$$

$$\sum_{i=1}^r \left(\int_a^b \left(\prod_{\substack{j=1 \\ j\neq i}}^r \|f_j(x)\|\right) dx\right), \tag{2.63}$$

and finally by (2.58) we derive that

$$\|\Phi(f_1,...,f_r)\| \leq \frac{(b-a)(g(b)-g(a))^{2\alpha-\frac{1}{q}} N_3(f_1,...,f_r)}{\Gamma(2\alpha)(p(2\alpha-1)+1)^{\frac{1}{p}}}$$

$$\sum_{i=1}^r \left(\int_a^b \left(\prod_{\substack{j=1 \\ j\neq i}}^r \|f_j(x)\|\right) dx\right), \tag{2.64}$$

all the above are valid under proper assumptions, respectively.

We have proved the following generalized Grüss-Cebysev type inequalities for several functions over a Banach algebra. A uniform estimate follows.

Theorem 2.24 *Let $(A, \|\cdot\|)$ be a Banach algebra, $[a,b]\subset\mathbb{R}$, $0<\alpha<1$, $f_i \in C^1([a,b],A)$, $i=1,...,r$. Let $g\in C^1([a,b])$ and strictly increasing, such that $g^{-1}\in C^1([g(a),g(b)])$. Assume that for every $x_0\in[a,b]$, $D^{\alpha}_{x_0-;g}f_i\in C^1([a,x_0],A)$ and $D^{2\alpha}_{x_0-;g}f_i\in C([a,x_0],A)$; $i=1,...,r$. Similarly, assume that for every $x_0\in[a,b]$, $D^{\alpha}_{x_0+;g}f_i\in C^1([x_0,b],A)$ and $D^{2\alpha}_{x_0+;g}f_i\in C([x_0,b],A)$;*

$i = 1, ..., r$. Here $N_1(f_1, ..., f_r)$ is as in (2.53). Then

$$\left\| \sum_{i=1}^{r} \left[(b-a) \left(\int_a^b \left(\prod_{\substack{j=1 \\ j \neq i}}^{r} f_j(x) \right) f_i(x)\, dx \right) - \left(\int_a^b \left(\prod_{\substack{j=1 \\ j \neq i}}^{r} f_j(x) \right) dx \right) \left(\int_a^b f_i(x)\, dx \right) \right] \right\| \leq$$

$$\frac{(b-a)(g(b)-g(a))^{2\alpha} N_1(f_1, ..., f_r)}{\Gamma(2\alpha+1)} \sum_{i=1}^{r} \left(\int_a^b \left(\prod_{\substack{j=1 \\ j \neq i}}^{r} \|f_j(x)\| \right) dx \right). \quad (2.65)$$

An L_1 estimate follows.

Theorem 2.25 *All as in Theorem 2.24, $\frac{1}{2} \leq \alpha < 1$. Here $N_2(f_1, ..., f_r)$ is as in (2.55). Then*

$$\left\| \sum_{i=1}^{r} \left[(b-a) \left(\int_a^b \left(\prod_{\substack{j=1 \\ j \neq i}}^{r} f_j(x) \right) f_i(x)\, dx \right) - \left(\int_a^b \left(\prod_{\substack{j=1 \\ j \neq i}}^{r} f_j(x) \right) dx \right) \left(\int_a^b f_i(x)\, dx \right) \right] \right\| \leq$$

$$\frac{\|g'\|_{\infty,[a,b]} (b-a)(g(b)-g(a))^{2\alpha-1} N_2(f_1, ..., f_r)}{\Gamma(2\alpha)} \sum_{i=1}^{r} \left(\int_a^b \left(\prod_{\substack{j=1 \\ j \neq i}}^{r} \|f_j(x)\| \right) dx \right). \quad (2.66)$$

An L_p estimate follows.

Theorem 2.26 *All as in Theorem 2.24, $p, q > 1 : \frac{1}{p} + \frac{1}{q} = 1$ and $\frac{1}{2q} < \alpha < 1$. Here $N_3(f_1, ..., f_r)$ is as in (2.57). Then*

$$\left\| \sum_{i=1}^{r} \left[(b-a) \left(\int_a^b \left(\prod_{\substack{j=1 \\ j \neq i}}^{r} f_j(x) \right) f_i(x)\, dx \right) - \right.\right.$$

$$\left(\int_a^b \left(\prod_{\substack{j=1\\j\neq i}}^r f_j(x)\right) dx\right)\left(\int_a^b f_i(x)\,dx\right)\Bigg]\Bigg\| \leq$$

$$\frac{(b-a)(g(b)-g(a))^{2\alpha-\frac{1}{q}} N_3(f_1,\ldots,f_r)}{\Gamma(2\alpha)(p(2\alpha-1)+1)^{\frac{1}{p}}} \sum_{i=1}^r \left(\int_a^b \left(\prod_{\substack{j=1\\j\neq i}}^r \|f_j(x)\|\right) dx\right). \tag{2.67}$$

2.5 Applications

We make

Remark 2.27 From now on we assume that $(A, \|\cdot\|)$ is a commutative Banach algebra. Then, we get that

$$\Omega(f_1,\ldots,f_r)(x_0) \overset{(2.28.)}{=} r\int_a^b \left(\prod_{j=1}^r f_j(x)\right) dx - \sum_{i=1}^r \left(\int_a^b \left(\prod_{\substack{j=1\\j\neq i}}^r f_j(x)\right) dx\right) f_i(x_0), \tag{2.68}$$

$x_0 \in [a,b]$.

Furthermore, it holds (case of $n=1$)

$$\Phi(f_1,\ldots,f_r) \overset{(2.59.)}{=} r(b-a)\int_a^b \left(\prod_{j=1}^r f_j(x)\right) dx -$$

$$\sum_{i=1}^r \left[\left(\int_a^b \left(\prod_{\substack{j=1\\j\neq i}}^r f_j(x)\right) dx\right)\left(\int_a^b f_i(x)\,dx\right)\right]. \tag{2.69}$$

When $r=2$, we get that

$$\Omega(f_1,f_2)(x_0) = 2\int_a^b f_1(x) f_2(x)\,dx - f_1(x_0)\int_a^b f_2(x)\,dx - f_2(x_0)\int_a^b f_1(x)\,dx, \tag{2.70}$$

$x_0 \in [a,b]$,

and

$$\Phi(f_1, f_2) = 2\left[(b-a)\int_a^b f_1(x) f_2(x)\,dx - \left(\int_a^b f_1(x)\,dx\right)\left(\int_a^b f_2(x)\,dx\right)\right], \tag{2.71}$$

case of $n = 1$.

We give

Corollary 2.28 *All as in Theorem 2.17, A is a commutative Banach algebra, $r = 2$. Then*

$$\|\Omega(f_1, f_2)(x_0)\| \leq \frac{1}{\Gamma((n+1)\alpha+1)} \sum_{i=1}^{2} \left[\left[\left\| \left(D_{x_0-;g}^{(n+1)\alpha} f_i\right)\right\|_{\infty,[a,x_0]} \right.\right.$$

$$\left.(g(x_0) - g(a))^{(n+1)\alpha} \left(\int_a^{x_0} \left(\prod_{\substack{j=1\\ j\neq i}}^{2} \|f_j(x)\|\right) dx\right)\right] + \tag{2.72}$$

$$\left.\left[\left\| \left(D_{x_0+;g}^{(n+1)\alpha} f_i\right)\right\|_{\infty,[x_0,b]} (g(b) - g(x_0))^{(n+1)\alpha} \left(\int_{x_0}^{b} \left(\prod_{\substack{j=1\\ j\neq i}}^{2} \|f_j(x)\|\right) dx\right)\right]\right].$$

Proof By Theorem 2.17. ■

We continue with

Corollary 2.29 *All as in Theorem 2.24, $r = 2$, $n = 1$, $N_1(f_1, f_2)$ as in (2.53); A is a commutative Banach algebra. Then*

$$\|\Phi(f_1, f_2)\| \leq \frac{(b-a)(g(b) - g(a))^{2\alpha} N_1(f_1, f_2)}{\Gamma(2\alpha+1)} \left(\int_a^b (\|f_1(x)\| + \|f_2(x)\|)\,dx\right). \tag{2.73}$$

Proof By Theorem 2.24. ■

We finish with

Corollary 2.30 (to Corollary 2.28) *All as in Corollary 2.28, for $g(t) = e^t$. Then*

$$\|\Omega(f_1, f_2)(x_0)\| \leq \frac{1}{\Gamma((n+1)\alpha+1)} \sum_{i=1}^{2} \left[\left[\left\| \left(D_{x_0-;e^t}^{(n+1)\alpha} f_i\right)\right\|_{\infty,[a,x_0]} \right.\right.$$

$$\left.\left(e^{x_0} - e^{a}\right)^{(n+1)\alpha} \left(\int_a^{x_0} \left(\prod_{\substack{j=1\\ j\neq i}}^{2} \|f_j(x)\|\right) dx\right)\right] +$$

$$\left[\left\| \left\| \left(D_{x_0+;e^t}^{(n+1)\alpha} f_i \right) \right\| \right\|_{\infty,[x_0,b]} \left(e^b - e^{x_0} \right)^{(n+1)\alpha} \left(\int_{x_0}^b \left(\prod_{\substack{j=1 \\ j \neq i}}^2 \| f_j(x) \| \right) dx \right) \right] \right]. \tag{2.74}$$

Proof By Corollary 2.28. ∎

References

1. Anastassiou, G.A.: Fractional Differentiation Inequalities. Research Monograph. Springer, New York (2009)
2. Anastassiou, G.A.: Advances on Fractional Inequalities. Research Monograph. Springer, New York (2011)
3. Anastassiou, G.A.: Intelligent Comparisons: Analytic Inequalities. Springer, Heidelberg, New York (2016)
4. Anastassiou, G.A.: Intelligent Computations: Abstract Fractional Calculus, Inequalities, Approximations. Springer, Heidelberg, New York (2018)
5. Anastassiou, G.A.: Sequential generalized fractional Ostrowski and Grüss type inequalities for several Banach algebra valued functions. Ann. Commun. Math. **4**(3), 207–225 (2021)
6. Čebyšev: Sur les expressions approximatives des intégrales définies par les aures prises entre les mêmes limites. Proc. Math. Soc. Charkov **2**, 93–98 (1882)
7. Dragomir, S.S.: Noncommutative Ostrowski type inequalities for functions in Banach algebras. RGMIA Res. Rep. Coll. **24**, Art. 10, 24 pp (2021)
8. Dragomir, S.S.: New Noncommuttive Grüss type inequalities for functions in Banach algebras. RGMIA Res. Rep. Coll. **24**, Art. 13, 19 pp (2021)
9. Mikusinski, J.: The Bochner Integral. Academic Press, New York (1978)
10. Ostrowski, A.: Über die Absolutabweichung einer differentiabaren Funcktion von ihrem Integralmittelwert. Comment. Math. Helv. **10**, 226–227 (1938)
11. Rudin, W.: Functional Analysis, 2nd edn. McGraw-Hill Inc, New York (1991)

Chapter 3
Generalized Canavati Fractional Ostrowski, Opial and Grüss Inequalities with Multiple Banach Algebra Valued Functions

Using generalized Canavati fractional left and right vectorial Taylor formulae we establish mixed fractional Ostrowski, Opial and Grüss inequalities involving several Banach algebra valued functions. The estimates are with respect to all norms $\|\cdot\|_p$, $1 \leq p \leq \infty$. We provide also applications. It follows [6].

3.1 Introduction

The following result motivates our chapter.

Theorem 3.1 (1938, Ostrowski [10]) *Let $f : [a, b] \to \mathbb{R}$ be continuous on $[a, b]$ and differentiable on (a, b) whose derivative $f' : (a, b) \to \mathbb{R}$ is bounded on (a, b), i.e., $\|f'\|_{\infty}^{\sup} := \sup_{t \in (a,b)} |f'(t)| < +\infty$. Then*

$$\left| \frac{1}{b-a} \int_a^b f(t)\, dt - f(x) \right| \leq \left[\frac{1}{4} + \frac{\left(x - \frac{a+b}{2}\right)^2}{(b-a)^2} \right] (b-a) \|f'\|_{\infty}^{\sup}, \tag{3.1}$$

for any $x \in [a, b]$. The constant $\frac{1}{4}$ is the best possible.

Ostrowski type inequalities have great applications to integral approximations in Numerical Analysis.

We presented also ([3], Chaps. 8, 9) mixed fractional Ostrowski and Grüss inequalities for several functions for various norms.

See also the monographs written by the author [1], Chaps. 24–26 and [2], Chaps. 2–6.

G. A. Anastassiou, *Fractional Inequalities In Banach Algebras*,
Studies in Systems, Decision and Control 441,
https://doi.org/10.1007/978-3-031-05148-7_3

In this chapter we generalize [3], Chaps. 8, 9 for several Banach algebra valued functions by employing their Canavati type generalized left and right fractional derivatives and our integrals are of Bochner type [8]. Motivation came also from [7].

We are also inspired by Z. Opial [9], 1960, famous inequality.

Theorem 3.2 *Let $x(t) \in C^1([0,h])$ be such that $x(0) = x(h) = 0$, and $x(t) > 0$ in $(0,h)$. Then*

$$\int_0^h |x(t)\, x'(t)|\, dt \leq \frac{h}{4} \int_0^h \left(x'(t)\right)^2 dt. \tag{3.2}$$

In (3.2), the constant $\frac{h}{4}$ is the best possible. Inequality (3.2) holds as equality for the optimal function

$$x(t) = \begin{cases} ct, & 0 \leq t \leq \frac{h}{2}, \\ c(h-t), & \frac{h}{2} \leq t \leq h, \end{cases} \tag{3.3}$$

where $c > 0$ is an arbitrary constant.

Opial-type inequalities are used a lot in proving uniqueness of solutions to differential equations and also to give upper bounds to their solutions.

For an extensive study about fractional Opial inequalities see the author's monograph [1].

In this chapter we also derive Opial type inequalities for Banach algebra valued functions with respect to their Canavati type generalized left and right fractional derivatives.

We include applications for Ostrowski and Opial inequalities. We finish the chapter with related Grüss type inequalities.

3.2 Background on Vectorial Generalized Canavati Fractional Calculus

All in this section come from [5], pp. 109–115 and [4].

Let $g : [a,b] \to \mathbb{R}$ be a strictly increasing function. such that $g \in C^1([a,b])$, and $g^{-1} \in C^n([g(a), g(b)])$, $n \in \mathbb{N}$, $(X, \|\cdot\|)$ is a Banach space. Let $f \in C^n([a,b], X)$, and call $l := f \circ g^{-1} : [g(a), g(b)] \to X$. It is clear that $l, l', ..., l^{(n)}$ are continuous functions from $[g(a), g(b)]$ into $f([a,b]) \subseteq X$.

Let $\nu \geq 1$ such that $[\nu] = n$, $n \in \mathbb{N}$ as above, where $[\cdot]$ is the integral part of the number.

Clearly when $0 < \nu < 1$, $[\nu] = 0$.

(I) Let $h \in C([g(a), g(b)], X)$, we define the left Riemann-Liouville Bochner fractional integral as

$$\left(J_\nu^{z_0} h\right)(z) := \frac{1}{\Gamma(\nu)} \int_{z_0}^{z} (z-t)^{\nu-1} h(t)\, dt, \tag{3.4}$$

for $g(a) \leq z_0 \leq z \leq g(b)$, where Γ is the gamma function; $\Gamma(\nu) = \int_0^\infty e^{-t} t^{\nu-1} dt$. We set $J_0^{z_0} h = h$.

Let $\alpha := \nu - [\nu]$ $(0 < \alpha < 1)$. We define the subspace $C_{g(x_0)}^{\nu}([g(a), g(b)], X)$ of $C^{[\nu]}([g(a), g(b)], X)$, where $x_0 \in [a, b]$ as:

$$C_{g(x_0)}^{\nu}([g(a), g(b)], X) =$$

$$\left\{ h \in C^{[\nu]}([g(a), g(b)], X) : J_{1-\alpha}^{g(x_0)} h^{([\nu])} \in C^1([g(x_0), g(b)], X) \right\}. \tag{3.5}$$

So let $h \in C_{g(x_0)}^{\nu}([g(a), g(b)], X)$, we define the left g-generalized X-valued fractional derivative of h of order ν, of Canavati type, over $[g(x_0), g(b)]$ as

$$D_{g(x_0)}^{\nu} h := \left(J_{1-\alpha}^{g(x_0)} h^{([\nu])} \right)'. \tag{3.6}$$

Clearly, for $h \in C_{g(x_0)}^{\nu}([g(a), g(b)], X)$, there exists

$$\left(D_{g(x_0)}^{\nu} h \right)(z) = \frac{1}{\Gamma(1-\alpha)} \frac{d}{dz} \int_{g(x_0)}^{z} (z-t)^{-\alpha} h^{([\nu])}(t) \, dt, \tag{3.7}$$

for all $g(x_0) \leq z \leq g(b)$.

In particular, when $f \circ g^{-1} \in C_{g(x_0)}^{\nu}([g(a), g(b)], X)$, we have that

$$\left(D_{g(x_0)}^{\nu} \left(f \circ g^{-1} \right) \right)(z) = \frac{1}{\Gamma(1-\alpha)} \frac{d}{dz} \int_{g(x_0)}^{z} (z-t)^{-\alpha} \left(f \circ g^{-1} \right)^{([\nu])}(t) \, dt, \tag{3.8}$$

for all $g(x_0) \leq z \leq g(b)$. We have that $D_{g(x_0)}^{n} \left(f \circ g^{-1} \right) = \left(f \circ g^{-1} \right)^{(n)}$ and $D_{g(x_0)}^{0} \left(f \circ g^{-1} \right) = f \circ g^{-1}$, see [4].

By [4], we have for $f \circ g^{-1} \in C_{g(x_0)}^{\nu}([g(a), g(b)], X)$, where $x_0 \in [a, b]$ the following left generalized g-fractional, of Canavati type,X-valued Taylor's formula:

Theorem 3.3 *Let $f \circ g^{-1} \in C_{g(x_0)}^{\nu}([g(a), g(b)], X)$, where $x_0 \in [a, b]$ is fixed.*

(i) If $\nu \geq 1$, then

$$f(x) - f(x_0) = \sum_{k=1}^{[\nu]-1} \frac{\left(f \circ g^{-1} \right)^{(k)}(g(x_0))}{k!} (g(x) - g(x_0))^k +$$

$$\frac{1}{\Gamma(\nu)} \int_{g(x_0)}^{g(x)} (g(x) - t)^{\nu-1} \left(D_{g(x_0)}^{\nu} \left(f \circ g^{-1} \right) \right)(t) \, dt, \tag{3.9}$$

for all $x_0 \leq x \leq b$.

(ii) If $0 < \nu < 1$, we get

$$f(x)=\frac{1}{\Gamma(\nu)}\int_{g(x_0)}^{g(x)}(g(x)-t)^{\nu-1}\left(D_{g(x_0)}^{\nu}\left(f\circ g^{-1}\right)\right)(t)\,dt, \tag{3.10}$$

for all $x_0\leq x\leq b$.

(II) Let $h\in C([g(a),g(b)],X)$, we define the right Riemann-Liouville Bochner fractional integral as

$$\left(J_{z_0-}^{\nu}h\right)(z):=\frac{1}{\Gamma(\nu)}\int_{z}^{z_0}(t-z)^{\nu-1}h(t)\,dt, \tag{3.11}$$

for $g(a)\leq z\leq z_0\leq g(b)$. We set $J_{z_0-}^{0}h=h$.

Let $\alpha:=\nu-[\nu]$ $(0<\alpha<1)$. We define the subspace $C_{g(x_0)-}^{\nu}([g(a),g(b)],X)$ of $C^{[\nu]}([g(a),g(b)],X)$, where $x_0\in[a,b]$ as:

$$C_{g(x_0)-}^{\nu}([g(a),g(b)],X):=$$

$$\left\{h\in C^{[\nu]}([g(a),g(b)],X):J_{g(x_0)-}^{1-\alpha}h^{([\nu])}\in C^{1}([g(a),g(x_0)],X)\right\}. \tag{3.12}$$

So let $h\in C_{g(x_0)-}^{\nu}([g(a),g(b)],X)$, we define the right g-generalized X-valued fractional derivative of h of order ν, of Canavati type, over $[g(a),g(x_0)]$ as

$$D_{g(x_0)-}^{\nu}h:=(-1)^{n-1}\left(J_{g(x_0)-}^{1-\alpha}h^{([\nu])}\right)'. \tag{3.13}$$

Clearly, for $h\in C_{g(x_0)-}^{\nu}([g(a),g(b)],X)$, there exists

$$\left(D_{g(x_0)-}^{\nu}h\right)(z)=\frac{(-1)^{n-1}}{\Gamma(1-\alpha)}\frac{d}{dz}\int_{z}^{g(x_0)}(t-z)^{-\alpha}h^{([\nu])}(t)\,dt, \tag{3.14}$$

for all $g(a)\leq z\leq g(x_0)\leq g(b)$.

In particular, when $f\circ g^{-1}\in C_{g(x_0)-}^{\nu}([g(a),g(b)],X)$, we have that

$$\left(D_{g(x_0)-}^{\nu}\left(f\circ g^{-1}\right)\right)(z)=\frac{(-1)^{n-1}}{\Gamma(1-\alpha)}\frac{d}{dz}\int_{z}^{g(x_0)}(t-z)^{-\alpha}\left(f\circ g^{-1}\right)^{([\nu])}(t)\,dt, \tag{3.15}$$

for all $g(a)\leq z\leq g(x_0)\leq g(b)$.

We get that

$$\left(D_{g(x_0)-}^{n}\left(f\circ g^{-1}\right)\right)(z)=(-1)^{n}\left(f\circ g^{-1}\right)^{(n)}(z) \tag{3.16}$$

and $\left(D_{g(x_0)-}^{0}\left(f\circ g^{-1}\right)\right)(z)=\left(f\circ g^{-1}\right)(z)$, all $z\in[g(a),g(b)]$, see [4].

By [4], we have for $f \circ g^{-1} \in C^{\nu}_{g(x_0)-}([g(a), g(b)], X)$, where $x_0 \in [a, b]$ is fixed, the following right generalized g-fractional, of Canavati type, X-valued Taylor's formula:

Theorem 3.4 *Let $f \circ g^{-1} \in C^{\nu}_{g(x_0)-}([g(a), g(b)], X)$, where $x_0 \in [a, b]$ is fixed.*
(i) If $\nu \geq 1$, then

$$f(x) - f(x_0) = \sum_{k=1}^{[\nu]-1} \frac{\left(f \circ g^{-1}\right)^{(k)}(g(x_0))}{k!} (g(x) - g(x_0))^k +$$

$$\frac{1}{\Gamma(\nu)} \int_{g(x)}^{g(x_0)} (t - g(x))^{\nu-1} \left(D^{\nu}_{g(x_0)-}\left(f \circ g^{-1}\right)\right)(t)\, dt, \tag{3.17}$$

for all $a \leq x \leq x_0$,
(ii) If $0 < \nu < 1$, we get

$$f(x) = \frac{1}{\Gamma(\nu)} \int_{g(x)}^{g(x_0)} (t - g(x))^{\nu-1} \left(D^{\nu}_{g(x_0)-}\left(f \circ g^{-1}\right)\right)(t)\, dt, \tag{3.18}$$

all $a \leq x \leq x_0$.

(III) Denote by

$$D^{m\nu}_{g(x_0)} = D^{\nu}_{g(x_0)} D^{\nu}_{g(x_0)} \ldots D^{\nu}_{g(x_0)} \ (m\text{ -times}),\ m \in \mathbb{N}. \tag{3.19}$$

We mention the following modified and generalized left X-valued fractional Taylor's formula of Canavati type:

Theorem 3.5 *Let $f \in C^1([a, b], X)$, $g \in C^1([a, b])$, strictly increasing: $g^{-1} \in C^1([g(a), g(b)])$. Assume that $\left(D^{i\nu}_{g(x_0)}\left(f \circ g^{-1}\right)\right) \in C^{\nu}_{g(x_0)}([g(a), g(b)], X)$, $0 < \nu < 1$, $x_0 \in [a, b]$, for $i = 0, 1, \ldots, m$. Then*

$$f(x) = \frac{1}{\Gamma((m+1)\nu)} \int_{g(x_0)}^{g(x)} (g(x) - z)^{(m+1)\nu-1} \left(D^{(m+1)\nu}_{g(x_0)}\left(f \circ g^{-1}\right)\right)(z)\, dz, \tag{3.20}$$

all $x_0 \leq x \leq b$.

(IV) Denote by

$$D^{m\nu}_{g(x_0)-} = D^{\nu}_{g(x_0)-} D^{\nu}_{g(x_0)-} \ldots D^{\nu}_{g(x_0)-} \ (m \text{ times}),\ m \in \mathbb{N}. \tag{3.21}$$

We mention the following modified and generalized right X-valued fractional Taylor's formula of Canavati type:

Theorem 3.6 *Let $f \in C^1([a,b], X)$, $g \in C^1([a,b])$, strictly increasing: $g^{-1} \in C^1([g(a), g(b)])$. Assume that $\left(D^{i\nu}_{g(x_0)-}\left(f \circ g^{-1}\right)\right) \in C^{\nu}_{g(x_0)-}([g(a), g(b)], X)$, $0 < \nu < 1$, $x_0 \in [a,b]$, for all $i = 0, 1, ..., m$. Then*

$$f(x) = \frac{1}{\Gamma((m+1)\nu)} \int_{g(x)}^{g(x_0)} (z - g(x))^{(m+1)\nu-1} \left(D^{(m+1)\nu}_{g(x_0)-}\left(f \circ g^{-1}\right)\right)(z)\, dz, \tag{3.22}$$

all $a \leq x \leq x_0 \leq b$.

3.3 Banach Algebras Background

All here come from [11].

We need

Definition 3.7 ([11], p. 245) A complex algebra is a vector space A over the complex filed $\mathbb{C}$ in which a multiplication is defined that satisfies

$$x(yz) = (xy)z, \tag{3.23}$$

$$(x+y)z = xz + yz,\ x(y+z) = xy + xz, \tag{3.24}$$

and

$$\alpha(xy) = (\alpha x)y = x(\alpha y), \tag{3.25}$$

for all x, y and z in A and for all scalars α.

Additionally if A is a Banach space with respect to a norm that satisfies the multiplicative inequality

$$\|xy\| \leq \|x\| \|y\| \quad (x \in A,\ y \in A) \tag{3.26}$$

and if A contains a unit element e such that

$$xe = ex = x \ (x \in A) \tag{3.27}$$

and

$$\|e\| = 1, \tag{3.28}$$

then A is called a Banach algebra.

A is commutative iff $xy = yx$ for all $x, y \in A$.

We make

Remark 3.8 Commutativity of A will be explicited stated when needed.

There exists at most one $e \in A$ that satisfies (3.27).

Inequality (3.26) makes multiplication to be continuous, more precisely left and right continuous, see [11], p. 246.

Multiplication in A is not necessarily the numerical multiplication, it is something more general and it is defined abstractly, that is for $x, y \in A$ we have $xy \in A$, e.g. composition or convolution, etc.

For nice examples about Banach algebras see [11], pp. 247–248, Sect. 10.3.

We also make

Remark 3.9 Next we mention about integration of A-valued functions, see [11], p. 259, Sect. 10.22:

If A is a Banach algebra and f is a continuous A-valued function on some compact Hausdorff space Q on which a complex Borel measure μ is defined, then $\int f d\mu$ exists and has all the properties that were discussed in Chap. 3 of [11], simply because A is a Banach space. However, an additional property can be added to these, namely: If $x \in A$, then

$$x \int_Q f \, d\mu = \int_Q x f(p) \, d\mu(p) \tag{3.29}$$

and

$$\left(\int_Q f \, d\mu \right) x = \int_Q f(p) \, x \, d\mu(p) . \tag{3.30}$$

The Bochner integrals we will involve in our chapter follow (3.29) and (3.30). Also, let $f \in C([a, b], X)$, where $[a, b] \subset \mathbb{R}$, $(X, \|\cdot\|)$ is a Banach space. By [5], p. 3, f is Bochner integrable.

3.4 Main Results

We start with mixed generalized Canavati type fractional Ostrowski type inequalities for several functions over a Banach algebra. A uniform estimate follows.

Theorem 3.10 *Let $(A, \|\cdot\|)$ be a Banach algebra, $x_0 \in [a, b] \subset \mathbb{R}$, $\nu \geq 1$, $n = [\nu]$, $f_i \in C^n([a, b], A)$, $i = 1, ..., r \in \mathbb{N} - \{1\}$; $g \in C^1([a, b])$ strictly increasing, such that $g^{-1} \in C^n([g(a), g(b)])$, with $\left(f_i \circ g^{-1}\right)^{(k)}(g(x_0)) = 0$, $k = 1, ..., n - 1$; $i = 1, ..., r$. Assume further that $f_i \circ g^{-1} \in C^{\nu}_{g(x_0)-}([g(a), g(b)], A) \cap C^{\nu}_{g(x_0)}([g(a), g(b)], A)$, i=1,...,r.*

Denote by

$$K(f_1, ..., f_r)(x_0) :=$$

$$\sum_{i=1}^{r} \left[\int_a^b \left(\prod_{\substack{j=1 \\ j \neq i}}^{r} f_j(x) \right) f_i(x) \, dx - \left(\int_a^b \left(\prod_{\substack{j=1 \\ j \neq i}}^{r} f_j(x) \right) dx \right) f_i(x_0) \right] . \tag{3.31}$$

Then

$$\|K(f_1, ..., f_r)(x_0)\| \leq \frac{1}{\Gamma(\nu+1)} \sum_{i=1}^{r} \Bigg[\Bigg[\left\| \left\| D_{g(x_0)-}^{\nu}\left(f_i \circ g^{-1}\right) \right\| \right\|_{\infty,[g(a),g(x_0)]}$$

$$(g(x_0) - g(a))^{\nu} \left(\int_a^{x_0} \left(\prod_{\substack{j=1 \\ j\neq i}}^{r} \|f_j(x)\| \right) dx \right) \Bigg] + \quad (3.32)$$

$$\left[\left\| \left\| D_{g(x_0)}^{\nu}\left(f_i \circ g^{-1}\right) \right\| \right\|_{\infty,[g(x_0),g(b)]} (g(b) - g(x_0))^{\nu} \left(\int_{x_0}^{b} \left(\prod_{\substack{j=1 \\ j\neq i}}^{r} \|f_j(x)\| \right) dx \right) \right] \Bigg].$$

Proof Since $\left(f_i \circ g^{-1}\right)^{(k)}(g(x_0)) = 0$, $k = 1, ..., [\nu] - 1$; $i = 1, ..., r$; we have by Theorem 3.3 that

$$f_i(x) - f_i(x_0) = \frac{1}{\Gamma(\nu)} \int_{g(x_0)}^{g(x)} (g(x) - t)^{\nu-1} \left(D_{g(x_0)}^{\nu}\left(f_i \circ g^{-1}\right)\right)(t)\, dt, \quad (3.33)$$

$\forall\, x \in [x_0, b]$,

and by Theorem 3.4 that

$$f_i(x) - f_i(x_0) = \frac{1}{\Gamma(\nu)} \int_{g(x)}^{g(x_0)} (t - g(x))^{\nu-1} \left(D_{g(x_0)-}^{\nu}\left(f_i \circ g^{-1}\right)\right)(t)\, dt, \quad (3.34)$$

$\forall\, x \in [a, x_0]$, for all $i = 1, ..., r$.

Left multiplying (3.33) and (3.34) with $\left(\prod_{\substack{j=1 \\ j\neq i}}^{r} f_j(x)\right)$ we get, respectively,

$$\left(\prod_{\substack{j=1 \\ j\neq i}}^{r} f_j(x) \right) f_i(x) - \left(\prod_{\substack{j=1 \\ j\neq i}}^{r} f_j(x) \right) f_i(x_0) =$$

$$\frac{\left(\prod_{\substack{j=1 \\ j\neq i}}^{r} f_j(x)\right)}{\Gamma(\nu)} \int_{g(x_0)}^{g(x)} (g(x) - t)^{\nu-1} \left(D_{g(x_0)}^{\nu}\left(f_i \circ g^{-1}\right)\right)(t)\, dt, \quad (3.35)$$

$\forall\, x \in [x_0, b]$,

and

$$\left(\prod_{\substack{j=1\\j\neq i}}^{r} f_j(x)\right) f_i(x) - \left(\prod_{\substack{j=1\\j\neq i}}^{r} f_j(x)\right) f_i(x_0) = \tag{3.36}$$

$$\frac{\left(\prod_{\substack{j=1\\j\neq i}}^{r} f_j(x)\right)}{\Gamma(\nu)} \int_{g(x)}^{g(x_0)} (t - g(x))^{\nu-1} \left(D_{g(x_0)-}^{\nu} \left(f_i \circ g^{-1}\right)\right)(t)\, dt,$$

$\forall\, x \in [a, x_0]$, for all $i = 1, ..., r$.

Adding (3.35) and (3.36) as separate groups, we obtain

$$\sum_{i=1}^{r} \left(\prod_{\substack{j=1\\j\neq i}}^{r} f_j(x)\right) f_i(x) - \sum_{i=1}^{r} \left(\prod_{\substack{j=1\\j\neq i}}^{r} f_j(x)\right) f_i(x_0) =$$

$$\frac{1}{\Gamma(\nu)} \sum_{i=1}^{r} \left(\prod_{\substack{j=1\\j\neq i}}^{r} f_j(x)\right) \int_{g(x_0)}^{g(x)} (g(x) - t)^{\nu-1} \left(D_{g(x_0)}^{\nu} \left(f_i \circ g^{-1}\right)\right)(t)\, dt, \tag{3.37}$$

$\forall\, x \in [x_0, b]$,

and

$$\sum_{i=1}^{r} \left(\prod_{\substack{j=1\\j\neq i}}^{r} f_j(x)\right) f_i(x) - \sum_{i=1}^{r} \left(\prod_{\substack{j=1\\j\neq i}}^{r} f_j(x)\right) f_i(x_0) =$$

$$\frac{1}{\Gamma(\nu)} \sum_{i=1}^{r} \left(\prod_{\substack{j=1\\j\neq i}}^{r} f_j(x)\right) \int_{g(x)}^{g(x_0)} (t - g(x))^{\nu-1} \left(D_{g(x_0)-}^{\nu} \left(f_i \circ g^{-1}\right)\right)(t)\, dt, \tag{3.38}$$

$\forall\, x \in [a, x_0]$.

Next, we integrate (3.37) and (3.38) with respect to $x \in [a, b]$. We have

$$\sum_{i=1}^{r} \int_{x_0}^{b} \left(\prod_{\substack{j=1\\j\neq i}}^{r} f_j(x)\right) f_i(x)\, dx - \sum_{i=1}^{r} \left(\int_{x_0}^{b} \left(\prod_{\substack{j=1\\j\neq i}}^{r} f_j(x)\right) dx\right) f_i(x_0) =$$

$$\frac{1}{\Gamma(\nu)}\sum_{i=1}^{r}\left[\int_{x_0}^{b}\left(\prod_{\substack{j=1\\ j\neq i}}^{r} f_j(x)\right)\left(\int_{g(x_0)}^{g(x)}(g(x)-t)^{\nu-1}\left(D_{g(x_0)}^{\nu}\left(f_i\circ g^{-1}\right)\right)(t)\,dt\right)dx\right], \tag{3.39}$$

and

$$\sum_{i=1}^{r}\int_{a}^{x_0}\left(\prod_{\substack{j=1\\ j\neq i}}^{r} f_j(x)\right) f_i(x)\,dx-\sum_{i=1}^{r}\left(\int_{a}^{x_0}\left(\prod_{\substack{j=1\\ j\neq i}}^{r} f_j(x)\right)dx\right) f_i(x_0)=$$

$$\frac{1}{\Gamma(\nu)}\sum_{i=1}^{r}\left[\int_{a}^{x_0}\left(\prod_{\substack{j=1\\ j\neq i}}^{r} f_j(x)\right)\left(\int_{g(x)}^{g(x_0)}(t-g(x))^{\nu-1}\left(D_{g(x_0)-}^{\nu}\left(f_i\circ g^{-1}\right)\right)(t)\,dt\right)dx\right], \tag{3.40}$$

Finally, adding (3.39) and (3.40) we obtain the useful identity

$$K(f_1,\ldots,f_r)(x_0):=$$

$$\sum_{i=1}^{r}\left[\int_{a}^{b}\left(\prod_{\substack{j=1\\ j\neq i}}^{r} f_j(x)\right) f_i(x)\,dx-\left(\int_{a}^{b}\left(\prod_{\substack{j=1\\ j\neq i}}^{r} f_j(x)\right)dx\right) f_i(x_0)\right]=$$

$$\frac{1}{\Gamma(\nu)}\sum_{i=1}^{r}\left[\left[\int_{a}^{x_0}\left(\prod_{\substack{j=1\\ j\neq i}}^{r} f_j(x)\right)\left(\int_{g(x)}^{g(x_0)}(t-g(x))^{\nu-1}\left(D_{g(x_0)-}^{\nu}\left(f_i\circ g^{-1}\right)\right)(t)\,dt\right)dx\right]\right.$$

$$\left.+\left[\int_{x_0}^{b}\left(\prod_{\substack{j=1\\ j\neq i}}^{r} f_j(x)\right)\left(\int_{g(x_0)}^{g(x)}(g(x)-t)^{\nu-1}\left(D_{g(x_0)}^{\nu}\left(f_i\circ g^{-1}\right)\right)(t)\,dt\right)dx\right]\right]. \tag{3.41}$$

Therefore, we get that

$$\|K(f_1,\ldots,f_r)(x_0)\|=$$

$$\left\|\sum_{i=1}^{r}\left[\int_{a}^{b}\left(\prod_{\substack{j=1\\ j\neq i}}^{r} f_j(x)\right) f_i(x)\,dx-\left(\int_{a}^{b}\left(\prod_{\substack{j=1\\ j\neq i}}^{r} f_j(x)\right)dx\right) f_i(x_0)\right]\right\|\leq\frac{1}{\Gamma(\nu)}$$

$$\sum_{i=1}^{r}\left[\left\|\left[\int_{a}^{x_0}\left(\prod_{\substack{j=1\\ j\neq i}}^{r} f_j(x)\right)\left(\int_{g(x)}^{g(x_0)}(t-g(x))^{\nu-1}\left(D_{g(x_0)-}^{\nu}\left(f_i\circ g^{-1}\right)\right)(t)\,dt\right)dx\right]\right\|\right. \tag{3.42}$$

$$\left.+\left\|\left[\int_{x_0}^{b}\left(\prod_{\substack{j=1\\ j\neq i}}^{r} f_j(x)\right)\left(\int_{g(x_0)}^{g(x)}(g(x)-t)^{\nu-1}\left(D_{g(x_0)}^{\nu}\left(f_i\circ g^{-1}\right)\right)(t)\,dt\right)dx\right]\right\|\right]\le$$

$$\frac{1}{\Gamma(\nu)}\sum_{i=1}^{r}\left[\left[\int_{a}^{x_0}\left\|\left(\prod_{\substack{j=1\\ j\neq i}}^{r} f_j(x)\right)\left(\int_{g(x)}^{g(x_0)}(t-g(x))^{\nu-1}\left(D_{g(x_0)-}^{\nu}\left(f_i\circ g^{-1}\right)\right)(t)\,dt\right)\right\|dx\right]\right. \tag{3.43}$$

$$\left.+\left[\int_{x_0}^{b}\left\|\left(\prod_{\substack{j=1\\ j\neq i}}^{r} f_j(x)\right)\left(\int_{g(x_0)}^{g(x)}(g(x)-t)^{\nu-1}\left(D_{g(x_0)}^{\nu}\left(f_i\circ g^{-1}\right)\right)(t)\,dt\right)\right\|dx\right]\right]\le$$

$$\frac{1}{\Gamma(\nu)}\sum_{i=1}^{r}\left[\left[\int_{a}^{x_0}\left(\prod_{\substack{j=1\\ j\neq i}}^{r}\|f_j(x)\|\right)\left(\int_{g(x)}^{g(x_0)}(t-g(x))^{\nu-1}\left\|\left(D_{g(x_0)-}^{\nu}\left(f_i\circ g^{-1}\right)\right)(t)\right\|dt\right)dx\right]\right.$$

$$\left.+\left[\int_{x_0}^{b}\left(\prod_{\substack{j=1\\ j\neq i}}^{r}\|f_j(x)\|\right)\left(\int_{g(x_0)}^{g(x)}(g(x)-t)^{\nu-1}\left\|\left(D_{g(x_0)}^{\nu}\left(f_i\circ g^{-1}\right)\right)(t)\right\|dt\right)dx\right]\right]=:(*). \tag{3.44}$$

Hence it holds

$$\|K(f_1,\ldots,f_r)(x_0)\|\le(*). \tag{3.45}$$

We have that

$$(*)\le\frac{1}{\Gamma(\nu+1)}$$

$$\sum_{i=1}^{r}\left[\left[\left\|\left\|\left(D_{g(x_0)-}^{\nu}\left(f_i\circ g^{-1}\right)\right)\right\|\right\|_{\infty,[g(a),g(x_0)]}\int_{a}^{x_0}\left(\prod_{\substack{j=1\\ j\neq i}}^{r}\|f_j(x)\|\right)(g(x_0)-g(x))^{\nu}\,dx\right]\right.$$

$$\left.+\left[\left\|\left\|\left(D_{g(x_0)}^{\nu}\left(f_i\circ g^{-1}\right)\right)\right\|\right\|_{\infty,[g(x_0),g(b)]}\int_{x_0}^{b}\left(\prod_{\substack{j=1\\ j\neq i}}^{r}\|f_j(x)\|\right)(g(x)-g(x_0))^{\nu}\,dx\right]\right]\le \tag{3.46}$$

$$\frac{1}{\Gamma(\nu+1)}\sum_{i=1}^{r}\Bigg[\Bigg[\left\|\left\|\left(D_{g(x_0)-}^{\nu}\left(f_i\circ g^{-1}\right)\right)\right\|\right\|_{\infty,[g(a),g(x_0)]}$$

$$(g(x_0)-g(a))^{\nu}\left(\int_a^{x_0}\left(\prod_{\substack{j=1\\j\neq i}}^{r}\|f_j(x)\|\right)dx\right)\Bigg]+ \tag{3.47}$$

$$\left[\left\|\left\|\left(D_{g(x_0)}^{\nu}\left(f_i\circ g^{-1}\right)\right)\right\|\right\|_{\infty,[g(x_0),g(b)]}(g(b)-g(x_0))^{\nu}\left(\int_{x_0}^{b}\left(\prod_{\substack{j=1\\j\neq i}}^{r}\|f_j(x)\|\right)dx\right)\right]\Bigg],$$

proving (3.32). ■

Next comes an L_1 estimate.

Theorem 3.11 *All as in Theorem 3.10. Then*

$$\|\|K(f_1,\ldots,f_r)(x_0)\|\leq\frac{1}{\Gamma(\nu)}$$

$$\sum_{i=1}^{r}\Bigg[\Bigg[\left\|\left\|\left(D_{g(x_0)-}^{\nu}\left(f_i\circ g^{-1}\right)\right)\right\|\right\|_{L_1([g(a),g(x_0)])}\int_a^{x_0}\left(\prod_{\substack{j=1\\j\neq i}}^{r}\|f_j(x)\|\right)(g(x_0)-g(x))^{\nu-1}dx\Bigg]$$

$$+\left[\left\|\left\|\left(D_{g(x_0)}^{\nu}\left(f_i\circ g^{-1}\right)\right)\right\|\right\|_{L_1([g(x_0),g(b)])}\int_{x_0}^{b}\left(\prod_{\substack{j=1\\j\neq i}}^{r}\|f_j(x)\|\right)(g(x)-g(x_0))^{\nu-1}dx\right]\Bigg]. \tag{3.48}$$

Proof We observe that (by (3.45)

$$(*)\leq\frac{1}{\Gamma(\nu)}$$

$$\sum_{i=1}^{r}\Bigg[\Bigg[\left\|\left\|\left(D_{g(x_0)-}^{\nu}\left(f_i\circ g^{-1}\right)\right)\right\|\right\|_{L_1([g(a),g(x_0)])}\int_a^{x_0}\left(\prod_{\substack{j=1\\j\neq i}}^{r}\|f_j(x)\|\right)(g(x_0)-g(x))^{\nu-1}dx\Bigg]$$

$$+\left[\left\|\left\|\left(D_{g(x_0)}^{\nu}\left(f_i\circ g^{-1}\right)\right)\right\|\right\|_{L_1([g(x_0),g(b)])}\int_{x_0}^{b}\left(\prod_{\substack{j=1\\j\neq i}}^{r}\|f_j(x)\|\right)(g(x)-g(x_0))^{\nu-1}dx\right]\Bigg], \tag{3.49}$$

proving (3.48). ■

An L_p estimate follows.

Theorem 3.12 *All as in Theorem 3.10. Let now $p, q > 1 : \frac{1}{p} + \frac{1}{q} = 1$. Then*

$$\|K(f_1, ..., f_r)(x_0)\| \leq \frac{1}{(p(\nu-1)+1)^{\frac{1}{p}}\Gamma(\nu)}$$

$$\sum_{i=1}^{r}\left[\left[\left\|\left(D^{\nu}_{g(x_0)-}\left(f_i \circ g^{-1}\right)\right)\right\|_{q,[g(a),g(x_0)]}\left(\int_a^{x_0}(g(x_0)-g(x))^{\nu-\frac{1}{q}}\left(\prod_{\substack{j=1\\j\neq i}}^{r}\|f_j(x)\|\right)dx\right)\right]\right.$$

$$\left.+\left[\left\|\left(D^{\nu}_{g(x_0)}\left(f_i \circ g^{-1}\right)\right)\right\|_{q,[g(x_0),g(b)]}\left(\int_{x_0}^{b}(g(x)-g(x_0))^{\nu-\frac{1}{q}}\left(\prod_{\substack{j=1\\j\neq i}}^{r}\|f_j(x)\|\right)dx\right)\right]\right]. \tag{3.50}$$

Proof We have that

$$(*) \overset{(3.45)}{\leq} \frac{1}{\Gamma(\nu)}\sum_{i=1}^{r}\left[\left[\int_a^{x_0}\left(\prod_{\substack{j=1\\j\neq i}}^{r}\|f_j(x)\|\right)\left(\int_{g(x)}^{g(x_0)}(t-g(x))^{p(\nu-1)}dt\right)^{\frac{1}{p}}\right.\right.$$

$$\left.\left(\int_{g(x)}^{g(x_0)}\left\|\left(D^{\nu}_{g(x_0)-}\left(f_i \circ g^{-1}\right)\right)(t)\right\|^q dt\right)^{\frac{1}{q}}dx\right]+$$

$$\left[\int_{x_0}^{b}\left(\prod_{\substack{j=1\\j\neq i}}^{r}\|f_j(x)\|\right)\left(\int_{g(x_0)}^{g(x)}(g(x)-t)^{p(\nu-1)}dt\right)^{\frac{1}{p}}\right.$$

$$\left.\left.\left(\int_{g(x)}^{g(x_0)}\left\|\left(D^{\nu}_{g(x_0)}\left(f_i \circ g^{-1}\right)\right)(t)\right\|^q dt\right)^{\frac{1}{q}}dx\right]\right] \leq \tag{3.51}$$

$$\frac{1}{\Gamma(\nu)}\sum_{i=1}^{r}\left[\left[\int_a^{x_0}\left(\prod_{\substack{j=1\\j\neq i}}^{r}\|f_j(x)\|\right)\frac{(g(x_0)-g(x))^{\nu-1+\frac{1}{p}}}{(p(\nu-1)+1)^{\frac{1}{p}}}\left\|\left(D^{\nu}_{g(x_0)-}\left(f_i \circ g^{-1}\right)\right)\right\|_{q,[g(a),g(x_0)]}dx\right]\right.$$

$$\left.+\left[\int_{x_0}^{b}\left(\prod_{\substack{j=1\\j\neq i}}^{r}\|f_j(x)\|\right)\frac{(g(x)-g(x_0))^{\nu-1+\frac{1}{p}}}{(p(\nu-1)+1)^{\frac{1}{p}}}\left\|\left(D^{\nu}_{g(x_0)}\left(f_i \circ g^{-1}\right)\right)(z)\right\|_{q,[g(x_0),g(b)]}dx\right]\right]$$

$$= \frac{1}{\left(p\left(\nu-1\right)+1\right)^{\frac{1}{p}}\Gamma\left(\nu\right)}$$

$$\sum_{i=1}^{r}\left[\left\|\left(D_{g(x_0)-}^{\nu}\left(f_i\circ g^{-1}\right)\right)\right\|_{q,[g(a),g(x_0)]}\left(\int_a^{x_0}\left(g\left(x_0\right)-g\left(x\right)\right)^{\nu-\frac{1}{q}}\left(\prod_{\substack{j=1\\ j\neq i}}^{r}\left\|f_j\left(x\right)\right\|\right)dx\right)\right.$$

$$\left.+\left\|\left(D_{g(x_0)}^{\nu}\left(f_i\circ g^{-1}\right)\right)\right\|_{q,[g(x_0),g(b)]}\left(\int_{x_0}^{b}\left(g\left(x\right)-g\left(x_0\right)\right)^{\nu-\frac{1}{q}}\left(\prod_{\substack{j=1\\ j\neq i}}^{r}\left\|f_j\left(x\right)\right\|\right)dx\right)\right], \tag{3.52}$$

proving (3.50). ■

Next we present a left generalized Canavati Opial type inequality:

Theorem 3.13 *Let $(A, \|\cdot\|)$ be a Banach algebra, $x_0 \in [a, b] \subset \mathbb{R}$, $\nu \geq 1$, $n = [\nu]$, $f \in C^n([a, b], A)$; $g \in C^1([a, b])$, strictly increasing, such that $g^{-1} \in C^n([g(a), g(b)])$, with $\left(f \circ g^{-1}\right)^{(k)}(g(x_0)) = 0$, $k = 0, 1, ..., n-1$. Assume further that $f \circ g^{-1} \in C_{g(x_0)}^{\nu}([g(a), g(b)], A)$. Let also $p, q > 1 : \frac{1}{p} + \frac{1}{q} = 1$. Then*

$$\int_{g(x_0)}^{z}\left\|\left(\left(f\circ g^{-1}\right)(w)\right)\left(\left(D_{g(x_0)}^{\nu}\left(f\circ g^{-1}\right)\right)(w)\right)\right\|dw\leq \tag{3.53}$$

$$\frac{2^{-\frac{1}{q}}\left(z-g\left(x_0\right)\right)^{\nu+\frac{1}{p}-\frac{1}{q}}}{\Gamma\left(\nu\right)\left[\left(p\left(\nu-1\right)+1\right)\left(p\left(\nu-1\right)+2\right)\right]^{\frac{1}{p}}}\left(\int_{g(x_0)}^{z}\left\|\left(D_{g(x_0)}^{\nu}\left(f\circ g^{-1}\right)\right)(t)\right\|^{q}dt\right)^{\frac{2}{q}},$$

for all $g(x_0) \leq z \leq g(b)$.

Proof By (3.9) and assumptions we get that

$$\left(f\circ g^{-1}\right)(z)=\frac{1}{\Gamma\left(\nu\right)}\int_{g(x_0)}^{z}(z-t)^{\nu-1}\left(D_{g(x_0)}^{\nu}\left(f\circ g^{-1}\right)\right)(t)\,dt, \tag{3.54}$$

for all $g(x_0) \leq z \leq g(b)$.

By Hölder's inequality we obtain

$$\left\|\left(f\circ g^{-1}\right)(z)\right\|\leq\frac{1}{\Gamma\left(\nu\right)}\int_{g(x_0)}^{z}(z-t)^{\nu-1}\left\|\left(D_{g(x_0)}^{\nu}\left(f\circ g^{-1}\right)\right)(t)\right\|dt\leq$$

$$\frac{1}{\Gamma\left(\nu\right)}\left(\int_{g(x_0)}^{z}(z-t)^{p(\nu-1)}dt\right)^{\frac{1}{p}}\left(\int_{g(x_0)}^{z}\left\|\left(D_{g(x_0)}^{\nu}\left(f\circ g^{-1}\right)\right)(t)\right\|^{q}dt\right)^{\frac{1}{q}}=$$

$$\frac{1}{\Gamma\left(\nu\right)}\frac{\left(z-g\left(x_0\right)\right)^{\frac{p(\nu-1)+1}{p}}}{\left(p\left(\nu-1\right)+1\right)^{\frac{1}{p}}}\left(\int_{g(x_0)}^{z}\left\|\left(D_{g(x_0)}^{\nu}\left(f\circ g^{-1}\right)\right)(t)\right\|^{q}dt\right)^{\frac{1}{q}}. \tag{3.55}$$

Call

$$\varphi(z) := \int_{g(x_0)}^{z} \left\| \left(D_{g(x_0)}^{\nu} \left(f \circ g^{-1} \right) \right)(t) \right\|^{q} dt, \tag{3.56}$$

$\varphi(g(x_0)) = 0.$

Thus

$$\varphi'(z) = \left\| \left(D_{g(x_0)}^{\nu} \left(f \circ g^{-1} \right) \right)(z) \right\|^{q} \geq 0, \tag{3.57}$$

and

$$\left(\varphi'(z) \right)^{\frac{1}{q}} = \left\| \left(D_{g(x_0)}^{\nu} \left(f \circ g^{-1} \right) \right)(z) \right\| \geq 0, \tag{3.58}$$

$\forall\, z \in [g(x_0), g(b)]$.

Consequently, we get

$$\left\| \left(\left(f \circ g^{-1} \right)(w) \right) \right\| \left\| \left(\left(D_{g(x_0)}^{\nu} \left(f \circ g^{-1} \right) \right)(w) \right) \right\| \leq$$

$$\frac{1}{\Gamma(\nu)} \frac{(w - g(x_0))^{\frac{p(\nu-1)+1}{p}}}{(p(\nu-1)+1)^{\frac{1}{p}}} \left(\varphi(w) \varphi'(w) \right)^{\frac{1}{q}}, \tag{3.59}$$

$\forall\, w \in [g(x_0), g(b)]$.

Then

$$\int_{g(x_0)}^{z} \left\| \left(\left(f \circ g^{-1} \right)(w) \right) \left(\left(D_{g(x_0)}^{\nu} \left(f \circ g^{-1} \right) \right)(w) \right) \right\| dw \overset{(3.26)}{\leq}$$

$$\int_{g(x_0)}^{z} \left\| \left(\left(f \circ g^{-1} \right)(w) \right) \right\| \left\| \left(\left(D_{g(x_0)}^{\nu} \left(f \circ g^{-1} \right) \right)(w) \right) \right\| dw \leq$$

$$\frac{1}{\Gamma(\nu)(p(\nu-1)+1)^{\frac{1}{p}}} \int_{g(x_0)}^{z} (w - g(x_0))^{\frac{p(\nu-1)+1}{p}} \left(\varphi(w) \varphi'(w) \right)^{\frac{1}{q}} dw \leq \tag{3.60}$$

$$\frac{1}{\Gamma(\nu)(p(\nu-1)+1)^{\frac{1}{p}}} \left(\int_{g(x_0)}^{z} (w - g(x_0))^{p(\nu-1)+1} dw \right)^{\frac{1}{p}} \left(\int_{g(x_0)}^{z} \varphi(w)\, d\varphi(w) \right)^{\frac{1}{q}} =$$

$$\frac{1}{\Gamma(\nu)(p(\nu-1)+1)^{\frac{1}{p}} (p(\nu-1)+2)^{\frac{1}{p}}} (z - g(x_0))^{\frac{p(\nu-1)+2}{p}} \left(\frac{\varphi^2(z)}{2} \right)^{\frac{1}{q}} =$$

$$\frac{2^{-\frac{1}{q}} (z - g(x_0))^{\nu + \frac{1}{p} - \frac{1}{q}}}{\Gamma(\nu)\left[(p(\nu-1)+1)(p(\nu-1)+2) \right]^{\frac{1}{p}}} \left(\int_{g(x_0)}^{z} \left\| \left(D_{g(x_0)}^{\nu} \left(f \circ g^{-1} \right) \right)(t) \right\|^{q} dt \right)^{\frac{2}{q}}, \tag{3.61}$$

for all $g(x_0) \leq z \leq g(b)$, proving (3.53). ∎

It follows the corresponding right side fractional Opial type inequality:

Theorem 3.14 *All as in Theorem 3.13, however now it is* $f \circ g^{-1} \in C^{\nu}_{g(x_0)-}([g(a), g(b)], A)$. *Then*

$$\int_{z}^{g(x_0)} \left\|\left(\left(f \circ g^{-1}\right)(w)\right)\left(\left(D^{\nu}_{g(x_0)-}\left(f \circ g^{-1}\right)\right)(w)\right)\right\| dw \leq$$

$$\frac{2^{-\frac{1}{q}}\left(g(x_0) - z\right)^{\nu+\frac{1}{p}-\frac{1}{q}}}{\Gamma(\nu)\left[\left(p(\nu-1)+1\right)\left(p(\nu-1)+2\right)\right]^{\frac{1}{p}}} \left(\int_{z}^{g(x_0)} \left\|\left(D^{\nu}_{g(x_0)-}\left(f \circ g^{-1}\right)\right)(t)\right\|^{q} dt\right)^{\frac{2}{q}}, \tag{3.62}$$

for all $g(a) \leq z \leq g(x_0)$.

Proof Based on (3.17), and as similar to the proof of Theorem 3.13 is omitted. ■

It follows the modified generalized left A-valued fractional Opial inequality:

Theorem 3.15 *All as in Theorem 3.5 and let* $p, q > 1 : \frac{1}{p} + \frac{1}{q} = 1$. *Here we assume that* $\frac{1}{(m+1)q} < \nu < 1$. *Then*

$$\int_{g(x_0)}^{z} \left\|\left(\left(f \circ g^{-1}\right)(w)\right)\left(\left(D^{(m+1)\nu}_{g(x_0)}\left(f \circ g^{-1}\right)\right)(w)\right)\right\| dw \leq \tag{3.63}$$

$$\frac{2^{-\frac{1}{q}}\left(z - g(x_0)\right)^{(m+1)\nu+\frac{1}{p}-\frac{1}{q}}}{\Gamma((m+1)\nu)\left[\left(p((m+1)\nu-1)+1\right)\left(p((m+1)\nu-1)+2\right)\right]^{\frac{1}{p}}}$$

$$\left(\int_{g(x_0)}^{z} \left\|\left(D^{(m+1)\nu}_{g(x_0)}\left(f \circ g^{-1}\right)\right)(t)\right\|^{q} dt\right)^{\frac{2}{q}},$$

for all $g(x_0) \leq z \leq g(b)$.

Proof As in Theorem 3.13. ■

The corresponding modified generalized right A-valued fractional Opial inequality comes next:

Theorem 3.16 *All as in Theorem 3.6 and let* $p, q > 1 : \frac{1}{p} + \frac{1}{q} = 1$. *Here we assume that* $\frac{1}{(m+1)q} < \nu < 1$. *Then*

$$\int_{z}^{g(x_0)} \left\|\left(\left(f \circ g^{-1}\right)(w)\right)\left(\left(D^{(m+1)\nu}_{g(x_0)-}\left(f \circ g^{-1}\right)\right)(w)\right)\right\| dw \leq \tag{3.64}$$

$$\frac{2^{-\frac{1}{q}}\left(g(x_0) - z\right)^{(m+1)\nu+\frac{1}{p}-\frac{1}{q}}}{\Gamma((m+1)\nu)\left[\left(p((m+1)\nu-1)+1\right)\left(p((m+1)\nu-1)+2\right)\right]^{\frac{1}{p}}}$$

$$\left(\int_z^{g(x_0)} \left\| \left(D_{g(x_0)-}^{(m+1)\nu} \left(f \circ g^{-1}\right)\right)(t)\right\|^q dt\right)^{\frac{2}{q}},$$

for all $g(a) \leq z \leq g(x_0)$.

Proof As in Theorem 3.13. ∎

3.5 Applications

We make

Remark 3.17 Assume in this section that $(A, \|\cdot\|)$ is a commutative Banach algebra. Then, we get that

$$K(f_1, ..., f_r)(x_0) \overset{(3.31)}{=} r \int_a^b \left(\prod_{j=1}^r f_j(x)\right) dx - \sum_{i=1}^r \left(\int_a^b \left(\prod_{\substack{j=1 \\ j \neq i}}^r f_j(x)\right) dx\right) f_i(x_0), \tag{3.65}$$

$x_0 \in [a, b]$.

When $r = 2$, we find

$$K(f_1, f_2)(x_0) = 2\int_a^b f_1(x) f_2(x) dx - f_1(x_0) \int_a^b f_2(x) dx - f_2(x_0) \int_a^b f_1(x) dx, \tag{3.66}$$

$x_0 \in [a, b]$.

We give

Corollary 3.18 *All as in Theorem 3.10, A is a commutative Banach algebra, r* $= 2$. *Then*

$$\|K(f_1, f_2)(x_0)\| \leq \frac{1}{\Gamma(\nu+1)} \sum_{i=1}^2 \left[\left[\left\|\left\|\left(D_{g(x_0)-}^{\nu}\left(f_i \circ g^{-1}\right)\right)\right\|\right\|_{\infty,[g(a),g(x_0)]}\right.\right.$$

$$\left.(g(x_0) - g(a))^\nu \left(\int_a^{x_0} \left(\prod_{\substack{j=1 \\ j \neq i}}^2 \|f_j(x)\|\right) dx\right)\right] +$$

$$\left.\left[\left\|\left\|\left(D_{g(x_0)}^{\nu}\left(f_i \circ g^{-1}\right)\right)\right\|\right\|_{\infty,[g(x_0),g(b)]} (g(b) - g(x_0))^\nu \left(\int_{x_0}^b \left(\prod_{\substack{j=1 \\ j \neq i}}^2 \|f_j(x)\|\right) dx\right)\right]\right]. \tag{3.67}$$

Proof By Theorem 3.10. ■

We also give

Corollary 3.19 *All as in Corollary 3.18, for $g(t) = e^t$. Then*

$$\|K(f_1, f_2)(x_0)\| \leq \frac{1}{\Gamma(\nu+1)} \sum_{i=1}^{2} \Bigg[\Bigg[\left\| \left\| D_{e^{x_0}-}^{\nu} (f_i \circ \log) \right\| \right\|_{\infty,[e^a, e^{x_0}]}$$

$$(e^{x_0} - e^a)^{\nu} \left(\int_a^{x_0} \left(\prod_{\substack{j=1 \\ j \neq i}}^{2} \|f_j(x)\| \right) dx \right) \Bigg] + \tag{3.68}$$

$$\left[\left\| \left\| D_{e^{x_0}}^{\nu} (f_i \circ \log) \right\| \right\|_{\infty,[e^{x_0}, e^b]} (e^b - e^{x_0})^{\nu} \left(\int_{x_0}^{b} \left(\prod_{\substack{j=1 \\ j \neq i}}^{2} \|f_j(x)\| \right) dx \right) \right] \Bigg].$$

Proof By Corollary 3.18. ■

We add also the following:

Corollary 3.20 *(to Theorem 3.13) All as in Theorem 3.13 for $g(t) = e^t$. Then*

$$\int_{e^{x_0}}^{z} \left\| ((f \circ \log)(w)) \left(\left(D_{e^{x_0}}^{\nu} (f \circ \log) \right)(w) \right) \right\| dw \leq \tag{3.69}$$

$$\frac{2^{-\frac{1}{q}} (z - e^{x_0})^{\nu + \frac{1}{p} - \frac{1}{q}}}{\Gamma(\nu) \left[(p(\nu-1)+1)(p(\nu-1)+2) \right]^{\frac{1}{p}}} \left(\int_{e^{x_0}}^{z} \left\| \left(D_{e^{x_0}}^{\nu} (f \circ \log) \right)(t) \right\|^q dt \right)^{\frac{2}{q}},$$

for all $e^{x_0} \leq z \leq e^b$.

Proof By Theorem 3.13. ■

3.6 Addendum

We give the following generalized Canavati type fractional Grüss type inequalities for several functions over a Banach algebra.

We start with a uniform estimate.

Theorem 3.21 *Let $(A, \|\cdot\|)$ be a Banach algebra, $x_0 \in [a,b] \subset \mathbb{R}$, $1 \leq \nu < 2$, $f_i \in C^1([a,b], A)$, $i = 1, ..., r \in \mathbb{N} - \{1\}$; $g \in C^1([a,b])$ strictly increasing, such that $g^{-1} \in C^1([g(a), g(b)])$. Assume further that $f_i \circ g^{-1} \in C_{g(x_0)-}^{\nu}([g(a), g(b)], A) \cap C_{g(x_0)}^{\nu}([g(a), g(b)], A)$, i=1,...,r.*

Assume that

$$M_1(f_1, \ldots, f_r)(x_0) := \max_{i=1,\ldots,r} \left\{ \sup_{x_0 \in [a,b]} \left\| \left\| D^{\nu}_{g(x_0)-} \left(f_i \circ g^{-1} \right) \right\| \right\|_{\infty,[g(a),g(x_0)]}, \right.$$

$$\left. \sup_{x_0 \in [a,b]} \left\| \left\| D^{\nu}_{g(x_0)} \left(f_i \circ g^{-1} \right) \right\| \right\|_{\infty,[g(x_0),g(b)]} \right\} < \infty. \quad (3.70)$$

Denote by

$$\Delta(f_1, \ldots, f_r) := \int_a^b K(f_1, \ldots, f_r)(x)\, dx =$$

$$\sum_{i=1}^{r} \left[(b-a) \left(\int_a^b \left(\prod_{\substack{j=1 \\ j \neq i}}^{r} f_j(x) \right) f_i(x)\, dx \right) - \left(\int_a^b \left(\prod_{\substack{j=1 \\ j \neq i}}^{r} f_j(x) \right) dx \right) \left(\int_a^b f_i(x)\, dx \right) \right]. \quad (3.71)$$

Then

$$\|\Delta(f_1, \ldots, f_r)\| =$$

$$\left\| \sum_{i=1}^{r} \left[(b-a) \left(\int_a^b \left(\prod_{\substack{j=1 \\ j \neq i}}^{r} f_j(x) \right) f_i(x)\, dx \right) - \left(\int_a^b \left(\prod_{\substack{j=1 \\ j \neq i}}^{r} f_j(x) \right) dx \right) \left(\int_a^b f_i(x)\, dx \right) \right] \right\|$$

$$\leq \frac{M_1(f_1, \ldots, f_r)(b-a)^2 (g(b) - g(a))^{\nu}}{\Gamma(\nu+1)} \sum_{i=1}^{r} \left(\left\| \prod_{\substack{j=1 \\ j \neq i}}^{r} \|f_j(x)\| \right\|_{\infty,[a,b]} \right). \quad (3.72)$$

Proof From (3.32) we get

$$R.H.S.(3.32) \leq \frac{M_1(f_1, \ldots, f_r)(g(b) - g(a))^{\nu}}{\Gamma(\nu+1)} \left(\sum_{i=1}^{r} \int_a^b \left(\prod_{\substack{j=1 \\ j \neq i}}^{r} \|f_j(x)\| \right) dx \right) \leq$$

$$\frac{M_1(f_1, \ldots, f_r)(g(b) - g(a))^{\nu}(b-a)}{\Gamma(\nu+1)} \left(\sum_{i=1}^{r} \left\| \prod_{\substack{j=1 \\ j \neq i}}^{r} \|f_j\| \right\|_{\infty,[a,b]} \right) =: \lambda_1. \quad (3.73)$$

We have that

$$\|\Delta(f_1,...,f_r)\| \leq \int_a^b \|K(f_1,...,f_r)(x_0)\| \, dx_0 \overset{((3.32),(3.73))}{\leq} \int_a^b \lambda_1 dx =$$

$$\frac{M_1(f_1,...,f_r)(g(b)-g(a))^{\nu}(b-a)^2}{\Gamma(\nu+1)} \left(\sum_{i=1}^{r} \left(\left\| \prod_{\substack{j=1 \\ j \neq i}}^{r} \|f_j(x)\| \right\|_{\infty,[a,b]} \right) \right), \tag{3.74}$$

proving (3.72). ■

Next comes an L_1-estimate.

Theorem 3.22 *All as in Theorem 3.21, however now we assume that*

$$M_2(f_1,...,f_r)(x_0) := \max_{i=1,...,r} \left\{ \sup_{x_0 \in [a,b]} \left\| \left\| D^{\nu}_{g(x_0)-}\left(f_i \circ g^{-1}\right) \right\| \right\|_{L_1([g(a),g(x_0)])}, \right.$$

$$\left. \sup_{x_0 \in [a,b]} \left\| \left\| D^{\nu}_{g(x_0)}\left(f_i \circ g^{-1}\right) \right\| \right\|_{L_1([g(x_0),g(b)])} \right\} < \infty. \tag{3.75}$$

Then

$$\|\Delta(f_1,...,f_r)\| \leq$$

$$\leq \frac{M_2(f_1,...,f_r)(b-a)^2(g(b)-g(a))^{\nu-1}}{\Gamma(\nu)} \sum_{i=1}^{r} \left(\left\| \prod_{\substack{j=1 \\ j \neq i}}^{r} \|f_j(x)\| \right\|_{\infty,[a,b]} \right). \tag{3.76}$$

Proof By (3.48) we get

$$R.H.S.(3.48) \leq \frac{M_2(f_1,...,f_r)(g(b)-g(a))^{\nu-1}}{\Gamma(\nu)} \left(\sum_{i=1}^{r} \int_a^b \left(\prod_{\substack{j=1 \\ j \neq i}}^{r} \|f_j(x)\| \right) dx \right) \leq$$

$$\frac{M_2(f_1,...,f_r)(g(b)-g(a))^{\nu-1}(b-a)}{\Gamma(\nu)} \left(\sum_{i=1}^{r} \left\| \prod_{\substack{j=1 \\ j \neq i}}^{r} \|f_j\| \right\|_{\infty,[a,b]} \right) =: \lambda_2. \tag{3.77}$$

We have that

$$\|\Delta(f_1,...,f_r)\| \leq \int_a^b \|K(f_1,...,f_r)(x_0)\| \, dx_0 \overset{((3.48),(3.76))}{\leq} \int_a^b \lambda_2 dx =$$

$$\frac{M_2(f_1,...,f_r)(g(b)-g(a))^{\nu-1}(b-a)^2}{\Gamma(\nu)}\left(\sum_{i=1}^{r}\left(\left\|\prod_{\substack{j=1\\j\neq i}}^{r}\|f_j(x)\|\right\|_{\infty,[a,b]}\right)\right), \tag{3.78}$$

proving (3.76). ■

An L_p-estimate follows:

Theorem 3.23 *All as in Theorem 3.21, however here we assume that*

$$M_3(f_1,...,f_r)(x_0) := \max_{i=1,...,r}\left\{\sup_{x_0\in[a,b]}\left\|\left\|D^{\nu}_{g(x_0)-}\left(f_i\circ g^{-1}\right)\right\|\right\|_{L_q([g(a),g(x_0)])},\right.$$

$$\left.\sup_{x_0\in[a,b]}\left\|\left\|D^{\nu}_{g(x_0)}\left(f_i\circ g^{-1}\right)\right\|\right\|_{L_q([g(x_0),g(b)])}\right\}<\infty. \tag{3.79}$$

Above it is $p,q>1:\frac{1}{p}+\frac{1}{q}=1$.
Then

$$\|\Delta(f_1,...,f_r)\|\leq$$

$$\leq\frac{M_3(f_1,...,f_r)(b-a)^2(g(b)-g(a))^{\nu-\frac{1}{q}}}{(p(\nu-1)+1)^{\frac{1}{p}}\Gamma(\nu)}\sum_{i=1}^{r}\left(\left\|\prod_{\substack{j=1\\j\neq i}}^{r}\|f_j(x)\|\right\|_{\infty,[a,b]}\right). \tag{3.80}$$

Proof By (3.50) we get

$$R.H.S.(3.50)\leq\frac{M_3(f_1,...,f_r)(g(b)-g(a))^{\nu-\frac{1}{q}}}{(p(\nu-1)+1)^{\frac{1}{p}}\Gamma(\nu)}\left(\sum_{i=1}^{r}\int_a^b\left(\prod_{\substack{j=1\\j\neq i}}^{r}\|f_j(x)\|\right)dx\right)\leq$$

$$\frac{M_3(f_1,...,f_r)(g(b)-g(a))^{\nu-\frac{1}{q}}(b-a)}{(p(\nu-1)+1)^{\frac{1}{p}}\Gamma(\nu)}\left(\sum_{i=1}^{r}\left\|\prod_{\substack{j=1\\j\neq i}}^{r}\|f_j\|\right\|_{\infty,[a,b]}\right)=:\lambda_3. \tag{3.81}$$

Hence it holds

$$\|\Delta(f_1,...,f_r)\|\leq\int_a^b\|K(f_1,...,f_r)(x_0)\|\,dx_0\overset{((3.50),(3.80))}{\leq}\int_a^b\lambda_3dx=$$

$$\frac{M_3 (f_1, ..., f_r) (g (b) - g (a))^{\nu - \frac{1}{q}} (b - a)^2}{(p (\nu - 1) + 1)^{\frac{1}{p}} \Gamma (\nu)} \left(\sum_{i=1}^{r} \left(\left\| \prod_{\substack{j=1 \\ j \neq i}}^{r} \| f_j (x) \| \right\|_{\infty, [a,b]} \right) \right), \tag{3.82}$$

proving (3.80). ■

References

1. Anastassiou, G.A.: Fractional Differentiation Inequalities. Research Monograph. Springer, New York (2009)
2. Anastassiou, G.A.: Advances on Fractional Inequalities. Research Monograph. Springer, New York (2011)
3. Anastassiou, G.A.: Intelligent Comparisons: Analytic Inequalities. Springer, Heidelberg, New York (2016)
4. Anastassiou, G.A.: Strong mixed and generalized fractional calculus for Banach space valued functions. Mat. Vesnik **69**(3), 176–191 (2017)
5. Anastassiou, G.A.: Intelligent Computations: Abstract Fractional Calculus, Inequalities, Approximations. Springer, Heidelberg, New York (2018)
6. Anastassiou, G.A.: Generalized Canavati Fractional Ostrowski, Opial and Grüss type inequalities for Banach algebra valued functions . Moroccan J. of Pure and Applied Analysis, accepted for publication (2022)
7. Dragomir, S.S.: Noncommutative Ostrowski type inequalities for functions in Banach algebras, RGMIA Res. Rep. Coll. **24**, Art. 10, 24 pp (2021)
8. Mikusinski, J.: The Bochner Integral. Academic Press, New York (1978)
9. Opial, Z.: Sur une inegalite. Ann. Polon. Math. **8**, 29–32 (1960)
10. Ostrowski, A.: Über die Absolutabweichung einer differentiabaren Funcktion von ihrem Integralmittelwert. Comment. Math. Helv. **10**, 226–227 (1938)
11. Rudin, W.: Functional Analysis, 2nd edn. McGraw-Hill Inc, New York (1991)

Chapter 4
Generalized Canavati Fractional Hilbert–Pachpatte Inequalities for Banach Algebra Valued Functions

Using generalized Canavati fractional left and right vectorial Taylor formulae we prove corresponding left and right fractional Hilbert–Pachpatte inequalities for Banach algebra valued functions. We cover also the sequential fractional case. We finish with applications. It follows [3].

4.1 Introduction

Motivation follows:

We need

Definition 4.1 (*see* [6]) A definition of the Hausdorff measure h_α goes as follows: if (T, d) is a metric space, $A \subseteq T$ and $\delta > 0$, let $\Lambda(A, \delta)$ be the set of all arbitrary collections $(C)_i$ of subsets of T, such that $A \subseteq \cup_i C_i$ and $diam(C_i) \leq \delta$ ($diam =$ diameter) for every i. Now, for every $\alpha > 0$ define

$$h_\alpha^\delta(A) := \inf\left\{\sum (diam C_i)^\alpha \mid (C_i)_i \in \Lambda(A, \delta)\right\}. \tag{4.1}$$

Then there exists $\lim_{\delta \to 0} h_\alpha^\delta(A) = \sup_{\delta > 0} h_\alpha^\delta(A)$, and $h_\alpha(A) := \lim_{\delta \to 0} h_\alpha^\delta(A)$ gives an outer measure on the power set $\mathcal{P}(T)$, which is countably additive on the σ-field of all Borel subsets of T. If $T = \mathbb{R}^n$, then the Hausdorff measure h_n, restricted to the σ-field of the Borel subsets of $\mathbb{R}^n$, equals the Lebesgue measure on $\mathbb{R}^n$ up to a constant multiple. In particular, $h_1(C) = \mu(C)$ for every Borel set $C \subseteq \mathbb{R}$, where μ is the Lebesgue measure.

G. A. Anastassiou, *Fractional Inequalities In Banach Algebras*, Studies in Systems, Decision and Control 441,
https://doi.org/10.1007/978-3-031-05148-7_4

We also need

Definition 4.2 ([2], *Chap. 1*) Let $[a,b] \subset \mathbb{R}$, X be a Banach space, $\nu > 0$; $n := \lceil \nu \rceil \in \mathbb{N}$, $\lceil \cdot \rceil$ is the ceiling of the number, $f : [a,b] \to X$. We assume that $f^{(n)} \in L_1([a,b], X)$. We call the Caputo-Bochner left fractional derivative of order ν:

$$\left(D_{*a}^{\nu} f\right)(x) := \frac{1}{\Gamma(n-\nu)} \int_a^x (x-t)^{n-\nu-1} f^{(n)}(t)\, dt, \ \forall\, x \in [a,b]. \tag{4.2}$$

If $\nu \in \mathbb{N}$, we set $D_{*a}^{\nu} f := f^{(\nu)}$ the ordinary X-valued derivative, and also set $D_{*a}^{0} f := f$. Here Γ is the gamma function and integrals are of Bochner type [4].

By [2], Chap. 1, $\left(D_{*a}^{\nu} f\right)(x)$ exists almost everywhere in $x \in [a,b]$ and $D_{*a}^{\nu} f \in L_1([a,b], X)$.

If $\left\| f^{(n)} \right\|_{L_\infty([a,b],X)} < \infty$, then by [2], Chap. 1, $D_{*a}^{\nu} f \in C([a,b], X)$.

We are motivated by a Hilbert–Pachpatte left fractional inequality:

Theorem 4.3 ([2], Chap. 1) *Let $p, q > 1 : \frac{1}{p} + \frac{1}{q} = 1$, and $\nu_1 > \frac{1}{q}$, $\nu_2 > \frac{1}{p}$, $n_i := \lceil \nu_i \rceil$, $i = 1, 2$. Here $[a_i, b_i] \subset \mathbb{R}$, $i = 1, 2$; X is a Banach space. Let $f_i \in C^{n_i - 1}([a_i, b_i], X)$, $i = 1, 2$. Set*

$$F_{x_i}(t_i) := \sum_{j_i=0}^{n_i-1} \frac{(x_i - t_i)^{j_i}}{j_i!} f_i^{(j_i)}(t_i), \tag{4.3}$$

$\forall\, t_i \in [a_i, x_i]$, where $x_i \in [a_i, b_i]$; $i = 1, 2$. Assume that $f_i^{(n_i)}$ exists outside a μ-null Borel set $B_{x_i} \subseteq [a_i, x_i]$, such that

$$h_1\left(F_{x_i}\left(B_{x_i}\right)\right) = 0, \ \forall\, x_i \in [a_i, b_i]; \ i = 1, 2. \tag{4.4}$$

We also assume that $f_i^{(n_i)} \in L_1([a_i, b_i], X)$, and

$$f_i^{(k_i)}(a_i) = 0, \ k_i = 0, 1, \ldots, n_i - 1; \ i = 1, 2, \tag{4.5}$$

and

$$\left(D_{*a_1}^{\nu_1} f_1\right) \in L_q([a_1, b_1], X), \ \left(D_{*a_2}^{\nu_2} f_2\right) \in L_p([a_2, b_2], X). \tag{4.6}$$

Then

$$\int_{a_1}^{b_1} \int_{a_2}^{b_2} \frac{\|f_1(x_1)\| \, \|f_2(x_2)\| \, dx_1 dx_2}{\left(\frac{(x_1-a_1)^{p(\nu_1-1)+1}}{p(p(\nu_1-1)+1)} + \frac{(x_2-a_2)^{q(\nu_2-1)+1}}{q(q(\nu_2-1)+1)} \right)} \leq$$

$$\frac{(b_1-a_1)(b_2-a_2)}{\Gamma(\nu_1)\Gamma(\nu_2)} \left\| D_{*a_1}^{\nu_1} f_1 \right\|_{L_q([a_1,b_1],X)} \left\| D_{*a_2}^{\nu_2} f_2 \right\|_{L_p([a_2,b_2],X)}. \tag{4.7}$$

We need

Definition 4.4 ([2], *Chap. 2*) Let $[a, b] \subset \mathbb{R}$, X be a Banach space, $\alpha > 0$, $m := \lceil \alpha \rceil$. We assume that $f^{(m)} \in L_1([a, b], X)$, where $f : [a, b] \to X$. We call the Caputo-Bochner right fractional derivative of order α:

$$\left(D_{b-}^{\alpha} f\right)(x) := \frac{(-1)^m}{\Gamma(m-\alpha)} \int_x^b (J-x)^{m-\alpha-1} f^{(m)}(J)\, dJ, \ \forall\, x \in [a, b]. \quad (4.8)$$

We observe that $D_{b-}^{m} f(x) = (-1)^m f^{(m)}(x)$, for $m \in \mathbb{N}$, and $D_{b-}^{0} f(x) = f(x)$.

By [2], Chap. 2, $\left(D_{b-}^{\alpha} f\right)(x)$ exists almost everywhere on $[a, b]$ and $\left(D_{b-}^{\alpha} f\right) \in L_1([a, b], X)$.

If $\left\| f^{(m)} \right\|_{L_\infty([a,b],X)} < \infty$, and $\alpha \notin \mathbb{N}$, then by [2], Chap. 2, $D_{b-}^{\alpha} f \in C([a, b], X)$, hence $\left\| D_{b-}^{\alpha} f \right\| \in C([a, b])$.

We are motivated also by the following Hilbert–Pachpatte right fractional inequality:

Theorem 4.5 ([2], Chap. 2) *Let* $p, q > 1 : \frac{1}{p} + \frac{1}{q} = 1$, *and* $\alpha_1 > \frac{1}{q}$, $\alpha_2 > \frac{1}{p}$, $m_i := \lceil \alpha_i \rceil$, $i = 1, 2$. *Here* $[a_i, b_i] \subset \mathbb{R}$, $i = 1, 2$; X *is a Banach space. Let* $f_i \in C^{m_i - 1}([a_i, b_i], X)$, $i = 1, 2$. *Set*

$$F_{x_i}(t_i) := \sum_{j_i=0}^{m_i-1} \frac{(x_i - t_i)^{j_i}}{j_i!} f_i^{(j_i)}(t_i), \quad (4.9)$$

$\forall\, t_i \in [x_i, b_i]$, *where* $x_i \in [a_i, b_i]$; $i = 1, 2$. *Assume that* $f_i^{(m_i)}$ *exists outside a* μ*-null Borel set* $B_{x_i} \subseteq [x_i, b_i]$, *such that*

$$h_1\left(F_{x_i}\left(B_{x_i}\right)\right) = 0, \ \forall\, x_i \in [a_i, b_i]; \ i = 1, 2. \quad (4.10)$$

We also assume that $f_i^{(m_i)} \in L_1([a_i, b_i], X)$, *and*

$$f_i^{(k_i)}(b_i) = 0, \ k_i = 0, 1, \ldots, , m_i - 1; \ i = 1, 2, \quad (4.11)$$

and

$$\left(D_{b_1-}^{\alpha_1} f_1\right) \in L_q([a_1, b_1], X), \ \left(D_{b_2-}^{\alpha_2} f_2\right) \in L_p([a_2, b_2], X). \quad (4.12)$$

Then

$$\int_{a_1}^{b_1} \int_{a_2}^{b_2} \frac{\| f_1(x_1) \| \, \| f_2(x_2) \| \, dx_1 dx_2}{\left(\frac{(b_1 - x_1)^{p(\alpha_1 - 1)+1}}{p(p(\alpha_1 - 1)+1)} + \frac{(b_2 - x_2)^{q(\alpha_2 - 1)+1}}{q(q(\alpha_2 - 1)+1)} \right)} \leq$$

$$\frac{(b_1 - a_1)(b_2 - a_2)}{\Gamma(\alpha_1)\Gamma(\alpha_2)} \left\| D_{b_1-}^{\alpha_1} f_1 \right\|_{L_q([a_1,b_1],X)} \left\| D_{b_2-}^{\alpha_2} f_2 \right\|_{L_p([a_2,b_2],X)}. \quad (4.13)$$

In this chapter we derive Hilbert–Pachpatte inequalities for Banach algebra valued functions with respect to their Canavati type generalized left and right fractional derivatives. We cover also the sequential fractional case. We finish with applications.

4.2 Background on Vectorial Generalized Canavati Fractional Calculus

All in this section come from [2], pp. 109–115 and [1].

Let $g : [a, b] \to \mathbb{R}$ be a strictly increasing function. such that $g \in C^1([a, b])$, and $g^{-1} \in C^n([g(a), g(b)])$, $n \in \mathbb{N}$, $(X, \|\cdot\|)$ is a Banach space. Let $f \in C^n([a, b], X)$, and call $l := f \circ g^{-1} : [g(a), g(b)] \to X$. It is clear that $l, l', ..., l^{(n)}$ are continuous functions from $[g(a), g(b)]$ into $f([a, b]) \subseteq X$.

Let $\nu \geq 1$ such that $[\nu] = n$, $n \in \mathbb{N}$ as above, where $[\cdot]$ is the integral part of the number.

Clearly when $0 < \nu < 1$, $[\nu] = 0$.

(I) Let $h \in C([g(a), g(b)], X)$, we define the left Riemann–Liouville Bochner fractional integral as

$$\left(J_{\nu}^{z_0} h\right)(z) := \frac{1}{\Gamma(\nu)} \int_{z_0}^{z} (z-t)^{\nu-1} h(t)\, dt, \tag{4.14}$$

for $g(a) \leq z_0 \leq z \leq g(b)$, where Γ is the gamma function; $\Gamma(\nu) = \int_0^\infty e^{-t} t^{\nu-1} dt$. We set $J_0^{z_0} h = h$.

Let $\alpha := \nu - [\nu]$ $(0 < \alpha < 1)$. We define the subspace $C_{g(x_0)}^{\nu}([g(a), g(b)], X)$ of $C^{[\nu]}([g(a), g(b)], X)$, where $x_0 \in [a, b]$ as:

$$C_{g(x_0)}^{\nu}([g(a), g(b)], X) =$$

$$\left\{h \in C^{[\nu]}([g(a), g(b)], X) : J_{1-\alpha}^{g(x_0)} h^{([\nu])} \in C^1([g(x_0), g(b)], X)\right\}. \tag{4.15}$$

So let $h \in C_{g(x_0)}^{\nu}([g(a), g(b)], X)$, we define the left g-generalized X -valued fractional derivative of h of order ν, of Canavati type, over $[g(x_0), g(b)]$ as

$$D_{g(x_0)}^{\nu} h := \left(J_{1-\alpha}^{g(x_0)} h^{([\nu])}\right)'. \tag{4.16}$$

Clearly, for $h \in C_{g(x_0)}^{\nu}([g(a), g(b)], X)$, there exists

$$\left(D_{g(x_0)}^{\nu} h\right)(z) = \frac{1}{\Gamma(1-\alpha)} \frac{d}{dz} \int_{g(x_0)}^{z} (z-t)^{-\alpha} h^{([\nu])}(t)\, dt, \tag{4.17}$$

for all $g(x_0) \leq z \leq g(b)$.

In particular, when $f \circ g^{-1} \in C^{\nu}_{g(x_0)}([g(a), g(b)], X)$, we have that

$$\left(D^{\nu}_{g(x_0)}\left(f \circ g^{-1}\right)\right)(z) = \frac{1}{\Gamma(1-\alpha)} \frac{d}{dz} \int_{g(x_0)}^{z} (z-t)^{-\alpha} \left(f \circ g^{-1}\right)^{([\nu])}(t)\, dt, \tag{4.18}$$

for all $g(x_0) \leq z \leq g(b)$. We have that $D^{n}_{g(x_0)}\left(f \circ g^{-1}\right) = \left(f \circ g^{-1}\right)^{(n)}$ and $D^{0}_{g(x_0)}\left(f \circ g^{-1}\right) = f \circ g^{-1}$, see [1].

By [1], we have for $f \circ g^{-1} \in C^{\nu}_{g(x_0)}([g(a), g(b)], X)$, where $x_0 \in [a, b]$ the following left generalized g-fractional, of Canavati type, X-valued Taylor's formula:

Theorem 4.6 *Let $f \circ g^{-1} \in C^{\nu}_{g(x_0)}([g(a), g(b)], X)$, where $x_0 \in [a, b]$ is fixed.*
(i) If $\nu \geq 1$, then

$$f(x) - f(x_0) = \sum_{k=1}^{[\nu]-1} \frac{\left(f \circ g^{-1}\right)^{(k)}(g(x_0))}{k!} (g(x) - g(x_0))^k +$$

$$\frac{1}{\Gamma(\nu)} \int_{g(x_0)}^{g(x)} (g(x) - t)^{\nu-1} \left(D^{\nu}_{g(x_0)}\left(f \circ g^{-1}\right)\right)(t)\, dt, \tag{4.19}$$

for all $x_0 \leq x \leq b$.
(ii) If $0 < \nu < 1$, we get

$$f(x) = \frac{1}{\Gamma(\nu)} \int_{g(x_0)}^{g(x)} (g(x) - t)^{\nu-1} \left(D^{\nu}_{g(x_0)}\left(f \circ g^{-1}\right)\right)(t)\, dt, \tag{4.20}$$

for all $x_0 \leq x \leq b$.

(II) Let $h \in C([g(a), g(b)], X)$, we define the right Riemann–Liouville Bochner fractional integral as

$$\left(J^{\nu}_{z_0-}h\right)(z) := \frac{1}{\Gamma(\nu)} \int_{z}^{z_0} (t-z)^{\nu-1} h(t)\, dt, \tag{4.21}$$

for $g(a) \leq z \leq z_0 \leq g(b)$. We set $J^{0}_{z_0-}h = h$.

Let $\alpha := \nu - [\nu]$ $(0 < \alpha < 1)$. We define the subspace $C^{\nu}_{g(x_0)-}([g(a), g(b)], X)$ of $C^{[\nu]}([g(a), g(b)], X)$, where $x_0 \in [a, b]$ as:

$$C^{\nu}_{g(x_0)-}([g(a), g(b)], X) :=$$

$$\left\{h \in C^{[\nu]}([g(a), g(b)], X) : J^{1-\alpha}_{g(x_0)-}h^{([\nu])} \in C^{1}([g(a), g(x_0)], X)\right\}. \tag{4.22}$$

So let $h \in C^{\nu}_{g(x_0)-}([g(a), g(b)], X)$, we define the right g-generalized X-valued fractional derivative of h of order ν, of Canavati type, over $[g(a), g(x_0)]$ as

$$D^{\nu}_{g(x_0)-}h := (-1)^{n-1}\left(J^{1-\alpha}_{g(x_0)-}h^{([\nu])}\right)'. \quad (4.23)$$

Clearly, for $h \in C^{\nu}_{g(x_0)-}([g(a), g(b)], X)$, there exists

$$\left(D^{\nu}_{g(x_0)-}h\right)(z) = \frac{(-1)^{n-1}}{\Gamma(1-\alpha)}\frac{d}{dz}\int_z^{g(x_0)}(t-z)^{-\alpha}h^{([\nu])}(t)\,dt, \quad (4.24)$$

for all $g(a) \leq z \leq g(x_0) \leq g(b)$.

In particular, when $f \circ g^{-1} \in C^{\nu}_{g(x_0)-}([g(a), g(b)], X)$, we have that

$$\left(D^{\nu}_{g(x_0)-}\left(f \circ g^{-1}\right)\right)(z) = \frac{(-1)^{n-1}}{\Gamma(1-\alpha)}\frac{d}{dz}\int_z^{g(x_0)}(t-z)^{-\alpha}\left(f \circ g^{-1}\right)^{([\nu])}(t)\,dt, \quad (4.25)$$

for all $g(a) \leq z \leq g(x_0) \leq g(b)$.

We get that

$$\left(D^{n}_{g(x_0)-}\left(f \circ g^{-1}\right)\right)(z) = (-1)^n\left(f \circ g^{-1}\right)^{(n)}(z) \quad (4.26)$$

and $\left(D^{0}_{g(x_0)-}\left(f \circ g^{-1}\right)\right)(z) = \left(f \circ g^{-1}\right)(z)$, all $z \in [g(a), g(b)]$, see [1].

By [1], we have for $f \circ g^{-1} \in C^{\nu}_{g(x_0)-}([g(a), g(b)], X)$, where $x_0 \in [a, b]$ is fixed, the following right generalized g-fractional, of Canavati type, X-valued Taylor's formula:

Theorem 4.7 *Let $f \circ g^{-1} \in C^{\nu}_{g(x_0)-}([g(a), g(b)], X)$, where $x_0 \in [a, b]$ is fixed.*

(i) If $\nu \geq 1$, then

$$f(x) - f(x_0) = \sum_{k=1}^{[\nu]-1}\frac{\left(f \circ g^{-1}\right)^{(k)}(g(x_0))}{k!}(g(x) - g(x_0))^k +$$

$$\frac{1}{\Gamma(\nu)}\int_{g(x)}^{g(x_0)}(t - g(x))^{\nu-1}\left(D^{\nu}_{g(x_0)-}\left(f \circ g^{-1}\right)\right)(t)\,dt, \quad (4.27)$$

for all $a \leq x \leq x_0$,

(ii) If $0 < \nu < 1$, we get

$$f(x) = \frac{1}{\Gamma(\nu)}\int_{g(x)}^{g(x_0)}(t - g(x))^{\nu-1}\left(D^{\nu}_{g(x_0)-}\left(f \circ g^{-1}\right)\right)(t)\,dt, \quad (4.28)$$

all $a \leq x \leq x_0$.

(III) Denote by

$$D^{m\nu}_{g(x_0)} = D^{\nu}_{g(x_0)}D^{\nu}_{g(x_0)}...D^{\nu}_{g(x_0)} \text{ (m-times)}, m \in \mathbb{N}. \quad (4.29)$$

We mention the following modified and generalized left X-valued fractional Taylor's formula of Canavati type:

Theorem 4.8 *Let $f \in C^1([a,b], X)$, $g \in C^1([a,b])$, strictly increasing: $g^{-1} \in C^1([g(a), g(b)])$. Assume that $\left(D_{g(x_0)}^{i\nu}(f \circ g^{-1})\right) \in C_{g(x_0)}^{\nu}([g(a), g(b)], X)$, $0 < \nu < 1$, $x_0 \in [a,b]$, for $i = 0, 1, ..., m$. Then*

$$f(x) = \frac{1}{\Gamma((m+1)\nu)} \int_{g(x_0)}^{g(x)} (g(x) - z)^{(m+1)\nu - 1} \left(D_{g(x_0)}^{(m+1)\nu}(f \circ g^{-1})\right)(z)\, dz, \tag{4.30}$$

all $x_0 \leq x \leq b$.

(IV) Denote by

$$D_{g(x_0)-}^{m\nu} = D_{g(x_0)-}^{\nu} D_{g(x_0)-}^{\nu} ... D_{g(x_0)-}^{\nu} \text{ (m times)}, m \in \mathbb{N}. \tag{4.31}$$

We mention the following modified and generalized right X-valued fractional Taylor's formula of Canavati type:

Theorem 4.9 *Let $f \in C^1([a,b], X)$, $g \in C^1([a,b])$, strictly increasing: $g^{-1} \in C^1([g(a), g(b)])$. Assume that $\left(D_{g(x_0)-}^{i\nu}(f \circ g^{-1})\right) \in C_{g(x_0)-}^{\nu}([g(a), g(b)], X)$, $0 < \nu < 1$, $x_0 \in [a,b]$, for all $i = 0, 1, ..., m$. Then*

$$f(x) = \frac{1}{\Gamma((m+1)\nu)} \int_{g(x)}^{g(x_0)} (z - g(x))^{(m+1)\nu - 1} \left(D_{g(x_0)-}^{(m+1)\nu}(f \circ g^{-1})\right)(z)\, dz, \tag{4.32}$$

all $a \leq x \leq x_0 \leq b$.

4.3 Banach Algebras Background

All here come from [5].

We need

Definition 4.10 ([5], *p. 245*) A complex algebra is a vector space A over the complex filed $\mathbb{C}$ in which a multiplication is defined that satisfies

$$x(yz) = (xy)z, \tag{4.33}$$

$$(x+y)z = xz + yz, \; x(y+z) = xy + xz, \tag{4.34}$$

and

$$\alpha(xy) = (\alpha x)y = x(\alpha y), \tag{4.35}$$

for all x, y and z in A and for all scalars α.

Additionally if A is a Banach space with respect to a norm that satisfies the multiplicative inequality

$$\|xy\| \leq \|x\| \, \|y\| \ (x \in A, y \in A) \tag{4.36}$$

and if A contains a unit element e such that

$$xe = ex = x \ (x \in A) \tag{4.37}$$

and

$$\|e\| = 1, \tag{4.38}$$

then A is called a Banach algebra.

A is commutative iff $xy = yx$ for all $x, y \in A$.

We make

Remark 4.11 Commutativity of A will be explicated stated when needed.

There exists at most one $e \in A$ that satisfies (4.37).

Inequality (4.36) makes multiplication to be continuous, more precisely left and right continuous, see [5], p. 246.

Multiplication in A is not necessarily the numerical multiplication, it is something more general and it is defined abstractly, that is for $x, y \in A$ we have $xy \in A$, e.g. composition or convolution, etc.

For nice examples about Banach algebras see [5], pp. 247–248, Sect. 10.3.

We also make

Remark 4.12 Next we mention about integration of A-valued functions, see [5], p. 259, Sect. 10.22:

If A is a Banach algebra and f is a continuous A-valued function on some compact Hausdorff space Q on which a complex Borel measure μ is defined, then $\int f d\mu$ exists and has all the properties that were discussed in Chap. 3 of [5], simply because A is a Banach space. However, an additional property can be added to these, namely: If $x \in A$, then

$$x \int_Q f \, d\mu = \int_Q xf(p) \, d\mu(p) \tag{4.39}$$

and

$$\left(\int_Q f \, d\mu \right) x = \int_Q f(p) \, x \, d\mu(p). \tag{4.40}$$

The Bochner integrals we will involve in our chapter follow (4.39) and (4.40). Also, let $f \in C([a, b], X)$, where $[a, b] \subset \mathbb{R}$, $(X, \|\cdot\|)$ is a Banach space. By [2], p. 3, f is Bochner integrable.

4.4 Main Results

We start with a left generalized Canavati fractional Hilbert–Pachpatte type inequality over a Banach algebra.

Theorem 4.13 *Let $p, q > 1$, such that $\frac{1}{p} + \frac{1}{q} = 1$, and $(A, \|\cdot\|)$ is a Banach algebra; and $i=1, 2$. Let also $x_{0i} \in [a_i, b_i] \subset \mathbb{R}$, $\nu_i \geq 1$, $n_i = [\nu_i]$, $f_i \in C^{n_i}([a_i, b_i], A)$; $g_i \in C^1([a_i, b_i])$, strictly increasing, such that $g_i^{-1} \in C^{n_i}([g_i(a_i), g_i(b_i)])$, with $\left(f_i \circ g_i^{-1}\right)^{(k_i)}(g_i(x_{0i})) = 0$, $k_i = 0, 1, ..., n_i - 1$. Assume further that $f_i \circ g_i^{-1} \in C_{g_i(x_{0i})}^{\nu_i}([g_i(a_i), g_i(b_i)], A)$. Then*

$$\int_{g_1(x_{01})}^{g_1(b_1)} \int_{g_2(x_{02})}^{g_2(b_2)} \frac{\left\| \left(f_1 \circ g_1^{-1}\right)(z_1) \left(f_2 \circ g_2^{-1}\right)(z_2) \right\| dz_1 dz_2}{\left(\frac{(z_1 - g_1(x_{01}))^{p(\nu_1 - 1)+1}}{p(p(\nu_1-1)+1)} + \frac{(z_2 - g_2(x_{02}))^{q(\nu_2-1)+1}}{q(q(\nu_2-1)+1)} \right)} \leq$$

$$\frac{(g_1(b_1) - g_1(x_{01}))(g_2(b_2) - g_2(x_{02}))}{\Gamma(\nu_1)\Gamma(\nu_2)} \tag{4.41}$$

$$\left\| \left\| D_{g_1(x_{01})}^{\nu_1} \left(f_1 \circ g_1^{-1}\right) \right\| \right\|_{L_q([g_1(x_{01}), g_1(b_1)], A)} \left\| \left\| D_{g_2(x_{02})}^{\nu_2} \left(f_2 \circ g_2^{-1}\right) \right\| \right\|_{L_p([g_2(x_{02}), g_2(b_2)], A)}.$$

Proof By (4.19) and assumptions we get that

$$\left(f_i \circ g_i^{-1}\right)(z_i) = \frac{1}{\Gamma(\nu_i)} \int_{g_i(x_{0i})}^{z_i} (z_i - t_i)^{\nu_i - 1} \left(D_{g_i(x_{0i})}^{\nu_i} \left(f_i \circ g_i^{-1}\right) \right)(t_i) \, dt_i, \tag{4.42}$$

for all $g_i(x_{0i}) \leq z_i \leq g_i(b_i)$; $i = 1, 2$.

By Hölder's inequality we obtain

$$\left\| \left(f_1 \circ g_1^{-1}\right)(z_1) \right\| \leq \frac{1}{\Gamma(\nu_1)} \int_{g_1(x_{01})}^{z_1} (z_1 - t_1)^{\nu_1 - 1} \left\| \left(D_{g_1(x_{01})}^{\nu_1} \left(f_1 \circ g_1^{-1}\right) \right)(t_1) \right\| dt_1 \leq$$

$$\frac{1}{\Gamma(\nu_1)} \left(\int_{g_1(x_{01})}^{z_1} (z_1 - t_1)^{p(\nu_1 - 1)} dt_1 \right)^{\frac{1}{p}} \left(\int_{g_1(x_{01})}^{z_1} \left\| \left(D_{g_1(x_{01})}^{\nu_1} \left(f_1 \circ g_1^{-1}\right) \right)(t_1) \right\|^q dt_1 \right)^{\frac{1}{q}} =$$

$$\frac{1}{\Gamma(\nu_1)} \frac{(z_1 - g_1(x_{01}))^{\frac{p(\nu_1-1)+1}{p}}}{(p(\nu_1 - 1) + 1)^{\frac{1}{p}}} \left(\int_{g_1(x_{01})}^{z_1} \left\| \left(D_{g_1(x_{01})}^{\nu_1} \left(f_1 \circ g_1^{-1}\right) \right)(t_1) \right\|^q dt_1 \right)^{\frac{1}{q}}. \tag{4.43}$$

That is

$$\left\| \left(f_1 \circ g_1^{-1}\right)(z_1) \right\| \leq \frac{1}{\Gamma(\nu_1)} \frac{(z_1 - g_1(x_{01}))^{\frac{p(\nu_1-1)+1}{p}}}{(p(\nu_1 - 1) + 1)^{\frac{1}{p}}}$$

$$\left(\int_{g_1(x_{01})}^{z_1} \left\| \left(D_{g_1(x_{01})}^{\nu_1} \left(f_1 \circ g_1^{-1}\right)\right)(t_1)\right\|^q dt_1\right)^{\frac{1}{q}}, \tag{4.44}$$

for all $g_1(x_{01}) \leq z_1 \leq g_1(b_1)$.

Similarly, we prove that

$$\left\| \left(f_2 \circ g_2^{-1}\right)(z_2)\right\| \leq \frac{1}{\Gamma(\nu_2)} \frac{(z_2 - g_2(x_{02}))^{\frac{q(\nu_2-1)+1}{q}}}{(q(\nu_2 - 1) + 1)^{\frac{1}{q}}}$$

$$\left(\int_{g_2(x_{02})}^{z_2} \left\| \left(D_{g_2(x_{02})}^{\nu_2} \left(f_2 \circ g_2^{-1}\right)\right)(t_2)\right\|^p dt_2\right)^{\frac{1}{p}}, \tag{4.45}$$

for all $g_2(x_{02}) \leq z_2 \leq g_2(b_2)$.

Therefore we have

$$\left\| \left(f_1 \circ g_1^{-1}\right)(z_1)\right\| \leq \frac{1}{\Gamma(\nu_1)} \frac{(z_1 - g_1(x_{01}))^{\frac{p(\nu_1-1)+1}{p}}}{(p(\nu_1 - 1) + 1)^{\frac{1}{p}}}$$

$$\left\| \left\| \left(D_{g_1(x_{01})}^{\nu_1} \left(f_1 \circ g_1^{-1}\right)\right)\right\| \right\|_{q,[g_1(x_{01}),g_1(b_1)]}, \tag{4.46}$$

for all $g_1(x_{01}) \leq z_1 \leq g_1(b_1)$;

and

$$\left\| \left(f_2 \circ g_2^{-1}\right)(z_2)\right\| \leq \frac{1}{\Gamma(\nu_2)} \frac{(z_2 - g_2(x_{02}))^{\frac{q(\nu_2-1)+1}{q}}}{(q(\nu_2 - 1) + 1)^{\frac{1}{q}}}$$

$$\left\| \left\| \left(D_{g_2(x_{02})}^{\nu_2} \left(f_2 \circ g_2^{-1}\right)\right)\right\| \right\|_{p,[g_2(x_{02}),g_2(b_2)]}, \tag{4.47}$$

for all $g_2(x_{02}) \leq z_2 \leq g_2(b_2)$.

Hence we get that

$$\left\| \left(f_1 \circ g_1^{-1}\right)(z_1)\right\| \left\| \left(f_2 \circ g_2^{-1}\right)(z_2)\right\| \leq \frac{1}{\Gamma(\nu_1)\Gamma(\nu_2)(p(\nu_1 - 1) + 1)^{\frac{1}{p}}(q(\nu_2 - 1) + 1)^{\frac{1}{q}}}$$

$$(z_1 - g_1(x_{01}))^{\frac{p(\nu_1-1)+1}{p}} (z_2 - g_2(x_{02}))^{\frac{q(\nu_2-1)+1}{q}} \tag{4.48}$$

$$\left\| \left\| \left(D_{g_1(x_{01})}^{\nu_1} \left(f_1 \circ g_1^{-1}\right)\right)\right\| \right\|_{q,[g_1(x_{01}),g_1(b_1)]} \left\| \left\| \left(D_{g_2(x_{02})}^{\nu_2} \left(f_2 \circ g_2^{-1}\right)\right)\right\| \right\|_{p,[g_2(x_{02}),g_2(b_2)]} \leq$$

(using Young's inequality for $a, b \geq 0$, $a^{\frac{1}{p}} b^{\frac{1}{q}} \leq \frac{a}{p} + \frac{b}{q}$)

$$\frac{1}{\Gamma(\nu_1)\Gamma(\nu_2)}\left(\frac{(z_1-g_1(x_{01}))^{p(\nu_1-1)+1}}{p(p(\nu_1-1)+1)}+\frac{(z_2-g_2(x_{02}))^{q(\nu_2-1)+1}}{q(q(\nu_2-1)+1)}\right)$$

$$\left\|\left(D_{g_1(x_{01})}^{\nu_1}\left(f_1\circ g_1^{-1}\right)\right)\right\|_{L_q([g_1(x_{01}),g_1(b_1)],A)}\left\|\left(D_{g_2(x_{02})}^{\nu_2}\left(f_2\circ g_2^{-1}\right)\right)\right\|_{L_p([g_2(x_{02}),g_2(b_2)],A)},$$

$\forall (z_1, z_2) \in [g_1(x_{01}), g_1(b_1)] \times [g_2(x_{02}), g_2(b_2)]$.

So far we have

$$\frac{\left\|\left(f_1\circ g_1^{-1}\right)(z_1)\left(f_2\circ g_2^{-1}\right)(z_2)\right\|}{\left(\frac{(z_1-g_1(x_{01}))^{p(\nu_1-1)+1}}{p(p(\nu_1-1)+1)}+\frac{(z_2-g_2(x_{02}))^{q(\nu_2-1)+1}}{q(q(\nu_2-1)+1)}\right)}\leq \tag{4.50}$$

$$\frac{\left\|\left(f_1\circ g_1^{-1}\right)(z_1)\right\|\left\|\left(f_2\circ g_2^{-1}\right)(z_2)\right\|}{\left(\frac{(z_1-g_1(x_{01}))^{p(\nu_1-1)+1}}{p(p(\nu_1-1)+1)}+\frac{(z_2-g_2(x_{02}))^{q(\nu_2-1)+1}}{q(q(\nu_2-1)+1)}\right)}\leq \tag{4.51}$$

$$\frac{1}{\Gamma(\nu_1)\Gamma(\nu_2)}\left\|\left(D_{g_1(x_{01})}^{\nu_1}\left(f_1\circ g_1^{-1}\right)\right)\right\|_{L_q([g_1(x_{01}),g_1(b_1)],A)}$$

$$\left\|\left(D_{g_2(x_{02})}^{\nu_2}\left(f_2\circ g_2^{-1}\right)\right)\right\|_{L_p([g_2(x_{02}),g_2(b_2)],A)},$$

$\forall (z_1, z_2) \in [g_1(x_{01}), g_1(b_1)] \times [g_2(x_{02}), g_2(b_2)]$.

The denominators in (4.50), (4.51) can be zero only when both $z_1 = g_1(x_{01})$ and $z_2 = g_2(x_{02})$.

Therefore we obtain (4.41), by integrating (4.50), (4.51) over $[g_1(x_{01}), g_1(b_1)] \times [g_2(x_{02}), g_2(b_2)]$. ■

We continue with a right generalized Canavati fractional Hilbert–Pachpatte type inequality over a Banach algebra.

Theorem 4.14 *All as in Theorem 4.13, however now it is* $f_i \circ g_i^{-1} \in C_{g_i(x_{0i})-}^{\nu_i}([g_i(a_i), g_i(b_i)], A)$, *for* $i = 1, 2$. *Then*

$$\int_{g_1(a_1)}^{g_1(x_{01})}\int_{g_2(a_2)}^{g_2(x_{02})}\frac{\left\|\left(f_1\circ g_1^{-1}\right)(z_1)\left(f_2\circ g_2^{-1}\right)(z_2)\right\|dz_1dz_2}{\left(\frac{(g_1(x_{01})-z_1)^{p(\nu_1-1)+1}}{p(p(\nu_1-1)+1)}+\frac{(g_2(x_{02})-z_2)^{q(\nu_2-1)+1}}{q(q(\nu_2-1)+1)}\right)}\leq$$

$$\frac{(g_1(x_{01})-g_1(a_1))(g_2(x_{02})-g_2(a_2))}{\Gamma(\nu_1)\Gamma(\nu_2)} \tag{4.52}$$

$$\left\|D_{g_1(x_{01})-}^{\nu_1}\left(f_1\circ g_1^{-1}\right)\right\|_{L_q([g_1(a_1),g_1(x_{01})],A)}\left\|D_{g_2(x_{02})-}^{\nu_2}\left(f_2\circ g_2^{-1}\right)\right\|_{L_p([g_2(a_2),g_2(x_{02})],A)}.$$

Proof Similar to Theorem 4.13, by using now (4.27). ∎

Next comes a sequential left generalized Canavati fractional Hilbert–Pachpatte type inequality over a Banach algebra.

Theorem 4.15 *Let* $p, q > 1$*, such that* $\frac{1}{p} + \frac{1}{q} = 1$*, and* $(A, \|\cdot\|)$ *is a Banach algebra; and* $i = 1, 2$*. Let also* $f_i \in C^1([a_i, b_i], A)$*;* $g_i \in C^1([a_i, b_i])$*, strictly increasing, such that* $g_i^{-1} \in C^1([g_i(a_i), g_i(b_i)])$*. Assume that* $\frac{1}{(m_i+1)q} < \nu_i < 1$*,* $x_{0i} \in [a_i, b_i]$*, and* $D_{g_i(x_{0i})}^{j_i \nu_i}(f_i \circ g_i^{-1}) \in C_{g_i(x_{0i})}^{\nu_i}([g_i(a_i), g_i(b_i)], A)$*, for* $j_i = 0, 1, \ldots, m_i \in \mathbb{N}$*. Then*

$$\int_{g_1(x_{01})}^{g_1(b_1)} \int_{g_2(x_{02})}^{g_2(b_2)} \frac{\left\| \left(f_1 \circ g_1^{-1}\right)(z_1) \left(f_2 \circ g_2^{-1}\right)(z_2) \right\| dz_1 dz_2}{\left(\frac{(z_1 - g_1(x_{01}))^{p((m_1+1)\nu_1 - 1)+1}}{p(p((m_1+1)\nu_1-1)+1)} + \frac{(z_2 - g_2(x_{02}))^{q((m_2+1)\nu_2-1)+1}}{q(q((m_2+1)\nu_2-1)+1)} \right)} \leq$$

$$\frac{(g_1(b_1) - g_1(x_{01}))(g_2(b_2) - g_2(x_{02}))}{\Gamma((m_1+1)\nu_1)\Gamma((m_2+1)\nu_2)} \tag{4.53}$$

$$\left\| \left\| D_{g_1(x_{01})}^{(m_1+1)\nu_1}\left(f_1 \circ g_1^{-1}\right) \right\| \right\|_{L_q([g_1(x_{01}), g_1(b_1)], A)} \left\| \left\| D_{g_2(x_{02})}^{(m_2+1)\nu_2}\left(f_2 \circ g_2^{-1}\right) \right\| \right\|_{L_p([g_2(x_{02}), g_2(b_2)], A)}.$$

Proof Using (4.30), as similar to Theorem 4.13 the proof is omitted. ∎

The right side analog of Theorem 4.15 follows:

Theorem 4.16 *Let* $p, q > 1$*, such that* $\frac{1}{p} + \frac{1}{q} = 1$*, and* $(A, \|\cdot\|)$ *is a Banach algebra; and* $i = 1, 2$*. Let also* $f_i \in C^1([a_i, b_i], A)$*;* $g_i \in C^1([a_i, b_i])$*, strictly increasing, such that* $g_i^{-1} \in C^1([g_i(a_i), g_i(b_i)])$*. Assume that* $\frac{1}{(m_i+1)q} < \nu_i < 1$*,* $x_{0i} \in [a_i, b_i]$*, and* $D_{g_i(x_{0i})-}^{j_i \nu_i}(f_i \circ g_i^{-1}) \in C_{g_i(x_{0i})-}^{\nu_i}([g_i(a_i), g_i(b_i)], A)$*, for* $j_i = 0, 1, \ldots, m_i \in \mathbb{N}$*. Then*

$$\int_{g_1(a_1)}^{g_1(x_{01})} \int_{g_2(a_2)}^{g_2(x_{02})} \frac{\left\| \left(f_1 \circ g_1^{-1}\right)(z_1) \left(f_2 \circ g_2^{-1}\right)(z_2) \right\| dz_1 dz_2}{\left(\frac{(g_1(x_{01}) - z_1)^{p((m_1+1)\nu_1 - 1)+1}}{p(p((m_1+1)\nu_1-1)+1)} + \frac{(g_2(x_{02}) - z_2)^{q((m_2+1)\nu_2-1)+1}}{q(q((m_2+1)\nu_2-1)+1)} \right)} \leq$$

$$\frac{(g_1(x_{01}) - g_1(a_1))(g_2(x_{02}) - g_2(a_2))}{\Gamma((m_1+1)\nu_1)\Gamma((m_2+1)\nu_2)} \tag{4.54}$$

$$\left\| \left\| D_{g_1(x_{01})-}^{(m_1+1)\nu_1}\left(f_1 \circ g_1^{-1}\right) \right\| \right\|_{L_q([g_1(a_1), g_1(x_{01})], A)} \left\| \left\| D_{g_2(x_{02})-}^{(m_2+1)\nu_2}\left(f_2 \circ g_2^{-1}\right) \right\| \right\|_{L_p([g_2(a_2), g_2(x_{02})], A)}.$$

Proof Using (4.32), as similar to Theorem 4.13 is omitted. ∎

4.5 Applications

We give

Corollary 4.17 (to Theorem 4.13) *All as in Theorem 4.13 for $g_i(t) = e^t$, $i = 1, 2$. Then*

$$\int_{e^{x_{01}}}^{e^{b_1}} \int_{e^{x_{02}}}^{e^{b_2}} \frac{\|(f_1 \circ \log)(z_1)(f_2 \circ \log)(z_2)\| \, dz_1 dz_2}{\left(\frac{(z_1 - e^{x_{01}})^{p(\nu_1 - 1)+1}}{p(p(\nu_1 - 1)+1)} + \frac{(z_2 - e^{x_{02}})^{q(\nu_2 - 1)+1}}{q(q(\nu_2 - 1)+1)} \right)} \leq$$

$$\frac{\left(e^{b_1} - e^{x_{01}}\right)\left(e^{b_2} - e^{x_{02}}\right)}{\Gamma(\nu_1)\Gamma(\nu_2)} \tag{4.55}$$

$$\left\| \left\| D_{e^{x_{01}}}^{\nu_1} (f_1 \circ \log) \right\| \right\|_{L_q([e^{x_{01}}, e^{b_1}], A)} \left\| \left\| D_{e^{x_{02}}}^{\nu_2} (f_2 \circ \log) \right\| \right\|_{L_p([e^{x_{02}}, e^{b_2}], A)}.$$

We finish with

Corollary 4.18 (to Theorem 4.15) *All as in Theorem 4.15 for $[a_1, b_1] \subset \mathbb{R}$, $[a_2, b_2] \subset (0, \infty)$, and $g_1(t) = e^t$ and $g_2(t) = \log t$. Then*

$$\int_{e^{x_{01}}}^{e^{b_1}} \int_{\log(x_{02})}^{\log(b_2)} \frac{\left\|(f_1 \circ \log)(z_1)\left(f_2 \circ e^t\right)(z_2)\right\| dz_1 dz_2}{\left(\frac{(z_1 - e^{x_{01}})^{p((m_1+1)\nu_1 - 1)+1}}{p(p((m_1+1)\nu_1 - 1)+1)} + \frac{(z_2 - \log(x_{02}))^{q((m_2+1)\nu_2 - 1)+1}}{q(q((m_2+1)\nu_2 - 1)+1)} \right)} \leq$$

$$\frac{\left(e^{b_1} - e^{x_{01}}\right)\log(b_2/x_{02})}{\Gamma((m_1+1)\nu_1)\Gamma((m_2+1)\nu_2)} \tag{4.56}$$

$$\left\| \left\| D_{e^{x_{01}}}^{(m_1+1)\nu_1} (f_1 \circ \log) \right\| \right\|_{L_q([e^{x_{01}}, e^{b_1}], A)} \left\| \left\| D_{\log(x_{02})}^{(m_2+1)\nu_2} \left(f_2 \circ e^t\right) \right\| \right\|_{L_p([\log(x_{02}), \log(b_2)], A)}.$$

References

1. Anastassiou, G.A.: Strong mixed and generalized fractional calculus for Banach space valued functions. Mat. Vesnik **69**(3), 176–191 (2017)
2. Anastassiou, G.A.: Intelligent Computations: Abstract Fractional Calculus, Inequalities, Approximations. Springer, Heidelberg (2018)
3. Anastassiou, G.A.: Generalized Canavati Fractional Hilbert-Pachpatte type inequalities for Banach algebra valued functions. J. Comput. Anal. Appl. **30**(1), 66–77 (2022)
4. Mikusinski, J.: The Bochner Integral. Academic, New York (1978)
5. Rudin, W.: Functional Analysis, 2nd edn. McGraw-Hill Inc, New York (1991)
6. Volintiru, C.: A proof of the fundamental theorem of Calculus using Hausdorff measures. Real Anal. Exch. **26**(1), 381–390 (2000/2001)

Chapter 5
Generalized Ostrowski, Opial and Hilbert-Pachpatte Inequalities for Banach Algebra Valued Functions Involving Integer Vectorial Derivatives

Using a generalized vectorial Taylor formula involving ordinary vector derivatives we establish mixed Ostrowski, Opial and Hilbert-Pachpatte inequalities for several Banach algebra valued functions. The estimates are with respect to all norms $\|\cdot\|_p$, $1 \leq p \leq \infty$. We finish with applications. It follows [3].

5.1 Introduction

The following result motivates this chapter.

Theorem 5.1 (1938, Ostrowski [7]) *Let $f : [a, b] \to \mathbb{R}$ be continuous on $[a, b]$ and differentiable on (a, b) whose derivative $f' : (a, b) \to \mathbb{R}$ is bounded on (a, b), i.e., $\|f'\|_{\infty}^{\sup} := \sup_{t \in (a,b)} |f'(t)| < +\infty$. Then*

$$\left| \frac{1}{b-a} \int_a^b f(t)\,dt - f(x) \right| \leq \left[\frac{1}{4} + \frac{\left(x - \frac{a+b}{2}\right)^2}{(b-a)^2} \right] (b-a) \|f'\|_{\infty}^{\sup}, \qquad (5.1)$$

for any $x \in [a, b]$. The constant $\frac{1}{4}$ is the best possible.

Ostrowski type inequalities have great applications to integral approximations in Numerical Analysis.

We present ([1], Chap. 8,9) mixed fractional Ostrowski inequalities for several functions for various norms.

In this chapter we generalize [1], Chap. 8,9 for several Banach algebra valued functions by using ordinary vector valued derivatives and our integrals here are of Bochner type [5]. Motivation comes also from [4].

G. A. Anastassiou, *Fractional Inequalities In Banach Algebras*,
Studies in Systems, Decision and Control 441,
https://doi.org/10.1007/978-3-031-05148-7_5

We are also inspired by Opial [6], 1960, famous inequality.

Theorem 5.2 *Let $x(t) \in C^1([0,h])$ be such that $x(0) = x(h) = 0$, and $x(t) > 0$ in $(0,h)$. Then*

$$\int_0^h |x(t)\, x'(t)|\, dt \leq \frac{h}{4} \int_0^h (x'(t))^2\, dt. \tag{5.2}$$

In (5.2), the constant $\frac{h}{4}$ is the best possible. Inequality (5.2) holds as equality for the optimal function

$$x(t) = \begin{cases} ct, & 0 \leq t \leq \frac{h}{2}, \\ c(h-t), & \frac{h}{2} \leq t \leq h, \end{cases} \tag{5.3}$$

where $c > 0$ is an arbitrary constant.

Opial-type inequalities are used a lot in proving uniqueness of solutions to differential equations and also to give upper bounds to their solutions.

In this chapter we also derive Opial type inequalities for Banach algebra valued functions with respect to ordinary vector valued derivatives.

Additionally we include in this chapter related Hilbert-Pachpatte type inequalities, [8]. We finish with selective applications to Ostrowski, Opial and Hilbert-Pachpatte inequalities.

5.2 About Banach Algebras

All here come from [9].

We need

Definition 5.3 ([9], p. 245) A complex algebra is a vector space A over the complex field $\mathbb{C}$ in which a multiplication is defined that satisfies

$$x(yz) = (xy)z, \tag{5.4}$$

$$(x+y)z = xz + yz,\ x(y+z) = xy + xz, \tag{5.5}$$

and

$$\alpha(xy) = (\alpha x)y = x(\alpha y), \tag{5.6}$$

for all x, y and z in A and for all scalars α.

Additionally if A is a Banach space with respect to a norm that satisfies the multiplicative inequality

$$\|xy\| \leq \|x\|\,\|y\|\ (x \in A,\ y \in A) \tag{5.7}$$

and if A contains a unit element e such that

$$xe = ex = x \ (x \in A) \tag{5.8}$$

and

$$\|e\| = 1, \tag{5.9}$$

then A is called a Banach algebra.

A is commutative iff $xy = yx$ for all $x, y \in A$.

We make

Remark 5.4 Commutativity of A will be explicited stated when needed.

There exists at most one $e \in A$ that satisfies (5.8).

Inequality (5.7) makes multiplication to be continuous, more precisely left and right continuous, see [9], p. 246.

Multiplication in A is not necessarily the numerical multiplication, it is something more general and it is defined abstractly, that is for $x, y \in A$ we have $xy \in A$, e.g. composition or convolution, etc.

For nice examples about Banach algebras see [9], p. 247-248, § 10.3.

We also make

Remark 5.5 Next we mention about integration of A-valued functions, see [9], p. 259, Sect. 10.22:

If A is a Banach algebra and f is a continuous A-valued function on some compact Hausdorff space Q on which a complex Borel measure μ is defined, then $\int f d\mu$ exists and has all the properties that were discussed in Chapter 3 of [9], simply because A is a Banach space. However, an additional property can be added to these, namely: If $x \in A$, then

$$x \int_Q f \, d\mu = \int_Q x f(p) \, d\mu(p) \tag{5.10}$$

and

$$\left(\int_Q f \, d\mu \right) x = \int_Q f(p) \, x \, d\mu(p). \tag{5.11}$$

The Bochner integrals we will involve in our chapter follow (5.10) and (5.11).

5.3 Background

We use the following generalized vector Taylor's formula:

Theorem 5.6 ([2], p. 97) *Let $n \in \mathbb{N}$ and $f \in C^n([a, b], X)$, where $[a, b] \subset \mathbb{R}$ and $(X, \|\cdot\|)$ is a Banach space. Let $g \in C^1([a, b])$, strictly increasing, such that $g^{-1} \in C^n([g(a), g(b)])$. Let any $x, y \in [a, b]$. Then*

$$f(x) = f(y) + \sum_{i=1}^{n-1} \frac{(g(x) - g(y))^i}{i!} \left(f \circ g^{-1}\right)^{(i)} (g(y)) \tag{5.12}$$

$$+\frac{1}{(n-1)!} \int_{g(y)}^{g(x)} (g(x) - z)^{n-1} \left(f \circ g^{-1}\right)^{(n)} (z)\, dz.$$

The derivatives here are defined similarly to the numerical ones, see [10], pp. 83-86.

The above integral is of Bochner type [5], and so are the integrals in this work. By [2], p. 3, if $f \in C([a, b], X)$ then f is Bochner integrable.

5.4 Main Results

We start with mixed generalized Ostrowski type inequalities for several functions that are Banach algebra valued. A uniform estimate follows.

Theorem 5.7 *Let $n \in \mathbb{N}$ and $f_i \in C^n([a, b], A)$, $i = 1, ..., r \in \mathbb{N} - \{1\}$; where $[a, b] \subset \mathbb{R}$ and $(A, \|\cdot\|)$ is a Banach algebra. Let $g \in C^1([a, b])$, strictly increasing, such that $g^{-1} \in C^n([g(a), g(b)])$. We assume that $\left(f_i \circ g^{-1}\right)^{(j)} (g(x_0)) = 0$, $j = 1, ..., n-1$; $i = 1, ..., r$; where $x_0 \in [a, b]$ be fixed. Denote by*

$$E(f_1, ..., f_r)(x_0) :=$$

$$\sum_{i=1}^{r} \left[\int_a^b \left(\prod_{\substack{j=1 \\ j \neq i}}^{r} f_j(x) \right) f_i(x)\, dx - \left(\int_a^b \left(\prod_{\substack{j=1 \\ j \neq i}}^{r} f_j(x) \right) dx \right) f_i(x_0) \right]. \tag{5.13}$$

Then

(1)

$$E(f_1, ..., f_r)(x_0) = \frac{1}{(n-1)!}$$

$$\sum_{i=1}^{r} \left[(-1)^n \left[\int_a^{x_0} \left(\prod_{\substack{j=1 \\ j \neq i}}^{r} f_j(x) \right) \left(\int_{g(x)}^{g(x_0)} (z - g(x))^{n-1} \left(f_i \circ g^{-1}\right)^{(n)} (z)\, dz \right) dx \right] + \right. \tag{5.14}$$

$$\left. \left[\int_{x_0}^{b} \left(\prod_{\substack{j=1 \\ j \neq i}}^{r} f_j(x) \right) \left(\int_{g(x_0)}^{g(x)} (g(x) - z)^{n-1} \left(f_i \circ g^{-1}\right)^{(n)} (z)\, dz \right) dx \right] \right],$$

and

(2)

$$
\|E(f_1,...,f_r)(x_0)\| \leq \frac{1}{n!}
$$

$$
\left\{\sum_{i=1}^{r}\left[\left[\left\|\left(f_i\circ g^{-1}\right)^{(n)}\right\|_{\infty,[g(a),g(x_0)]}(g(x_0)-g(a))^n\left(\int_a^{x_0}\left(\prod_{\substack{j=1\\j\neq i}}^{r}\|f_j(x)\|\right)dx\right)\right]+\right.\right.
$$

$$
\left.\left.\left[\left\|\left(f_i\circ g^{-1}\right)^{(n)}\right\|_{\infty,[g(x_0),g(b)]}(g(b)-g(x_0))^n\left(\int_{x_0}^{b}\left(\prod_{\substack{j=1\\j\neq i}}^{r}\|f_j(x)\|\right)dx\right)\right]\right]\right\}. \tag{5.15}
$$

Proof Let $x_0 \in [a,b]$ such that $\left(f_i \circ g^{-1}\right)^{(j)}(g(x_0)) = 0$, $j = 1,...,n-1$; $i = 1,...,r$. Let $x \in [a, x_0]$, then by Theorem 5.6 we have

$$
f_i(x) - f_i(x_0) = \frac{1}{(n-1)!}\int_{g(x_0)}^{g(x)}(g(x)-z)^{n-1}\left(f_i\circ g^{-1}\right)^{(n)}(z)\,dz \tag{5.16}
$$

$$
= \frac{(-1)^n}{(n-1)!}\int_{g(x)}^{g(x_0)}(z-g(x))^{n-1}\left(f_i\circ g^{-1}\right)^{(n)}(z)\,dz,
$$

for $i = 1,...,r$.

And for $x \in [x_0, b]$, then again by Theorem 5.6 we get

$$
f_i(x) - f_i(x_0) = \frac{1}{(n-1)!}\int_{g(x_0)}^{g(x)}(g(x)-z)^{n-1}\left(f_i\circ g^{-1}\right)^{(n)}(z)\,dz, \tag{5.17}
$$

for $i = 1,...,r$.

We multiply (5.16) by $\left(\prod_{\substack{j=1\\j\neq i}}^{r} f_j(x)\right)$ to get:

$$
\left(\prod_{\substack{j=1\\j\neq i}}^{r} f_j(x)\right)f_i(x) - \left(\prod_{\substack{j=1\\j\neq i}}^{r} f_j(x)\right)f_i(x_0) =
$$

$$
\frac{\left(\prod_{\substack{j=1\\j\neq i}}^{r} f_j(x)\right)(-1)^n}{(n-1)!}\int_{g(x)}^{g(x_0)}(z-g(x))^{n-1}\left(f_i\circ g^{-1}\right)^{(n)}(z)\,dz, \tag{5.18}
$$

$\forall\, x \in [a, x_0]$; for $i = 1,...,r$.

Similarly, we get (by (5.17))

$$\left(\prod_{\substack{j=1\\j\neq i}}^{r} f_j(x)\right) f_i(x) - \left(\prod_{\substack{j=1\\j\neq i}}^{r} f_j(x)\right) f_i(x_0) =$$

$$\frac{\left(\prod_{\substack{j=1\\j\neq i}}^{r} f_j(x)\right)}{(n-1)!} \int_{g(x_0)}^{g(x)} (g(x) - z)^{n-1} \left(f_i \circ g^{-1}\right)^{(n)}(z)\, dz, \tag{5.19}$$

$\forall\, x \in [x_0, b]$; for $i = 1, ..., r$.

Adding (5.18) and (5.19) as separate groups, we obtain

$$\sum_{i=1}^{r} \left(\prod_{\substack{j=1\\j\neq i}}^{r} f_j(x)\right) f_i(x) - \sum_{i=1}^{r} \left(\prod_{\substack{j=1\\j\neq i}}^{r} f_j(x)\right) f_i(x_0) =$$

$$\frac{(-1)^n}{(n-1)!} \sum_{i=1}^{r} \left(\prod_{\substack{j=1\\j\neq i}}^{r} f_j(x)\right) \int_{g(x)}^{g(x_0)} (z - g(x))^{n-1} \left(f_i \circ g^{-1}\right)^{(n)}(z)\, dz, \tag{5.20}$$

$\forall\, x \in [a, x_0]$,

and

$$\sum_{i=1}^{r} \left(\prod_{\substack{j=1\\j\neq i}}^{r} f_j(x)\right) f_i(x) - \sum_{i=1}^{r} \left(\prod_{\substack{j=1\\j\neq i}}^{r} f_j(x)\right) f_i(x_0) =$$

$$\frac{1}{(n-1)!} \sum_{i=1}^{r} \left(\prod_{\substack{j=1\\j\neq i}}^{r} f_j(x)\right) \int_{g(x_0)}^{g(x)} (g(x) - z)^{n-1} \left(f_i \circ g^{-1}\right)^{(n)}(z)\, dz, \tag{5.21}$$

$\forall\, x \in [x_0, b]$.

Next, we integrate (5.20) and (5.21) with respect to $x \in [a, b]$. We have

$$\sum_{i=1}^{r} \int_{a}^{x_0} \left(\prod_{\substack{j=1\\j\neq i}}^{r} f_j(x)\right) f_i(x)\, dx - \sum_{i=1}^{r} \left(\int_{a}^{x_0} \left(\prod_{\substack{j=1\\j\neq i}}^{r} f_j(x)\right) dx\right) f_i(x_0) = \tag{5.22}$$

$$\frac{(-1)^n}{(n-1)!}\sum_{i=1}^{r}\left[\int_a^{x_0}\left(\prod_{\substack{j=1\\j\neq i}}^{r} f_j(x)\right)\left(\int_{g(x)}^{g(x_0)}(z-g(x))^{n-1}\left(f_i\circ g^{-1}\right)^{(n)}(z)\,dz\right)dx\right],$$

and

$$\sum_{i=1}^{r}\int_{x_0}^{b}\left(\prod_{\substack{j=1\\j\neq i}}^{r} f_j(x)\right)f_i(x)\,dx-\sum_{i=1}^{r}\left(\int_{x_0}^{b}\left(\prod_{\substack{j=1\\j\neq i}}^{r} f_j(x)\right)dx\right)f_i(x_0)= \tag{5.23}$$

$$\frac{1}{(n-1)!}\sum_{i=1}^{r}\left[\int_{x_0}^{b}\left(\prod_{\substack{j=1\\j\neq i}}^{r} f_j(x)\right)\left(\int_{g(x_0)}^{g(x)}(g(x)-z)^{n-1}\left(f_i\circ g^{-1}\right)^{(n)}(z)\,dz\right)dx\right].$$

Finally, adding (5.22) and (5.23) we obtain the useful identity

$$E(f_1,\ldots,f_r)(x_0):=$$

$$\sum_{i=1}^{r}\left[\int_a^{b}\left(\prod_{\substack{j=1\\j\neq i}}^{r} f_j(x)\right)f_i(x)\,dx-\left(\int_a^{b}\left(\prod_{\substack{j=1\\j\neq i}}^{r} f_j(x)\right)dx\right)f_i(x_0)\right]=\frac{1}{(n-1)!}$$

$$\sum_{i=1}^{r}\left[(-1)^n\left[\int_a^{x_0}\left(\prod_{\substack{j=1\\j\neq i}}^{r} f_j(x)\right)\left(\int_{g(x)}^{g(x_0)}(z-g(x))^{n-1}\left(f_i\circ g^{-1}\right)^{(n)}(z)\,dz\right)dx\right]+\right.$$

$$\left.\left[\int_{x_0}^{b}\left(\prod_{\substack{j=1\\j\neq i}}^{r} f_j(x)\right)\left(\int_{g(x_0)}^{g(x)}(g(x)-z)^{n-1}\left(f_i\circ g^{-1}\right)^{(n)}(z)\,dz\right)dx\right]\right], \tag{5.24}$$

proving (5.14).

Therefore, we get that

$$\|E(f_1,\ldots,f_r)(x_0)\|=$$

$$\left\|\sum_{i=1}^{r}\left[\int_a^{b}\left(\prod_{\substack{j=1\\j\neq i}}^{r} f_j(x)\right)f_i(x)\,dx-\left(\int_a^{b}\left(\prod_{\substack{j=1\\j\neq i}}^{r} f_j(x)\right)dx\right)f_i(x_0)\right]\right\|\leq\frac{1}{(n-1)!}$$

$$\left\{\sum_{i=1}^{r}\left\|\left[\int_{a}^{x_0}\left(\prod_{\substack{j=1\\j\neq i}}^{r} f_j(x)\right)\left(\int_{g(x)}^{g(x_0)}(z-g(x))^{n-1}\left(f_i\circ g^{-1}\right)^{(n)}(z)\,dz\right)dx\right]\right\|\right.$$

$$\left.+\left\|\left[\int_{x_0}^{b}\left(\prod_{\substack{j=1\\j\neq i}}^{r} f_j(x)\right)\left(\int_{g(x_0)}^{g(x)}(g(x)-z)^{n-1}\left(f_i\circ g^{-1}\right)^{(n)}(z)\,dz\right)dx\right]\right\|\right\}\leq$$

$$\frac{1}{(n-1)!}\left\{\sum_{i=1}^{r}\left[\left[\int_{a}^{x_0}\left\|\left(\prod_{\substack{j=1\\j\neq i}}^{r} f_j(x)\right)\left(\int_{g(x)}^{g(x_0)}(z-g(x))^{n-1}\left(f_i\circ g^{-1}\right)^{(n)}(z)\,dz\right)\right\|dx\right]\right.\right.$$

$$\left.\left.+\left[\int_{x_0}^{b}\left\|\left(\prod_{\substack{j=1\\j\neq i}}^{r} f_j(x)\right)\left(\int_{g(x_0)}^{g(x)}(g(x)-z)^{n-1}\left(f_i\circ g^{-1}\right)^{(n)}(z)\,dz\right)\right\|dx\right]\right]\right\}\leq \tag{5.25}$$

$$\frac{1}{(n-1)!}\left\{\sum_{i=1}^{r}\left[\left[\int_{a}^{x_0}\left(\prod_{\substack{j=1\\j\neq i}}^{r}\|f_j(x)\|\right)\left(\int_{g(x)}^{g(x_0)}(z-g(x))^{n-1}\left\|\left(f_i\circ g^{-1}\right)^{(n)}(z)\right\|dz\right)dx\right]\right.\right.$$

$$\left.\left.+\left[\int_{x_0}^{b}\left(\prod_{\substack{j=1\\j\neq i}}^{r}\|f_j(x)\|\right)\left(\int_{g(x_0)}^{g(x)}(g(x)-z)^{n-1}\left\|\left(f_i\circ g^{-1}\right)^{(n)}(z)\right\|dz\right)dx\right]\right]\right\}=:(\xi). \tag{5.26}$$

Hence it holds

$$\|E(f_1,\ldots,f_r)(x_0)\|\leq(\xi). \tag{5.27}$$

We have that

$$(\xi)\leq\frac{1}{n!}\left\{\sum_{i=1}^{r}\left[\left[\left\|\left\|\left(f_i\circ g^{-1}\right)^{(n)}\right\|\right\|_{\infty,[g(a),g(x_0)]}\int_{a}^{x_0}\left(\prod_{\substack{j=1\\j\neq i}}^{r}\|f_j(x)\|\right)(g(x_0)-g(x))^{n}\,dx\right]\right.\right.$$

$$\left.\left.+\left[\left\|\left\|\left(f_i\circ g^{-1}\right)^{(n)}\right\|\right\|_{\infty,[g(x_0),g(b)]}\int_{x_0}^{b}\left(\prod_{\substack{j=1\\j\neq i}}^{r}\|f_j(x)\|\right)(g(x)-g(x_0))^{n}\,dx\right]\right]\right\}\leq \tag{5.28}$$

$$\frac{1}{n!}\left\{\sum_{i=1}^{r}\left[\left[\left\|\left(f_i\circ g^{-1}\right)^{(n)}\right\|_{\infty,[g(a),g(x_0)]}(g(x_0)-g(a))^n\left(\int_a^{x_0}\left(\prod_{\substack{j=1\\ j\neq i}}^{r}\|f_j(x)\|\right)dx\right)\right]\right.\right.$$

$$\left.\left.+\left[\left\|\left(f_i\circ g^{-1}\right)^{(n)}\right\|_{\infty,[g(x_0),g(b)]}(g(b)-g(x_0))^n\left(\int_{x_0}^{b}\left(\prod_{\substack{j=1\\ j\neq i}}^{r}\|f_j(x)\|\right)dx\right)\right]\right]\right\}, \quad (5.29)$$

proving (5.15). ∎

Next comes an L_1 estimate.

Theorem 5.8 *All as in Theorem 5.7. Then*

$$\|E(f_1,\ldots,f_r)(x_0)\|\leq\frac{1}{(n-1)!}$$

$$\left\{\sum_{i=1}^{r}\left[\left[\left\|\left(f_i\circ g^{-1}\right)^{(n)}\right\|_{L_1([g(a),g(x_0)])}\int_a^{x_0}\left(\prod_{\substack{j=1\\ j\neq i}}^{r}\|f_j(x)\|\right)(g(x_0)-g(x))^{n-1}dx\right]\right.\right.$$

$$\left.\left.+\left[\left\|\left(f_i\circ g^{-1}\right)^{(n)}\right\|_{L_1([g(x_0),g(b)])}\int_{x_0}^{b}\left(\prod_{\substack{j=1\\ j\neq i}}^{r}\|f_j(x)\|\right)(g(x)-g(x_0))^{n-1}dx\right]\right]\right\}. \quad (5.30)$$

Proof By (5.26), (5.27), we get that

$$\|E(f_1,\ldots,f_r)(x_0)\|\leq(\xi)\leq\frac{1}{(n-1)!}$$

$$\left\{\sum_{i=1}^{r}\left[\left[\left\|\left(f_i\circ g^{-1}\right)^{(n)}\right\|_{L_1([g(a),g(x_0)])}\int_a^{x_0}\left(\prod_{\substack{j=1\\ j\neq i}}^{r}\|f_j(x)\|\right)(g(x_0)-g(x))^{n-1}dx\right]\right.\right.$$

$$\left.\left.+\left[\left\|\left(f_i\circ g^{-1}\right)^{(n)}\right\|_{L_1([g(x_0),g(b)])}\int_{x_0}^{b}\left(\prod_{\substack{j=1\\ j\neq i}}^{r}\|f_j(x)\|\right)(g(x)-g(x_0))^{n-1}dx\right]\right]\right\}, \quad (5.31)$$

proving (5.30). ∎

An L_p estimate follows.

Theorem 5.9 *All as in Theorem 5.7, and let* $p, q > 1 : \frac{1}{p} + \frac{1}{q} = 1$. *Then*

$$\|E(f_1, ..., f_r)(x_0)\| \leq \frac{1}{(n-1)!\,(p(n-1)+1)^{\frac{1}{p}}}$$

$$\sum_{i=1}^{r}\left[\left\|\left(f_i \circ g^{-1}\right)^{(n)}\right\|_{L_q([g(a),g(x_0)])}\left(\int_a^{x_0}(g(x_0)-g(x))^{n-\frac{1}{q}}\left(\prod_{\substack{j=1\\ j\neq i}}^{r}\|f_j(x)\|\right)dx\right)\right.$$

$$\left.+\left\|\left(f_i \circ g^{-1}\right)^{(n)}\right\|_{L_q([g(x_0),g(b)])}\left(\int_{x_0}^{b}(g(x)-g(x_0))^{n-\frac{1}{q}}\left(\prod_{\substack{j=1\\ j\neq i}}^{r}\|f_j(x)\|\right)dx\right)\right]. \tag{5.32}$$

Proof By (5.26), (5.27), we get that

$$\|E(f_1, ..., f_r)(x_0)\| \leq (\xi) \leq \frac{1}{(n-1)!}$$

$$\left\{\sum_{i=1}^{r}\left[\left[\int_a^{x_0}\left(\prod_{\substack{j=1\\ j\neq i}}^{r}\|f_j(x)\|\right)\left(\int_{g(x)}^{g(x_0)}(z-g(x))^{p(n-1)}dz\right)^{\frac{1}{p}}\right.\right.\right.$$

$$\left.\left(\int_{g(x)}^{g(x_0)}\left\|\left(f_i \circ g^{-1}\right)^{(n)}(z)\right\|^q dz\right)^{\frac{1}{q}}dx\right]+$$

$$\left[\int_{x_0}^{b}\left(\prod_{\substack{j=1\\ j\neq i}}^{r}\|f_j(x)\|\right)\left(\int_{g(x_0)}^{g(x)}(g(x)-z)^{p(n-1)}dz\right)^{\frac{1}{p}}\right. \tag{5.33}$$

$$\left.\left.\left.\left(\int_{g(x_0)}^{g(x)}\left\|\left(f_i \circ g^{-1}\right)^{(n)}(z)\right\|^q dz\right)^{\frac{1}{q}}dx\right]\right]\right\} = \frac{1}{(n-1)!}$$

$$\left\{\sum_{i=1}^{r}\left[\left[\int_a^{x_0}\left(\prod_{\substack{j=1\\ j\neq i}}^{r}\|f_j(x)\|\right)\frac{(g(x_0)-g(x))^{\frac{p(n-1)]+1}{p}}}{(p(n-1)+1)^{\frac{1}{p}}}\left\|\left(f_i \circ g^{-1}\right)^{(n)}\right\|_{L_q([g(a),g(x_0)])}dx\right]\right.\right.$$

$$+\left[\int_{x_0}^{b}\left(\prod_{\substack{j=1\\j\neq i}}^{r}\|f_j(x)\|\right)\frac{(g(x)-g(x_0))^{\frac{p(n-1)|+1}{p}}}{(p(n-1)+1)^{\frac{1}{p}}}\left\|\left(f_i\circ g^{-1}\right)^{(n)}\right\|_{L_q([g(x_0),g(b)])}dx\right]\Bigg]\Bigg\}$$

$$=\frac{1}{(n-1)!\,(p(n-1)+1)^{\frac{1}{p}}}$$

$$\left\{\sum_{i=1}^{r}\left[\left\|\left(f_i\circ g^{-1}\right)^{(n)}\right\|_{L_q([g(a),g(x_0)])}\left(\int_a^{x_0}(g(x_0)-g(x))^{n-\frac{1}{q}}\left(\prod_{\substack{j=1\\j\neq i}}^{r}\|f_j(x)\|\right)dx\right)\right.\right.$$

$$\left.\left.+\left\|\left(f_i\circ g^{-1}\right)^{(n)}\right\|_{L_q([g(x_0),g(b)])}\left(\int_{x_0}^{b}(g(x)-g(x_0))^{n-\frac{1}{q}}\left(\prod_{\substack{j=1\\j\neq i}}^{r}\|f_j(x)\|\right)dx\right)\right]\right\},\tag{5.34}$$

proving (5.32). ■

Next we present a left generalized Opial type inequality for ordinary derivatives:

Theorem 5.10 *Let $p,q>1:\frac{1}{p}+\frac{1}{q}=1$, and $n\in\mathbb{N}$, $f\in C^n([a,b],A)$; where $[a,b]\subset\mathbb{R}$ and $(A,\|\cdot\|)$ is a Banach algebra. Let $g\in C^1([a,b])$, strictly increasing, such that $g^{-1}\in C^n([g(a),g(b)])$. We assume that $\left(f\circ g^{-1}\right)^{(j)}(g(x_0))=0$, $j=0,1,...,n-1$; where $x_0\in[a,b]$ be fixed. Then*

$$\int_{g(x_0)}^{g(x)}\left\|\left(f\circ g^{-1}\right)(z)\left(f\circ g^{-1}\right)^{(n)}(z)\right\|dz\leq$$

$$\frac{(g(x)-g(x_0))^{n+\frac{1}{p}-\frac{1}{q}}}{2^{\frac{1}{q}}(n-1)!\,[(p(n-1)+1)(p(n-1)+2)]^{\frac{1}{p}}}\left(\int_{g(x_0)}^{g(x)}\left\|\left(f\circ g^{-1}\right)^{(n)}(z)\right\|^q dz\right)^{\frac{2}{q}},\tag{5.35}$$

for all $x_0\leq x\leq b$.

Proof Let $x_0\in[a,b]$ such that $\left(f\circ g^{-1}\right)^{(j)}(g(x_0))=0$, $j=0,1,...,n-1$.

For $x\in[x_0,b]$ by Theorem 5.6 we have

$$\left(f\circ g^{-1}\right)(g(z))=\frac{1}{(n-1)!}\int_{g(x_0)}^{g(x)}(g(x)-z)^{n-1}\left(f\circ g^{-1}\right)^{(n)}(z)\,dz.\tag{5.36}$$

By Hölder's inequality we obtain

$$\left\|\left(f\circ g^{-1}\right)(g(x))\right\|\leq\frac{1}{(n-1)!}\int_{g(x_0)}^{g(x)}(g(x)-z)^{n-1}\left\|\left(f\circ g^{-1}\right)^{(n)}(z)\right\|dz\leq\tag{5.37}$$

$$\frac{1}{(n-1)!}\left(\int_{g(x_0)}^{g(x)}(g(x)-z)^{p(n-1)}dt\right)^{\frac{1}{p}}\left(\int_{g(x_0)}^{g(x)}\left\|\left(f\circ g^{-1}\right)^{(n)}(z)\right\|^q dz\right)^{\frac{1}{q}}=$$

$$\frac{1}{(n-1)!}\frac{(g(x)-g(x_0))^{\frac{p(n-1)+1}{p}}}{(p(n-1)+1)^{\frac{1}{p}}}\left(\int_{g(x_0)}^{g(x)}\left\|\left(f\circ g^{-1}\right)^{(n)}(z)\right\|^q dz\right)^{\frac{1}{q}}.$$

Call

$$\varphi(g(x)):=\int_{g(x_0)}^{g(x)}\left\|\left(f\circ g^{-1}\right)^{(n)}(z)\right\|^q dz, \tag{5.38}$$

$\varphi(g(x_0))=0$.

Thus

$$\frac{d\varphi(g(x))}{dg(x)}=\left\|\left(f\circ g^{-1}\right)^{(n)}(g(x))\right\|^q\geq 0, \tag{5.39}$$

and

$$\left(\frac{d\varphi(g(x))}{dg(x)}\right)^{\frac{1}{q}}=\left\|\left(f\circ g^{-1}\right)^{(n)}(g(x))\right\|\geq 0, \tag{5.40}$$

$\forall\, g(x)\in[g(x_0),g(b)]$.

Consequently, we get

$$\left\|\left(f\circ g^{-1}\right)(g(w))\right\|\left\|\left(f\circ g^{-1}\right)^{(n)}(g(w))\right\|\leq$$

$$\frac{(g(w)-g(x_0))^{\frac{p(n-1)+1}{p}}}{(n-1)!\,(p(n-1)+1)^{\frac{1}{p}}}\left(\varphi(g(w))\frac{d\varphi(g(w))}{dg(w)}\right)^{\frac{1}{q}}, \tag{5.41}$$

$\forall\, g(w)\in[g(x_0),g(b)]$.

Then we observe that

$$\int_{g(x_0)}^{g(x)}\left\|\left(f\circ g^{-1}\right)(g(w))\left(f\circ g^{-1}\right)^{(n)}(g(w))\right\|dg(w)\overset{(5.7)}{\leq}$$

$$\int_{g(x_0)}^{g(x)}\left\|\left(f\circ g^{-1}\right)(g(w))\right\|\left\|\left(f\circ g^{-1}\right)^{(n)}(g(w))\right\|dg(w)\leq$$

$$\frac{1}{(n-1)!\,(p(n-1)+1)^{\frac{1}{p}}}$$

$$\int_{g(x_0)}^{g(x)}(g(w)-g(x_0))^{\frac{p(n-1)+1}{p}}\left(\varphi(g(w))\frac{d\varphi(g(w))}{dg(w)}\right)^{\frac{1}{q}}dg(w)\leq \tag{5.42}$$

$$\frac{1}{(n-1)!\,(p\,(n-1)+1)^{\frac{1}{p}}}$$

$$\left(\int_{g(x_0)}^{g(x)} (g\,(w)-g\,(x_0))^{p(n-1)+1}\,dg\,(w)\right)^{\frac{1}{p}} \left(\int_{g(x_0)}^{g(x)} \varphi\,(g\,(w))\,\frac{d\varphi\,(g\,(w))}{dg\,(w)}dg\,(w)\right)^{\frac{1}{q}} =$$

$$\frac{1}{(n-1)!\,(p\,(n-1)+1)^{\frac{1}{p}}\,(p\,(n-1)+2)^{\frac{1}{p}}}$$

$$(g\,(x)-g\,(x_0))^{\frac{p(n-1)+2}{p}} \left(\int_{g(x_0)}^{g(x)} \varphi\,(g\,(w))\,d\varphi\,(g\,(w))\right)^{\frac{1}{q}} = \tag{5.43}$$

$$\frac{(g\,(x)-g\,(x_0))^{n+\frac{1}{p}-\frac{1}{q}}}{(n-1)!\,(p\,(n-1)+1)^{\frac{1}{p}}\,(p\,(n-1)+2)^{\frac{1}{p}}} \left(\frac{\varphi^2\,(g\,(x))}{2}\right)^{\frac{1}{q}} =$$

$$\frac{(g\,(x)-g\,(x_0))^{n+\frac{1}{p}-\frac{1}{q}}}{2^{\frac{1}{q}}\,(n-1)!\,((p\,(n-1)+1)\,(p\,(n-1)+2))^{\frac{1}{p}}} \left(\int_{g(x_0)}^{g(x)} \left\|\left(f\circ g^{-1}\right)^{(n)}(z)\right\|^q dz\right)^{\frac{2}{q}}, \tag{5.44}$$

for all $g\,(x_0) \le g\,(x) \le g\,(b)$, proving (5.35). ■

The corresponding right generalized Opial type inequality follows:

Theorem 5.11 *All as in Theorem 5.10. Then*

$$\int_{g(x)}^{g(x_0)} \left\|\left(f\circ g^{-1}\right)(z)\left(f\circ g^{-1}\right)^{(n)}(z)\right\| dz \le$$

$$\frac{(g\,(x_0)-g\,(x))^{n+\frac{1}{p}-\frac{1}{q}}}{2^{\frac{1}{q}}\,(n-1)!\,((p\,(n-1)+1)\,(p\,(n-1)+2))^{\frac{1}{p}}} \left(\int_{g(x)}^{g(x_0)} \left\|\left(f\circ g^{-1}\right)^{(n)}(z)\right\|^q dz\right)^{\frac{2}{q}}, \tag{5.45}$$

for all $a \le x \le x_0$.

Proof As similar to Theorem 5.10 is omitted. ■

Next we present a left generalized Hilbert-Pachpatte inequality for ordinary derivatives.

Theorem 5.12 *Let* $i = 1, 2;$ $p, q > 1 : \frac{1}{p} + \frac{1}{q} = 1,$ *and* $n_i \in \mathbb{N},$ $f_i \in C^{n_i}$ $([a_i, b_i], A)$*; where* $[a_i, b_i] \subset \mathbb{R}$ *and* $(A, \|\cdot\|)$ *is a Banach algebra. Let* $g_i \in C^1$ $([a_i, b_i])$*, strictly increasing, such that* $g_i^{-1} \in C^{n_i}\,([g_i\,(a_i)\,, g_i\,(b_i)])$*. We assume that* $\left(f_i \circ g_i^{-1}\right)^{(j_i)}(g_i\,(x_{0i})) = 0,$ $j_i = 0, 1, ..., n_i - 1;$ *where* $x_{0i} \in [a_i, b_i]$ *be fixed. Then*

$$\int_{g_1(x_{01})}^{g_1(b_1)} \int_{g_2(x_{02})}^{g_2(b_2)} \frac{\left\| \left(f_1 \circ g_1^{-1}\right)(z_1) \left(f_2 \circ g_2^{-1}\right)(z_2) \right\| dz_1 dz_2}{\left(\frac{(z_1 - g_1(x_{01}))^{p(n_1-1)+1}}{p(p(n_1-1)+1)} + \frac{(z_2 - g_2(x_{02}))^{q(n_2-1)+1}}{q(q(n_2-1)+1)} \right)} \leq$$

$$\frac{(g_1(b_1) - g_1(x_{01}))(g_2(b_2) - g_2(x_{02}))}{(n_1-1)!(n_2-1)!} \tag{5.46}$$

$$\left\| \left\| \left(f_1 \circ g_1^{-1}\right)^{(n_1)} \right\| \right\|_{L_q([g_1(x_{01}), g_1(b_1)], A)} \left\| \left\| \left(f_2 \circ g_2^{-1}\right)^{(n_2)} \right\| \right\|_{L_p([g_2(x_{02}), g_2(b_2)], A)}.$$

Proof Let $i = 1, 2$; $x_0 \in [a_i, b_i]$, such that $\left(f_i \circ g_i^{-1}\right)^{(j_i)}(g_i(x_{0i})) = 0$, $j_i = 0, 1, ..., n_i - 1$.

For $x_i \in [x_{0i}, b_i]$ by Theorem 5.6 we have

$$\left(f_i \circ g_i^{-1}\right)(g_i(x_i)) = \frac{1}{(n_i-1)!} \int_{g_i(x_{0i})}^{g_i(x_i)} (g_i(x_i) - z_i)^{n_i-1} \left(f_i \circ g_i^{-1}\right)^{(n_i)}(z_i) \, dz_i. \tag{5.47}$$

As in (5.37) we have

$$\left\| \left(f_1 \circ g_1^{-1}\right)(g_1(x_1)) \right\| \leq \frac{1}{(n_1-1)!} \frac{(g_1(x_1) - g_1(x_{01}))^{\frac{p(n_1-1)+1}{p}}}{(p(n_1-1)+1)^{\frac{1}{p}}}$$

$$\left(\int_{g_1(x_{01})}^{g_1(x_1)} \left\| \left(f_1 \circ g_1^{-1}\right)^{(n_1)}(z) \right\|^q dz \right)^{\frac{1}{q}} \leq$$

$$\frac{1}{(n_1-1)!} \frac{(g_1(x_1) - g_1(x_{01}))^{\frac{p(n_1-1)+1}{p}}}{(p(n_1-1)+1)^{\frac{1}{p}}} \left\| \left\| \left(f_1 \circ g_1^{-1}\right)^{(n_1)} \right\| \right\|_{L_q([g_1(x_{01}), g_1(b_1)])}, \tag{5.48}$$

for all $x_1 \in [x_{01}, b_1]$.

Similarly, we obtain that

$$\left\| \left(f_2 \circ g_2^{-1}\right)(g_2(x_2)) \right\| \leq \frac{1}{(n_2-1)!} \frac{(g_2(x_2) - g_2(x_{02}))^{\frac{q(n_2-1)+1}{q}}}{(q(n_2-1)+1)^{\frac{1}{q}}}$$

$$\left\| \left\| \left(f_2 \circ g_2^{-1}\right)^{(n_2)} \right\| \right\|_{L_p([g_2(x_{02}), g_2(b_2)])}, \tag{5.49}$$

for all $x_2 \in [x_{02}, b_2]$.

By (5.48) and (5.49) we get

$$\left\| \left(f_1 \circ g_1^{-1}\right)(g_1(x_1)) \left(f_2 \circ g_2^{-1}\right)(g_2(x_2)) \right\| \leq$$

$$\left\|\left(f_1 \circ g_1^{-1}\right)\left(g_1\left(x_1\right)\right)\right\| \left\|\left(f_2 \circ g_2^{-1}\right)\left(g_2\left(x_2\right)\right)\right\| \leq \frac{1}{\left(n_1-1\right)!\left(n_2-1\right)!}$$

$$\frac{\left(g_1\left(x_1\right)-g_1\left(x_{01}\right)\right)^{\frac{p\left(n_1-1\right)+1}{p}}}{\left(p\left(n_1-1\right)+1\right)^{\frac{1}{p}}} \frac{\left(g_2\left(x_2\right)-g_2\left(x_{02}\right)\right)^{\frac{q\left(n_2-1\right)+1}{q}}}{\left(q\left(n_2-1\right)+1\right)^{\frac{1}{q}}} \tag{5.50}$$

$$\left\|\left(f_1 \circ g_1^{-1}\right)^{(n_1)}\right\|_{L_q([g_1(x_{01}), g_1(b_1)])} \left\|\left(f_2 \circ g_2^{-1}\right)^{(n_2)}\right\|_{L_p([g_2(x_{02}), g_2(b_2)])} \leq$$

(using Young's inequality for $a, b \geq 0$, $a^{\frac{1}{p}} b^{\frac{1}{q}} \leq \frac{a}{p}+\frac{b}{q}$)

$$\frac{1}{\left(n_1-1\right)!\left(n_2-1\right)!}\left(\frac{\left(g_1\left(x_1\right)-g_1\left(x_{01}\right)\right)^{p\left(n_1-1\right)+1}}{p\left(p\left(n_1-1\right)+1\right)}+\frac{\left(g_2\left(x_2\right)-g_2\left(x_{02}\right)\right)^{q\left(n_2-1\right)+1}}{q\left(q\left(n_2-1\right)+1\right)}\right) \tag{5.51}$$

$$\left\|\left(f_1 \circ g_1^{-1}\right)^{(n_1)}\right\|_{L_q([g_1(x_{01}), g_1(b_1)])} \left\|\left(f_2 \circ g_2^{-1}\right)^{(n_2)}\right\|_{L_p([g_2(x_{02}), g_2(b_2)])},$$

$\forall\left(x_1, x_2\right) \in\left[x_{01}, b_1\right] \times\left[x_{02}, b_2\right]$.

So far we have

$$\frac{\left\|\left(f_1 \circ g_1^{-1}\right)\left(g_1\left(x_1\right)\right)\left(f_2 \circ g_2^{-1}\right)\left(g_2\left(x_2\right)\right)\right\|}{\left(\frac{\left(g_1(x_1)-g_1(x_{01})\right)^{p(n_1-1)+1}}{p(p(n_1-1)+1)}+\frac{\left(g_2(x_2)-g_2(x_{02})\right)^{q(n_2-1)+1}}{q(q(n_2-1)+1)}\right)} \leq \tag{5.52}$$

$$\frac{1}{\left(n_1-1\right)!\left(n_2-1\right)!}\left\|\left(f_1 \circ g_1^{-1}\right)^{(n_1)}\right\|_{L_q([g_1(x_{01}), g_1(b_1)], A)}$$

$$\left\|\left(f_2 \circ g_2^{-1}\right)^{(n_2)}\right\|_{L_p([g_2(x_{02}), g_2(b_2)], A)},$$

$\forall\left(x_1, x_2\right) \in\left[x_{01}, b_1\right] \times\left[x_{02}, b_2\right]$.

The denominator in (5.52) can be zero, only when both $g_1\left(x_1\right)=g_1\left(x_{01}\right)$ and $g_2\left(x_2\right)=g_2\left(x_{02}\right)$.

Therefore we obtain (5.46), by integrating (5.52) over $\left[g_1\left(x_{01}\right), g_1\left(b_1\right)\right] \times \left[g_2\left(x_{02}\right), g_2\left(b_2\right)\right]$. ■

It follows the right generalized Hilbert-Pachpate inequality for ordinary derivatives.

Theorem 5.13 *All as in Theorem 5.12. Then*

$$\int_{g_1(a_1)}^{g_1(x_{01})} \int_{g_2(a_2)}^{g_2(x_{02})} \frac{\left\|\left(f_1 \circ g_1^{-1}\right)\left(z_1\right)\left(f_2 \circ g_2^{-1}\right)\left(z_2\right)\right\| dz_1 dz_2}{\left(\frac{\left(g_1(x_{01})-z_1\right)^{p(n_1-1)+1}}{p(p(n_1-1)+1)}+\frac{\left(g_2(x_{02})-z_2\right)^{q(n_2-1)+1}}{q(q(n_2-1)+1)}\right)} \leq$$

$$\frac{\left(g_1\left(x_{01}\right)-g_1\left(a_1\right)\right)\left(g_2\left(x_{02}\right)-g_2\left(a_2\right)\right)}{\left(n_1-1\right)!\left(n_2-1\right)!} \tag{5.53}$$

$$\left\|\left\|\left(f_1 \circ g_1^{-1}\right)^{(n_1)}\right\|\right\|_{L_q([g_1(a_1),g_1(x_{01})],A)} \left\|\left\|\left(f_2 \circ g_2^{-1}\right)^{(n_2)}\right\|\right\|_{L_p([g_2(a_2),g_2(x_{02})],A)}.$$

Proof As similar to Theorem 5.12 is omitted. ■

5.5 Applications

We make

Remark 5.14 Assume next that $(A, \|\cdot\|)$ is a commutative Banach algebra. Then, we get that

$$E\left(f_1,...,f_r\right)\left(x_0\right) \overset{(5.13)}{=} r\int_a^b \left(\prod_{j=1}^r f_j(x)\right)dx - \sum_{i=1}^r \left(\int_a^b \left(\prod_{\substack{j=1\\ j\neq i}}^r f_j(x)\right)dx\right) f_i\left(x_0\right), \tag{5.54}$$

$x_0 \in [a, b]$.

When $r = 2$, we have that

$$E\left(f_1, f_2\right)\left(x_0\right) = 2\int_a^b f_1(x)\, f_2(x)\, dx - f_1\left(x_0\right)\int_a^b f_2(x)\, dx - f_2\left(x_0\right)\int_a^b f_1(x)\, dx, \tag{5.55}$$

$x_0 \in [a, b]$.

We give

Corollary 5.15 (to Theorem 5.7) *All as in Theorem 5.7, $(A, \|\cdot\|)$ is a commutative Banach algebra, $r = 2$. Then*

$$\left\|E\left(f_1, f_2\right)\left(x_0\right)\right\| \leq \frac{1}{n!}\sum_{i=1}^2 \left[\left[\left\|\left\|\left(f_i \circ g^{-1}\right)^{(n)}\right\|\right\|_{\infty,[g(a),g(x_0)]}\right.\right.$$

$$\left.\left(g\left(x_0\right)-g(a)\right)^n \left(\int_a^{x_0}\left(\prod_{\substack{j=1\\ j\neq i}}^2 \left\|f_j(x)\right\|\right)dx\right)\right]+$$

$$\left.\left[\left\|\left\|\left(f_i \circ g^{-1}\right)^{(n)}\right\|\right\|_{\infty,[g(x_0),g(b)]}\left(g(b)-g\left(x_0\right)\right)^n\left(\int_{x_0}^b\left(\prod_{\substack{j=1\\ j\neq i}}^2 \left\|f_j(x)\right\|\right)dx\right)\right]\right]. \tag{5.56}$$

It follows

Corollary 5.16 (to Corollary 5.15) *All as in Corollary 5.15, with $g(t) = e^t$. Then*

$$\|E(f_1, f_2)(x_0)\| \leq \frac{1}{n!} \sum_{i=1}^{2} \Bigg[\Bigg[\left\| \|(f_i \circ \log)^{(n)}\| \right\|_{\infty,[e^a, e^{x_0}]} (e^{x_0} - e^a)^n \left(\int_a^{x_0} \left(\prod_{\substack{j=1 \\ j \neq i}}^{2} \|f_j(x)\| \right) dx \right) \Bigg] +$$

$$\Bigg[\left\| \|(f_i \circ \log)^{(n)}\| \right\|_{\infty,[e^{x_0}, e^b]} (e^b - e^{x_0})^n \left(\int_{x_0}^{b} \left(\prod_{\substack{j=1 \\ j \neq i}}^{2} \|f_j(x)\| \right) dx \right) \Bigg] \Bigg]. \quad (5.57)$$

We continue with

Corollary 5.17 (to Theorem 5.10) *All as in Theorem 5.10 for $g(t) = e^t$. Then*

$$\int_{e^{x_0}}^{e^x} \left\| ((f \circ \log)(z)) (f \circ \log)^{(n)}(z) \right\| dz \leq$$

$$\frac{(e^x - e^{x_0})^{n + \frac{1}{p} - \frac{1}{q}}}{2^{\frac{1}{q}} (n-1)! [(p(n-1)+1)(p(n-1)+2)]^{\frac{1}{p}}} \left(\int_{e^{x_0}}^{z} \left\| (f \circ \log)^{(n)}(z) \right\|^q dz \right)^{\frac{2}{q}}, \quad (5.58)$$

for all $x_0 \leq x \leq b$.

We finish with

Corollary 5.18 (to Theorem 5.12) *All as in Theorem 5.12 for $g_i(t) = e^t$, $i = 1, 2$. Then*

$$\int_{e^{x_{01}}}^{e^{b_1}} \int_{e^{x_{02}}}^{e^{b_2}} \frac{\|(f_1 \circ \log)(z_1)(f_2 \circ \log)(z_2)\| \, dz_1 dz_2}{\left(\frac{(z_1 - e^{x_{01}})^{p(n_1 - 1)+1}}{p(p(n_1-1)+1)} + \frac{(z_2 - e^{x_{02}})^{q(n_2-1)+1}}{q(q(n_2-1)+1)} \right)} \leq$$

$$\frac{\left(e^{b_1} - e^{x_{01}}\right)\left(e^{b_2} - e^{x_{02}}\right)}{(n_1 - 1)!(n_2 - 1)!} \quad (5.59)$$

$$\left\| \|(f_1 \circ \log)^{(n_1)}\| \right\|_{L_q([e^{x_{01}}, e^{b_1}], A)} \left\| \|(f_2 \circ \log)^{(n_2)}\| \right\|_{L_p([e^{x_{02}}, e^{b_2}], A)}.$$

The simplest applications derive when $g(t) = t$ and $A = \mathbb{R}$, leading to basic known results.

References

1. Anastassiou, G.A.: Intelligent Comparisons: Analytic Inequalities. Springer, Heidelberg, New York (2016)
2. Anastassiou, G.A.: Intelligent Computations: Abstract Fractional Calculus, Inequalities, Approximations. Springer, Heidelberg, New York (2018)
3. Anastassiou, G.A.: Generalized Ostrowski, Opial and Hilbert-Pachpatte type inequalities for Banach algebra valued functions involving integer vectorial derivatives. J. Comput. Anal. Appli. **30**(1), 78–94 (2022)
4. Dragomir, S.S.: Noncommutative Ostrowski type inequalities for functions in Banach algebras. RGMIA Res. Rep. Coll. **24**, Art. 10, 24 pp (2021)
5. Mikusinski, J.: The Bochner Integral. Academic Press, New York (1978)
6. Opial, Z.: Sur une inegalite. Ann. Polon. Math. **8**, 29–32 (1960)
7. Ostrowski, A.: Über die Absolutabweichung einer differentiabaren Funcktion von ihrem Integralmittelwert. Comment. Math. Helv. **10**, 226–227 (1938)
8. Pachpatte, B.G.: Inequalities similar to the integral analogue of Hilbert's inequalities. Tamkang J. Math. **30**(1), 139–146 (1999)
9. Rudin, W.: Functional Analysis, 2nd edn. McGraw-Hill Inc, New York (1991)
10. Shilov, G.E.: Elementary Functional Analysis. Dover Publications Inc., New York (1996)

Chapter 6
Multivariate Ostrowski Inequalities for Several Banach Algebra Valued Functions

Here we are dealing with several smooth functions from a compact convex set of $\mathbb{R}^k$, $k \geq 2$ to a Banach algebra. For these we prove general multivariate Ostrowski inequalities with estimates. It follows [2].

6.1 Introduction

In 1938, A Ostrowski [6] proved the following famous inequality:

Theorem 6.1 (1938, Ostrowski [7]) *Let $f : [a, b] \to \mathbb{R}$ be continuous on $[a, b]$ and differentiable on (a, b) whose derivative $f' : (a, b) \to \mathbb{R}$ is bounded on (a, b), i.e., $\|f'\|_\infty^{\sup} := \sup_{t \in (a,b)} |f'(t)| < +\infty$. Then*

$$\left| \frac{1}{b-a} \int_a^b f(t)\, dt - f(x) \right| \leq \left[\frac{1}{4} + \frac{\left(x - \frac{a+b}{2}\right)^2}{(b-a)^2} \right] (b-a) \|f'\|_\infty^{\sup}, \tag{6.1}$$

for any $x \in [a, b]$. The constant $\frac{1}{4}$ is the best possible.

Since then there has been a lot of activity around these inequalities with important applications to Numerical Analysis and Probability.

This chapter is also greatly motivated by the following result:

Theorem 6.2 (see [1]) *Let $f \in C^1\left(\prod_{i=1}^k [a_i, b_i]\right)$, where $a_i < b_i$; $a_i, b_i \in \mathbb{R}$, $i = 1, \ldots, k$, ans let $\overrightarrow{x_0} := (x_{01}, \ldots, x_{0k}) \in \prod_{i=1}^k [a_i, b_i]$ be fixed. Then*

G. A. Anastassiou, *Fractional Inequalities In Banach Algebras*,
Studies in Systems, Decision and Control 441,
https://doi.org/10.1007/978-3-031-05148-7_6

$$\left| \frac{1}{\prod_{i=1}^{k} (b_i - a_i)} \int_{a_1}^{b_1} \cdots \int_{a_i}^{b_i} \cdots \int_{a_k}^{b_k} f(z_1, \ldots, z_k) \, dz_1 \cdots dz_k - f(\overrightarrow{x_0}) \right| \leq \tag{6.2}$$

$$\sum_{i=1}^{k} \left(\frac{(x_{0i} - a_i)^2 + (b_i - x_{0i})^2}{2(b_i - a_i)} \right) \left\| \frac{\partial f}{\partial z_i} \right\|_{\infty}.$$

Inequality (6.2) is sharp, here the optimal function is

$$f^*(z_1, \ldots, z_k) := \sum_{i=1}^{k} |z_i - x_{0i}|^{\alpha_i}, \ \alpha_i > 1.$$

Clearly inequality (6.2) generalizes inequality (6.1) to multidimension.

We are inspired also by [3].

In this chapter we establish multivariate Ostrowski type inequalities for several smooth functions from a compact convex subset of $\mathbb{R}^k$, $k \geq 2$, to a Banach algebra. These involve the norms $\|\cdot\|_p$, $1 \leq p \leq \infty$.

6.2 About Banach Algebras

All here come from [7].

We need

Definition 6.3 ([7], *p. 245*) A complex algebra is a vector space A over the complex field $\mathbb{C}$ in which a multiplication is defined that satisfies

$$x(yz) = (xy)z, \tag{6.3}$$

$$(x + y)z = xz + yz, \ x(y + z) = xy + xz, \tag{6.4}$$

and

$$\alpha(xy) = (\alpha x)y = x(\alpha y), \tag{6.5}$$

for all x, y and z in A and for all scalars α.

Additionally if A is a Banach space with respect to a norm that satisfies the multiplicative inequality

$$\|xy\| \leq \|x\| \|y\| \ (x \in A, y \in A) \tag{6.6}$$

and if A contains a unit element e such that

$$xe = ex = x \ (x \in A) \tag{6.7}$$

and

$$\|e\| = 1, \tag{6.8}$$

then A is called a Banach algebra.

A is commutative iff $xy = yx$ for all $x, y \in A$.

We make

Remark 6.4 Commutativity of A will be explicited stated when needed.

There exists at most one $e \in A$ that satisfies (6.7).

Inequality (6.6) makes multiplication to be continuous, more precisely left and right continuous, see [7], p. 246.

Multiplication in A is not necessarily the numerical multiplication, it is something more general and it is defined abstractly, that is for $x, y \in A$ we have $xy \in A$, e.g. composition or convolution, etc.

For nice examples about Banach algebras see [7], pp. 247–248, Sect. 10.3.

We also make

Remark 6.5 Next we mention about integration of A-valued functions, see [7], p. 259, Sect. 10.22:

If A is a Banach algebra and f is a continuous A-valued function on some compact Hausdorff space Q on which a complex Borel measure μ is defined, then $\int f d\mu$ exists and has all the properties that were discussed in Chap. 3 of [7], simply because A is a Banach space. However, an additional property can be added to these, namely: If $x \in A$, then

$$x \int_Q f \, d\mu = \int_Q xf(p) \, d\mu(p) \tag{6.9}$$

and

$$\left(\int_Q f \, d\mu \right) x = \int_Q f(p) \, x \, d\mu(p). \tag{6.10}$$

The vector integrals we will involve in our chapter follow (6.9) and (6.10).

6.3 Vector Analysis Background

(see [9], pp. 83–94)

Let $f(t)$ be a function defined on $[a, b] \subseteq \mathbb{R}$ taking values in a real or complex normed linear space $(X, \|\cdot\|)$, Then $f(t)$ is said to be differentiable at a point $t_0 \in [a, b]$ if the limit

$$f'(t_0) = \lim_{h \to 0} \frac{f(t_0 + h) - f(t_0)}{h} \tag{6.11}$$

exists in X, the convergence is in $\|\cdot\|$. This is called the derivative of $f(t)$ at $t = t_0$.

We call $f(t)$ differentiable on $[a, b]$, iff there exists $f'(t) \in X$ for all $t \in [a, b]$.

Similarly and inductively are defined higher order derivatives of f, denoted f'', $f^{(3)}, \ldots, f^{(k)}$, $k \in \mathbb{N}$, just as for numerical functions.

For all the properties of derivatives see [9], pp. 83–86.

Let now $(X, \|\cdot\|)$ be a Banach space, and $f : [a, b] \to X$.

We define the vector valued Riemann integral $\int_a^b f(t)\, dt \in X$ as the limit of the vector valued Riemann sums in X, convergence is in $\|\cdot\|$. The definition is as for the numerical valued functions.

If $\int_a^b f(t)\, dt \in X$ we call f integrable on $[a, b]$. If $f \in C([a, b], X)$, then f is integrable, [9], p. 87.

For all the properties of vector valued Riemann integrals see [9], pp. 86–91.

We define the space $C^n([a, b], X)$, $n \in \mathbb{N}$, of n-times continuousky differentiable functions from $[a, b]$ into X; here continuity is with respect to $\|\cdot\|$ and defined in the usual way as for numerical functions·.

Let $(X, \|\cdot\|)$ be a Banach space and $f \in C^n([a, b], X)$, then we have the vector valued Taylor's formula, see [9], pp. 93–94, and also [8], (IV, 9; 47).

It holds

$$f(y) - f(x) - f'(x)(y - x) - \frac{1}{2} f''(x)(y - x)^2 - \cdots - \frac{1}{(n-1)!} f^{(n-1)}(x)(y - x)^{n-1} \tag{6.12}$$

$$= \frac{1}{(n-1)!} \int_x^y (y - t)^{n-1} f^{(n)}(t)\, dt, \ \forall\, x, y \in [a, b].$$

In particular (6.12) is true when $X = \mathbb{R}^m, \mathbb{C}^m$, $m \in \mathbb{N}$, etc.

A function $f(t)$ with values in a normed linear space X is said to be piecewise continuous (see [9], p. 85) on the interval $a \leq t \leq b$ if there exists a partition $a = t_0 < t_1 < t_2 < \cdots < t_n = b$ such that $f(t)$ is continuous on every open interval $t_k < t < t_{k+1}$ and has finite limits $f(t_0 + 0)$, $f(t_1 - 0)$, $f(t_1 + 0)$, $f(t_2 - 0)$, $f(t_2 + 0), \ldots, f(t_n - 0)$.

Here $f(t_k - 0) = \lim_{t \uparrow t_k} f(t)$, $f(t_k + 0) = \lim_{t \downarrow t_k} f(t)$.

The values of $f(t)$ at the points t_k can be arbitrary or even undefined.

A function $f(t)$ with values in normed linear space X is said to be piecewise smooth on $[a, b]$, if it is continuous on $[a, b]$ and has a derivative $f'(t)$ at all but a finite number of points of $[a, b]$, and if $f'(t)$ is piecewise continuous on $[a, b]$ (see [9], p. 85).

Let $u(t)$ and $v(t)$ be two piecewise smooth functions on $[a, b]$, one a numerical function and the other a vector function with values in Banach space X. Then we have the following integration by parts formula

$$\int_a^b u(t)\, dv(t) = u(t)\, v(t) \big|_a^b - \int_a^b v(t)\, du(t), \tag{6.13}$$

see [9], p. 93.

We mention also the mean value theorem for Banach space valued functions.

Theorem 6.6 (see [5], p. 3) *Let $f \in C([a,b], X)$, where X is a Banach space. Assume f' exists on $[a,b]$ and $\|f'(t)\| \leq K$, $a < t < b$, then*

$$\|f(b) - f(a)\| \leq K(b-a). \tag{6.14}$$

Here the multiple Riemann integral of a function from a real box or a real compact and convex subset to a Banach space is defined similarly to numerical one however convergence is with respect to $\|\cdot\|$. Similarly are defined the vector valued partial derivatives as in the numerical case.

We mention the equality of vector valued mixed partiasl derivatives.

Proposition 6.7 (see Proposition 4.11 of [4], p. 90) *Let $Q = (a,b) \times (c,d) \subseteq \mathbb{R}^2$ and $f \in C(Q, X)$, where $(X, \|\cdot\|)$ is a Banach space. Assume that $\frac{\partial}{\partial t} f(s,t)$, $\frac{\partial}{\partial s} f(s,t)$ and $\frac{\partial^2}{\partial t \partial s} f(s,t)$ exist and are continuous for $(s,t) \in Q$, then $\frac{\partial^2}{\partial s \partial t} f(s,t)$ exists for $(s,t) \in Q$ and*

$$\frac{\partial^2}{\partial s \partial t} f(s,t) = \frac{\partial^2}{\partial t \partial s} f(s,t), \text{ for } (s,t) \in Q. \tag{6.15}$$

6.4 Main Results

We present general Ostrowski type inequalities results regarding several Banach algebra valued functions.

Theorem 6.8 *Let $p, q > 1 : \frac{1}{p} + \frac{1}{q} = 1$; $(A, \|\cdot\|)$ a Banach algebra and $f_i \in C^{n+1}(Q, A)$, $i = 1, \ldots, r$; $r \in \mathbb{N}$, $n \in \mathbb{Z}_+$, and fixed $\overrightarrow{x_0} \in Q \subset \mathbb{R}^k$, $k \geq 2$, where Q is a compact and convex subset. Here all vector partial derivatives $f_{i\alpha} := \frac{\partial^\alpha f_i}{\partial z^\alpha}$, where $\alpha = (\alpha_1, \ldots, \alpha_k)$, $\alpha_\lambda \in \mathbb{Z}^+$, $\lambda = 1, \ldots, k$, $|\alpha| = \sum_{\lambda=1}^{k} \alpha_\lambda = j$, $j = 1, \ldots, n$, fulfill $f_{i\alpha}(\overrightarrow{x_0}) = 0$, $i = 1, \ldots, r$.*

Denote

$$D_{n+1}(f_i) := \max_{\alpha : |\alpha| = n+1} \| \|f_{i\alpha}\| \|_{\infty, Q}, \tag{6.16}$$

$i = 1, \ldots, r$, and

$$\left\| \overrightarrow{z} - \overrightarrow{x_0} \right\|_{l_1} := \sum_{\lambda=1}^{k} |z_\lambda - x_{0\lambda}|. \tag{6.17}$$

Then

$$\left\| \sum_{i=1}^{r} \int_{Q} \left(\prod_{\substack{\rho=1 \\ \rho \neq i}}^{r} f_{\rho}\left(\overrightarrow{z}\right) \right) f_{i}\left(\overrightarrow{z}\right) d\overrightarrow{z} - \sum_{i=1}^{r} \left(\int_{Q} \left(\prod_{\substack{\rho=1 \\ \rho \neq i}}^{r} f_{\rho}\left(\overrightarrow{z}\right) \right) d\overrightarrow{z} \right) f_{i}\left(\overrightarrow{x_0}\right) \right\| \leq \tag{6.18}$$

$$\frac{\max_{i \in \{1,\ldots,r\}} D_{n+1}(f_i)}{(n+1)!} \sum_{i=1}^{r} \left(\int_{Q} \left(\prod_{\substack{\rho=1 \\ \rho \neq i}}^{r} \left\| f_{\rho}\left(\overrightarrow{z}\right) \right\| \right) \left\| \overrightarrow{z} - \overrightarrow{x_0} \right\|_{l_1}^{n+1} d\overrightarrow{z} \right) \leq$$

$$\frac{\max_{i \in \{1,\ldots,r\}} D_{n+1}(f_i)}{(n+1)!} \min \left\{ \left(\int_{Q} \left\| \overrightarrow{z} - \overrightarrow{x_0} \right\|_{l_1}^{n+1} d\overrightarrow{z} \right) \left[\sum_{i=1}^{r} \left(\prod_{\substack{\rho=1 \\ \rho \neq i}}^{r} \left\| \left\| f_{\rho} \right\| \right\|_{\infty,Q} \right) \right], \right.$$

$$\left\| \left\| \cdot - \overrightarrow{x_0} \right\|_{l_1}^{n+1} \right\|_{\infty,Q} \left[\sum_{i=1}^{r} \left\| \left(\prod_{\substack{\rho=1 \\ \rho \neq i}}^{r} \left\| f_{\rho} \right\| \right) \right\|_{L_1(Q,A)} \right],$$

$$\left. \left\| \left\| \cdot - \overrightarrow{x_0} \right\|_{l_1}^{n+1} \right\|_{L_p(Q,A)} \left[\sum_{i=1}^{r} \left[\left\| \left(\prod_{\substack{\rho=1 \\ \rho \neq i}}^{r} \left\| f_{\rho} \right\| \right) \right\|_{L_q(Q,A)} \right] \right] \right\}. \tag{6.19}$$

Proof Take $g_{i\overrightarrow{z}}(t) := f_i\left(\overrightarrow{x_0} + t\left(\overrightarrow{z} - \overrightarrow{x_0}\right)\right), 0 \leq t \leq 1; i = 1, \ldots, r$. Notice that $g_{i\overrightarrow{z}}(0) = f_i\left(\overrightarrow{x_0}\right)$ and $g_{i\overrightarrow{z}}(1) = f_i\left(\overrightarrow{z}\right)$. The jth derivative of $g_{i\overrightarrow{z}}(t)$, based on Proposition 6.7, is given by

$$g_{i\overrightarrow{z}}^{(j)}(t) = \left[\left(\sum_{\lambda=1}^{k} (z_\lambda - x_{0\lambda}) \frac{\partial}{\partial z_\lambda} \right)^{j} f_i \right] (x_{01} + t(z_1 - x_{01}), \ldots, x_{0k} + t(z_k - x_{0k})) \tag{6.20}$$

and

$$g_{i\overrightarrow{z}}^{(j)}(0) = \left[\left(\sum_{\lambda=1}^{k} (z_\lambda - x_{0\lambda}) \frac{\partial}{\partial z_\lambda} \right)^{j} f_i \right] \left(\overrightarrow{x_0}\right), \tag{6.21}$$

for $j = 1, \ldots, n+1; i = 1, \ldots, r$.

Let $f_{i\alpha}$ be a partial derivative of $f_i \in C^{n+1}(Q, A)$. Because by assumption of the theorem we have $f_{i\alpha}\left(\overrightarrow{x_0}\right) = 0$ for all $\alpha : |\alpha| = j, j = 1, \ldots, n$, we find that

$$g_{i\overrightarrow{z}}^{(j)}(0) = 0, \quad j = 1, \ldots, n; \quad i = 1, \ldots, r.$$

Hence by vector Taylor's theorem (6.12) we see that

$$f_i\left(\overrightarrow{z}\right) - f_i\left(\overrightarrow{x_0}\right) = \sum_{j=1}^{n} \frac{g_{i\overrightarrow{z}}^{(j)}(0)}{j!} + R_{in}\left(\overrightarrow{z}, 0\right) = R_{in}\left(\overrightarrow{z}, 0\right), \tag{6.22}$$

where

$$R_{in}\left(\overrightarrow{z}, 0\right) := \int_0^1 \left(\int_0^{t_1} \cdots \left(\int_0^{t_{n-1}} \left(g_{i\overrightarrow{z}}^{(n)}(t_n) - g_{i\overrightarrow{z}}^{(n)}(0)\right) dt_n\right) \cdots \right) dt_1, \tag{6.23}$$

$i = 1, \ldots, r$.

Therefore,

$$\left\| R_{in}\left(\overrightarrow{z}, 0\right)\right\| \leq \int_0^1 \left(\int_0^{t_1} \cdots \left(\int_0^{t_{n-1}} \left\| g_{i\overrightarrow{z}}^{(n+1)}(\xi(t_n))\right\|_\infty t_n dt_n\right) \cdots \right) dt_1, \tag{6.24}$$

by the vector mean value Theorem 6.6 applied on $g_{i\overrightarrow{z}}^{(n)}$ over $(0, t_n)$. Moreover, we get

$$\left\| R_{in}\left(\overrightarrow{z}, 0\right)\right\| \leq \left\| \left\| g_{i\overrightarrow{z}}^{(n+1)} \right\| \right\|_{\infty,[0,1]} \int_0^1 \int_0^{t_1} \cdots \left(\int_0^{t_{n-1}} t_n dt_n\right) \cdots dt_1$$

$$= \frac{\left\| \left\| g_{i\overrightarrow{z}}^{(n+1)} \right\| \right\|_{\infty,[0,1]}}{(n+1)!}. \tag{6.25}$$

However, there exists a $t_{i0} \in [0, 1]$ such that $\left\| \left\| g_{i\overrightarrow{z}}^{(n+1)} \right\| \right\|_{\infty,[0,1]} = \left\| g_{i\overrightarrow{z}}^{(n+1)}(t_{i0})\right\|$.

That is

$$\left\| \left\| g_{i\overrightarrow{z}}^{(n+1)} \right\| \right\|_{\infty,[0,1]} = \left\| \left[\left(\sum_{\lambda=1}^{k} (z_\lambda - x_{0\lambda}) \frac{\partial}{\partial z_\lambda} \right)^{n+1} f_i \right] \left(\overrightarrow{x_0} + t_{i0}\left(\overrightarrow{z} - \overrightarrow{z_{0i}}\right)\right) \right\|$$

$$\leq \left[\left(\sum_{\lambda=1}^{k} |z_\lambda - x_{0\lambda}| \left\| \frac{\partial}{\partial z_\lambda} \right\| \right)^{n+1} f_i \right] \left(\overrightarrow{x_0} + t_{i0}\left(\overrightarrow{z} - \overrightarrow{z_{0i}}\right)\right).$$

I.e.,

$$\left\| \left\| g_{i\overrightarrow{z}}^{(n+1)} \right\| \right\|_{\infty,[0,1]} \leq \left[\left(\sum_{\lambda=1}^{k} |z_\lambda - x_{0\lambda}| \left\| \left\| \frac{\partial}{\partial z_\lambda} \right\| \right\|_\infty \right)^{n+1} f_i \right], \tag{6.26}$$

$i = 1, \ldots, r$.

Hence by (6.26) we get

$$\left\| R_{in}\left(\overrightarrow{z},0\right)\right\| \leq \frac{\left[\left(\sum_{\lambda=1}^{k}\left|z_{\lambda}-x_{0\lambda}\right|\left\|\left\|\frac{\partial}{\partial z_{\lambda}}\right\|\right\|_{\infty}\right)^{n+1} f_i\right]}{(n+1)!} \leq$$

$$\frac{D_{n+1}\left(f_i\right)}{(n+1)!}\left(\sum_{\lambda=1}^{k}\left|z_{\lambda}-x_{0\lambda}\right|\right)^{n+1} = \frac{D_{n+1}\left(f_i\right)}{(n+1)!}\left\|\overrightarrow{z}-\overrightarrow{x_0}\right\|_{l_1}^{n+1}, \tag{6.27}$$

$i = 1, \ldots, r$.

Therefore it holds

$$\left\| R_{in}\left(\overrightarrow{z},0\right)\right\| \leq \frac{\max_{i\in\{1,\ldots,r\}} D_{n+1}\left(f_i\right)}{(n+1)!}\left\|\overrightarrow{z}-\overrightarrow{x_0}\right\|_{l_1}^{n+1}, \tag{6.28}$$

for $i = 1, \ldots, r$.

By (6.22) we get that

$$\left(\prod_{\substack{\rho=1\\ \rho\neq i}}^{r} f_{\rho}\left(\overrightarrow{z}\right)\right) f_i\left(\overrightarrow{z}\right) - \left(\prod_{\substack{\rho=1\\ \rho\neq i}}^{r} f_{\rho}\left(\overrightarrow{z}\right)\right) f_i\left(\overrightarrow{x_0}\right) = \left(\prod_{\substack{\rho=1\\ \rho\neq i}}^{r} f_{\rho}\left(\overrightarrow{z}\right)\right) R_{in}\left(\overrightarrow{z},0\right), \tag{6.29}$$

for all $i = 1, \ldots, r$.

Hence

$$\sum_{i=1}^{r}\left(\prod_{\substack{\rho=1\\ \rho\neq i}}^{r} f_{\rho}\left(\overrightarrow{z}\right)\right) f_i\left(\overrightarrow{z}\right) - \sum_{i=1}^{r}\left(\prod_{\substack{\rho=1\\ \rho\neq i}}^{r} f_{\rho}\left(\overrightarrow{z}\right)\right) f_i\left(\overrightarrow{x_0}\right)$$

$$= \sum_{i=1}^{r}\left(\prod_{\substack{\rho=1\\ \rho\neq i}}^{r} f_{\rho}\left(\overrightarrow{z}\right)\right) R_{in}\left(\overrightarrow{z},0\right). \tag{6.30}$$

Therefore we find

$$E\left(f_1,\ldots,f_r\right)\left(x_0\right) :=$$

$$\sum_{i=1}^{r}\int_{Q}\left(\prod_{\substack{\rho=1\\ \rho\neq i}}^{r} f_{\rho}\left(\overrightarrow{z}\right)\right) f_i\left(\overrightarrow{z}\right) d\overrightarrow{z} - \sum_{i=1}^{r}\left(\int_{Q}\left(\prod_{\substack{\rho=1\\ \rho\neq i}}^{r} f_{\rho}\left(\overrightarrow{z}\right)\right) d\overrightarrow{z}\right) f_i\left(\overrightarrow{x_0}\right) =$$

$$\sum_{i=1}^{r}\int_{Q}\left(\prod_{\substack{\rho=1\\ \rho\neq i}}^{r}f_{\rho}\left(\overrightarrow{z}\right)\right)R_{in}\left(\overrightarrow{z},0\right)d\overrightarrow{z}. \tag{6.31}$$

Consequently, we have that

$$\left\|E\left(f_{1},\ldots,f_{r}\right)\left(x_{0}\right)\right\| =$$

$$\left\|\sum_{i=1}^{r}\int_{Q}\left(\prod_{\substack{\rho=1\\ \rho\neq i}}^{r}f_{\rho}\left(\overrightarrow{z}\right)\right)f_{i}\left(\overrightarrow{z}\right)d\overrightarrow{z}-\sum_{i=1}^{r}\left(\int_{Q}\left(\prod_{\substack{\rho=1\\ \rho\neq i}}^{r}f_{\rho}\left(\overrightarrow{z}\right)\right)d\overrightarrow{z}\right)f_{i}\left(\overrightarrow{x_{0}}\right)\right\| =$$

$$\left\|\sum_{i=1}^{r}\int_{Q}\left(\prod_{\substack{\rho=1\\ \rho\neq i}}^{r}f_{\rho}\left(\overrightarrow{z}\right)\right)R_{in}\left(\overrightarrow{z},0\right)d\overrightarrow{z}\right\| \leq \tag{6.32}$$

$$\sum_{i=1}^{r}\left\|\int_{Q}\left(\prod_{\substack{\rho=1\\ \rho\neq i}}^{r}f_{\rho}\left(\overrightarrow{z}\right)\right)R_{in}\left(\overrightarrow{z},0\right)d\overrightarrow{z}\right\| \leq$$

$$\sum_{i=1}^{r}\left(\int_{Q}\left\|\left(\prod_{\substack{\rho=1\\ \rho\neq i}}^{r}f_{\rho}\left(\overrightarrow{z}\right)\right)R_{in}\left(\overrightarrow{z},0\right)\right\|d\overrightarrow{z}\right)\overset{(6.6)}{\leq}$$

$$\sum_{i=1}^{r}\left(\int_{Q}\left(\prod_{\substack{\rho=1\\ \rho\neq i}}^{r}\left\|f_{\rho}\left(\overrightarrow{z}\right)\right\|\right)\left\|R_{in}\left(\overrightarrow{z},0\right)\right\|d\overrightarrow{z}\right)\overset{(6.28)}{\leq} \tag{6.33}$$

$$\frac{\max\limits_{i\in\{1,\ldots,r\}}D_{n+1}\left(f_{i}\right)}{(n+1)!}\sum_{i=1}^{r}\left(\int_{Q}\left(\prod_{\substack{\rho=1\\ \rho\neq i}}^{r}\left\|f_{\rho}\left(\overrightarrow{z}\right)\right\|\right)\left\|\overrightarrow{z}-\overrightarrow{x_{0}}\right\|_{l_{1}}^{n+1}d\overrightarrow{z}\right).$$

So far we have proved

$$\left\|E\left(f_{1},\ldots,f_{r}\right)\left(x_{0}\right)\right\| \leq$$

$$\frac{\max_{i\in\{1,\dots,r\}} D_{n+1}(f_i)}{(n+1)!}\sum_{i=1}^{r}\left(\int_Q\left(\prod_{\substack{\rho=1\\ \rho\neq i}}^{r}\left\|f_\rho\left(\overrightarrow{z}\right)\right\|\right)\left\|\overrightarrow{z}-\overrightarrow{x_0}\right\|_{l_1}^{n+1}d\overrightarrow{z}\right)=:(\xi). \tag{6.34}$$

Furthermore it holds

$$(\xi)\leq\frac{\max_{i\in\{1,\dots,r\}} D_{n+1}(f_i)}{(n+1)!}\left(\int_Q\left\|\overrightarrow{z}-\overrightarrow{x_0}\right\|_{l_1}^{n+1}d\overrightarrow{z}\right)\left[\sum_{i=1}^{r}\left(\prod_{\substack{\rho=1\\ \rho\neq i}}^{r}\left\|\left\|f_\rho\right\|\right\|_{\infty,Q}\right)\right], \tag{6.35}$$

and

$$(\xi)\leq\frac{\max_{i\in\{1,\dots,r\}} D_{n+1}(f_i)}{(n+1)!}\left\|\left\|\cdot-\overrightarrow{x_0}\right\|_{l_1}^{n+1}\right\|_{\infty,Q}\left[\sum_{i=1}^{r}\left\|\left(\prod_{\substack{\rho=1\\ \rho\neq i}}^{r}\left\|f_\rho\right\|\right)\right\|_{L_1(Q,A)}\right], \tag{6.36}$$

and finally

$$(\xi)\leq\frac{\max_{i\in\{1,\dots,r\}} D_{n+1}(f_i)}{(n+1)!}\left[\sum_{i=1}^{r}\left[\left\|\left(\prod_{\substack{\rho=1\\ \rho\neq i}}^{r}\left\|f_\rho\right\|\right)\right\|_{L_q(Q,A)}\right]\right]\left\|\left\|\cdot-\overrightarrow{x_0}\right\|_{l_1}^{n+1}\right\|_{L_p(Q,A)}, \tag{6.37}$$

proving (6.18), (6.19). ∎

We give

Corollary 6.9 (to Theorem 6.8) *All as in Theorem 6.8, with* $f_1=\cdots=f_r=f$, $r\in\mathbb{N}$. *Then*

$$\left\|\int_Q f^r\left(\overrightarrow{z}\right)d\overrightarrow{z}-\left(\int_Q f^{r-1}\left(\overrightarrow{z}\right)d\overrightarrow{z}\right)f\left(\overrightarrow{x_0}\right)\right\|\leq$$

$$\frac{D_{n+1}(f)}{(n+1)!}\left(\int_Q\left\|f\left(\overrightarrow{z}\right)\right\|^{r-1}\left\|\overrightarrow{z}-\overrightarrow{x_0}\right\|_{l_1}^{n+1}d\overrightarrow{z}\right)\leq \tag{6.38}$$

$$\frac{D_{n+1}(f)}{(n+1)!}\min\left\{\left(\int_Q\left\|\overrightarrow{z}-\overrightarrow{x_0}\right\|_{l_1}^{n+1}d\overrightarrow{z}\right)\left(\left\|\left\|f\right\|\right\|_{\infty,Q}\right)^{r-1},\right.$$

$$\left.\left\|\left\|\cdot-\overrightarrow{x_0}\right\|_{l_1}^{n+1}\right\|_{\infty,Q}\left\|\left\|f\right\|^{r-1}\right\|_{L_1(Q,A)},\left\|\left\|\cdot-\overrightarrow{x_0}\right\|_{l_1}^{n+1}\right\|_{L_p(Q,A)}\left\|\left\|f\right\|^{r-1}\right\|_{L_q(Q,A)}\right\}. \tag{6.39}$$

We also give

Corollary 6.10 (to Theorem 6.8) *All as in Theorem 6.8, with $(A, \|\cdot\|)$ being a commutative Banach algebra. Then*

$$\left\| r\int_Q \left(\prod_{\rho=1}^{r} f_\rho(\overrightarrow{z})\right) d\overrightarrow{z} - \sum_{i=1}^{r}\left(\int_Q \left(\prod_{\substack{\rho=1\\ \rho\neq i}}^{r} f_\rho(\overrightarrow{z})\right) d\overrightarrow{z}\right) f_i(\overrightarrow{x_0})\right\| \leq$$

$$\textit{Right hand side of (6.18)} \leq \textit{Right hand side of (6.19)}. \tag{6.40}$$

We make

Remark 6.11 Of great interest are applications of Theorem 6.8 when $Q = \prod_{\lambda=1}^{k} [a_\lambda, b_\lambda]$, where $[a_\lambda, b_\lambda] \subset \mathbb{R}$, $\lambda = 1, \ldots, k$.

We observe that by the multinomial theorem we get:

$$\int_{\prod_{\lambda=1}^{k}[a_\lambda,b_\lambda]} \left(\sum_{\lambda=1}^{k} |z_\lambda - x_{0\lambda}|\right)^{n+1} dz_1 \cdots dz_k = \sum_{\rho_1+\rho_2+\cdots\rho_k=n+1} \frac{(n+1)!}{\rho_1!\rho_2!\cdots\rho_k!}$$

$$\int_{\prod_{\lambda=1}^{k}[a_\lambda,b_\lambda]} |z_1 - x_{01}|^{\rho_1} |z_2 - x_{02}|^{\rho_2} \cdots |z_k - x_{0k}|^{\rho_k} dz_1 \cdots dz_k = \tag{6.41}$$

$$\sum_{\rho_1+\rho_2+\cdots\rho_k=n+1} \frac{(n+1)!}{\rho_1!\rho_2!\cdots\rho_k!} \prod_{\lambda=1}^{k}\left(\int_{a_\lambda}^{b_\lambda} |z_\lambda - x_{0\lambda}|^{\rho_\lambda} dz_\lambda\right) =$$

$$\sum_{\sum_{\lambda=1}^{k}\rho_\lambda=n+1} \frac{(n+1)!}{\prod_{\lambda=1}^{k}\rho_\lambda!} \prod_{\lambda=1}^{k}\left(\int_{a_\lambda}^{x_{0\lambda}} (x_{0\lambda} - z_\lambda)^{\rho_\lambda} dz_\lambda + \int_{x_{0\lambda}}^{b_\lambda} (z_\lambda - x_{0\lambda})^{\rho_\lambda} dz_\lambda\right) =$$

$$\sum_{\sum_{\lambda=1}^{k}\rho_\lambda=n+1} \frac{(n+1)!}{\prod_{\lambda=1}^{k}\rho_\lambda!} \prod_{\lambda=1}^{k}\left(\frac{(x_{0\lambda} - a_\lambda)^{\rho_\lambda+1} + (b_\lambda - x_{0\lambda})^{\rho_\lambda+1}}{\rho_\lambda + 1}\right). \tag{6.42}$$

We have found that

$$\int_{\prod_{\lambda=1}^{k}[a_\lambda,b_\lambda]} \|\overrightarrow{z} - \overrightarrow{x_0}\|_{l_1}^{n+1} d\overrightarrow{z} = \tag{6.43}$$

$$\sum_{\sum_{\lambda=1}^{k}\rho_\lambda=n+1} \frac{(n+1)!}{\prod_{\lambda=1}^{k}\rho_\lambda!} \prod_{\lambda=1}^{k}\left(\frac{(b_\lambda - x_{0\lambda})^{\rho_\lambda+1} + (x_{0\lambda} - a_\lambda)^{\rho_\lambda+1}}{\rho_\lambda + 1}\right).$$

Based on (6.18), (6.19) and (6.43) we conclude:

Theorem 6.12 *Let* $(A, \|\cdot\|)$ *a Banach algebra and* $f_i \in C^{n+1}\left(\prod_{\lambda=1}^{k}[a_\lambda, b_\lambda], A\right)$, $i = 1, \ldots, r$; $r \in \mathbb{N}$, $n \in \mathbb{Z}_+$, *and fixed* $\overrightarrow{x_0} \in \prod_{\lambda=1}^{k}[a_\lambda, b_\lambda] \subset \mathbb{R}^k$, $k \geq 2$. *Here all vector partial derivatives* $f_{i\alpha} := \frac{\partial^\alpha f_i}{\partial z^\alpha}$, *where* $\alpha = (\alpha_1, \ldots, \alpha_k)$, $\alpha_\lambda \in \mathbb{Z}^+$, $\lambda = 1, \ldots, k$, $|\alpha| = \sum_{\lambda=1}^{k}\alpha_\lambda = j$, $j = 1, \ldots, n$, *fulfill* $f_{i\alpha}\left(\overrightarrow{x_0}\right) = 0$, $i = 1, \ldots, r$.

Denote

$$D_{n+1}(f_i) := \max_{\alpha:|\alpha|=n+1} \left\|\left\|f_{i\alpha}\right\|\right\|_{\infty, \prod_{\lambda=1}^{k}[a_\lambda, b_\lambda]}, \tag{6.44}$$

$i = 1, \ldots, r$.

Then

$$\left\| \sum_{i=1}^{r} \int_{\prod_{\lambda=1}^{k}[a_\lambda, b_\lambda]} \left(\prod_{\substack{\rho=1\\ \rho\neq i}}^{r} f_\rho\left(\overrightarrow{z}\right)\right) f_i\left(\overrightarrow{z}\right) d\overrightarrow{z} - \right.$$

$$\left. \sum_{i=1}^{r} \left(\int_{\prod_{\lambda=1}^{k}[a_\lambda, b_\lambda]} \left(\prod_{\substack{\rho=1\\ \rho\neq i}}^{r} f_\rho\left(\overrightarrow{z}\right)\right) d\overrightarrow{z}\right) f_i\left(\overrightarrow{x_0}\right)\right\| \leq \tag{6.45}$$

$$\left(\max_{i\in\{1,\ldots,r\}} D_{n+1}(f_i)\right)\left[\sum_{i=1}^{r}\left(\prod_{\substack{\rho=1\\ \rho\neq i}}^{r} \left\|\left\|f_\rho\right\|\right\|_{\infty, \prod_{\lambda=1}^{k}[a_\lambda, b_\lambda]}\right)\right]$$

$$\left[\sum_{\sum_{\lambda=1}^{k}\rho_\lambda = n+1} \frac{1}{\prod_{\lambda=1}^{k}\rho_\lambda! \prod_{\lambda=1}^{k}(\rho_\lambda+1)} \prod_{\lambda=1}^{k}\left((b_\lambda - x_{0\lambda})^{\rho_\lambda+1} + (x_{0\lambda} - a_\lambda)^{\rho_\lambda+1}\right)\right].$$

References

1. Anastassiou, G.A.: Multivariate Ostrowski type inequalities. Acta Math. Hung. **76**(4), 267–278 (1997)
2. Anastassiou, G.A.: Multivariate Ostrowski type inequalities for several Banach algebra valued functions. J. Comput. Anal. Appl. **30**(1), 95–106 (2022)

3. Dragomir, S.S.: Noncommutative Ostrowski type inequalities for functions in Banach algebras. RGMIA Res. Rep. Coll. **24**(10), (2021)
4. Driver, B.: Analysis Tools with Applications. Springer, N.Y., Heidelberg (2003)
5. Ladas, G., Laksmikantham, V.: Differential Equations in Abstract Spaces. Academic Press, New York, London (1972)
6. Ostrowski, A.: Über die Absolutabweichung einer differentiabaren Funcktion von ihrem Integralmittelwert. Comment. Math. Helv. **10**, 226–227 (1938)
7. Rudin, W.: Functional Analysis, 2nd edn. McGraw-Hill Inc, New York (1991)
8. Schwartz, L.: Analyse Mathematique. Hermann, Paris (1967)
9. Shilov, G.E.: Elementary Functional Analysis. The MIT Press Cambridge, Massachusetts (1974)

Chapter 7
p-Schatten Norm Generalized Fractional Ostrowski and Grüss Inequalities for Multiple Functions

Employing generalized Caputo fractional left and right vectorial Taylor formulae we establish generalized fractional Ostrowski and Grüss inequalities involving several functions that take values in the von Neumann-Schatten class $\mathcal{B}_p(H)$, $1 \leq p < \infty$. The estimates are with respect to all p-Schatten norms, $1 \leq p < \infty$. We finish with applications. It follows [6].

7.1 Introduction

The following results motivate this chapter.

Theorem 7.1 (1938, Ostrowski [15]) *Let $f : [a, b] \to \mathbb{R}$ be continuous on $[a, b]$ and differentiable on (a, b) whose derivative $f' : (a, b) \to \mathbb{R}$ is bounded on (a, b), i.e., $\|f'\|_\infty^{\sup} := \sup_{t \in (a,b)} |f'(t)| < +\infty$. Then*

$$\left| \frac{1}{b-a} \int_a^b f(t)\, dt - f(x) \right| \leq \left[\frac{1}{4} + \frac{\left(x - \frac{a+b}{2}\right)^2}{(b-a)^2} \right] (b-a) \|f'\|_\infty^{\sup}, \tag{7.1}$$

for any $x \in [a, b]$. The constant $\frac{1}{4}$ is the best possible.

Ostrowski type inequalities have great applications to integral approximations in Numerical Analysis.

Theorem 7.2 (1882, Čebyšev [8]) *Let $f, g : [a, b] \to \mathbb{R}$ be absolutely continuous functions with $f', g' \in L_\infty([a, b])$. Then*

G. A. Anastassiou, *Fractional Inequalities In Banach Algebras*,
Studies in Systems, Decision and Control 441,
https://doi.org/10.1007/978-3-031-05148-7_7

$$\left|\frac{1}{b-a}\int_a^b f(x)g(x)\,dx - \left(\frac{1}{b-a}\int_a^b f(x)\,dx\right)\left(\frac{1}{b-a}\int_a^b g(x)\,dx\right)\right|$$

$$\leq \frac{1}{12}(b-a)^2 \left\|f'\right\|_\infty \left\|g'\right\|_\infty. \tag{7.2}$$

The above integrals are assumed to exist.

The related Grüss type inequalities have many applications to Probability Theory. We presented also ([4], Chaps. 8, 9) mixed fractional Ostrowski and Grüss-Cebysev type inequalities for several functions, acting to all possible directions. The estimates involve the left and right Caputo fractional derivatives. See also the monographs written by the author [2], Chaps. 24–26 and [3], Chaps. 2–6.

We are motivated also by S. Dragomir [11] recent work:

An operator $A \in \mathcal{B}(H)$ is said to belong to the von Neumann-Schatten class $\mathcal{B}_p(H)$, $1 \leq p < \infty$ if the p-Schatten norm is finite

$$\|A\|_p := \left[tr\left(|A|^p\right)\right]^{\frac{1}{p}} < \infty.$$

Assume that $A : [a,b] \to \mathcal{B}_p(H)$, $B : [a,b] \to \mathcal{B}_q(H)$, $p, q > 1$ with $\frac{1}{p} + \frac{1}{q} = 1$, are continuous and B is strongly differentiable on (a,b), then

$$\left\|\int_a^b A(t)B(t)\,dt - \left(\int_a^b A(s)\,ds\right)B(u)\right\|_1 \leq$$

$$\sup_{t\in[a,b]}\left\|B'(t)\right\|_q \times \begin{cases} \left[\frac{1}{2}(b-a) + \left|u - \frac{a+b}{2}\right|\right]\int_a^b \|A(t)\|_p\,dt, \\ \left[\frac{(u-a)^{\beta+1}+(b-u)^{\beta+1}}{\beta+1}\right]^{\frac{1}{\beta}}\left(\int_a^b \|A(t)\|_p^\alpha\right)^{\frac{1}{\alpha}}, \\ \text{for } \alpha, \beta > 1 \text{ with } \frac{1}{\alpha} + \frac{1}{\beta} = 1 \\ \left[\frac{1}{4}(b-a)^2 + \left(u - \frac{a+b}{2}\right)^2\right]\sup_{t\in[a,b]}\|A(t)\|_p, \end{cases} \tag{7.3}$$

for all $u \in [a,b]$, an Ostrowski type inequality.

Further inspiration comes from S. Dragomir [12] recent work on Grüss inequalities:

For two continuous functions $A, B : [a,b] \to \mathcal{B}(H)$ we define the noncommutative Cebysev fractional

$$D(A,B) := (b-a)\int_a^b A(t)B(t)\,dt - \int_a^b A(t)\,dt \int_a^b B(t)\,dt. \tag{7.4}$$

If $p, q > 1$ with $\frac{1}{p} + \frac{1}{q} = 1$, let $A : [a,b] \to \mathcal{B}_p(H)$, $B : [a,b] \to \mathcal{B}_q(H)$ be strongly differentiable functions on the interval (a,b), then

$$\|D(A,B)\|_1 \leq D\left(\int_a^b \|A'(u)\|_p du, \int_a^b \|B'(u)\|_q du\right) \leq$$

$$\frac{1}{4}(b-a)^2 \int_a^b \|A'(u)\|_p du \int_a^b \|B'(u)\|_q du.$$

In this chapter we generalize [4], Chaps. 8, 9 for several Banach algebra $\mathcal{B}_p(H)$ valued functions, in the sense of developing fractional Ostrowski and Grüss type inequalities. Now our left and right generalized Caputo fractional derivatives are for Banach space valued functions and our integrals are of Bochner type [13]. Applications finish the chapter.

7.2 Vectorial Background Fractional Calculus

Here all come from [5].

We need

Definition 7.3 ([5], *p. 106*) Let $\alpha > 0$, $\lceil \alpha \rceil = n$, $\lceil \cdot \rceil$ the ceiling of the number. Let $f \in C^n([a,b], X)$, where $[a,b] \subset \mathbb{R}$, and $(X, \|\cdot\|)$ is a Banach space. Let $g \in C^1([a,b])$, strictly increasing, such that $g^{-1} \in C^n([g(a), g(b)])$.

We define the left generalized g-fractional derivative X-valued of f of order α as follows:

$$\left(D_{a+;g}^{\alpha} f\right)(x) := \frac{1}{\Gamma(n-\alpha)} \int_a^x (g(x) - g(t))^{n-\alpha-1} g'(t) \left(f \circ g^{-1}\right)^{(n)} (g(t))\, dt, \tag{7.5}$$

$\forall\, x \in [a,b]$, where Γ is the gamma function. The last integral is of Bochner type ([1], pp. 422–428; [13]).

If $\alpha \notin \mathbb{N}$, by Theorem 4.10 ([5], p. 98), we have that $\left(D_{a+;g}^{\alpha} f\right) \in C([a,b], X)$.

We set

$$D_{a+;g}^{n} f(x) := \left(\left(f \circ g^{-1}\right)^{n} \circ g\right)(x) \in C([a,b], X), \ n \in \mathbb{N}, \tag{7.6}$$

$$D_{a+;g}^{0} f(x) = f(x), \ \forall x \in [a,b].$$

When $g = id$, then

$$D_{a+;g}^{\alpha} f = D_{a+;id}^{\alpha} f = D_{*a}^{\alpha} f, \tag{7.7}$$

the usual left X-valued Caputo fractional derivative, see [5], Chap. 1.

We also need

Definition 7.4 ([5], *p. 107*) Let $\alpha > 0$, $\lceil \alpha \rceil = n$, $\lceil \cdot \rceil$ the ceiling of the number. Let $f \in C^n([a,b], X)$, where $[a,b] \subset \mathbb{R}$, and $(X, \|\cdot\|)$ is a Banach space. Let $g \in C^1([a,b])$, strictly increasing, such that $g^{-1} \in C^n([g(a), g(b)])$.

We define the right generalized g-fractional derivative X-valued of f of order α as follows:

$$\left(D_{b-;g}^{\alpha} f\right)(x) := \frac{(-1)^n}{\Gamma(n-\alpha)} \int_x^b (g(t) - g(x))^{n-\alpha-1} g'(t) \left(f \circ g^{-1}\right)^{(n)} (g(t))\, dt, \tag{7.8}$$

$\forall\, x \in [a,b]$. The last integral is of Bochner type.

If $\alpha \notin \mathbb{N}$, by Theorem 4.11 ([5], p. 101), we have that $\left(D_{b-;g}^{\alpha} f\right) \in C([a,b], X)$.

We set

$$D_{b-;g}^{n} f(x) := (-1)^n \left(\left(f \circ g^{-1}\right)^n \circ g\right)(x) \in C([a,b], X), \; n \in \mathbb{N}, \tag{7.9}$$

$$D_{b-;g}^{0} f(x) := f(x), \; \forall x \in [a,b].$$

When $g = id$, then

$$D_{b-;g}^{\alpha} f(x) = D_{b-;id}^{\alpha} f(x) = D_{b-}^{\alpha} f, \tag{7.10}$$

the usual right X-valued Caputo fractional derivative, see [5], Chap. 2.

We mention the following generalized fractional Taylor formulae with integral remainders over Banach spaces.

Theorem 7.5 ([5], p. 107) *Let $\alpha > 0$, $n = \lceil \alpha \rceil$, and $f \in C^n([a,b], X)$, where $[a,b] \subset \mathbb{R}$ and $(X, \|\cdot\|)$ is a Banach space. Let $g \in C^1([a,b])$, strictly increasing, such that $g^{-1} \in C^n([g(a), g(b)])$, $a \leq x \leq b$. Then*

$$f(x) = f(a) + \sum_{i=1}^{n-1} \frac{(g(x) - g(a))^i}{i!} \left(f \circ g^{-1}\right)^{(i)} (g(a)) +$$

$$\frac{1}{\Gamma(\alpha)} \int_a^x (g(x) - g(t))^{\alpha-1} g'(t) \left(D_{a+;g}^{\alpha} f\right)(t)\, dt =$$

$$f(a) + \sum_{i=1}^{n-1} \frac{(g(x) - g(a))^i}{i!} \left(f \circ g^{-1}\right)^{(i)} (g(a)) + \tag{7.11}$$

$$\frac{1}{\Gamma(\alpha)} \int_{g(a)}^{g(x)} (g(x) - z)^{\alpha-1} \left(\left(D_{a+;g}^{\alpha} f\right) \circ g^{-1}\right)(z)\, dz.$$

We also mention

Theorem 7.6 ([5], p. 108) *Let $\alpha > 0$, $n = \lceil \alpha \rceil$, and $f \in C^n([a,b], X)$, where $[a,b] \subset \mathbb{R}$ and $(X, \|\cdot\|)$ is a Banach space. Let $g \in C^1([a,b])$, strictly increasing, such that $g^{-1} \in C^n([g(a), g(b)])$, $a \leq x \leq b$. Then*

$$f(x) = f(b) + \sum_{i=1}^{n-1} \frac{(g(x) - g(b))^i}{i!} \left(f \circ g^{-1}\right)^{(i)} (g(b)) +$$

$$\frac{1}{\Gamma(\alpha)} \int_x^b (g(t) - g(x))^{\alpha-1} g'(t) \left(D_{b-;g}^{\alpha} f\right)(t)\, dt =$$

$$f(b) + \sum_{i=1}^{n-1} \frac{(g(x) - g(b))^i}{i!} \left(f \circ g^{-1}\right)^{(i)} (g(b)) + \tag{7.12}$$

$$\frac{1}{\Gamma(\alpha)} \int_{g(x)}^{g(b)} (z - g(x))^{\alpha-1} \left(\left(D_{b-;g}^{\alpha} f\right) \circ g^{-1}\right)(z)\, dz.$$

If $0 < \alpha \leq 1$, then the sums in (7.11), (7.12) disappear.

Also in (7.11), (7.12), we have that $\left\|\left\|\left(D_{a+;g}^{\alpha} f\right) \circ g^{-1}\right\|\right\|_{\infty,[g(a),g(b)]}$, $\left\|\left\|\left(D_{b-;g}^{\alpha} f\right) \circ g^{-1}\right\|\right\|_{\infty,[g(a),g(b)]} < \infty$.

7.3 Banach Algebras Background

All here come from [16].

We need

Definition 7.7 ([16], *p. 245*) A complex algebra is a vector space A over the complex filed $\mathbb{C}$ in which a multiplication is defined that satisfies

$$x(yz) = (xy)z, \tag{7.13}$$

$$(x + y)z = xz + yz, \ x(y + z) = xy + xz, \tag{7.14}$$

and

$$\alpha(xy) = (\alpha x)y = x(\alpha y), \tag{7.15}$$

for all x, y and z in A and for all scalars α.

Additionally if A is a Banach space with respect to a norm that satisfies the multiplicative inequality

$$\|xy\| \leq \|x\| \|y\| \ (x \in A, y \in A) \tag{7.16}$$

and if A contains a unit element e such that

$$xe = ex = x \ (x \in A) \tag{7.17}$$

and

$$\|e\| = 1, \tag{7.18}$$

then A is called a Banach algebra.

A is commutative iff $xy = yx$ for all $x, y \in A$.

We make

Remark 7.8 Commutativity of A will be explicited stated when needed.

There exists at most one $e \in A$ that satisfies (7.17).

Inequality (7.16) makes multiplication to be continuous, more precisely left and right continuous, see [16], p. 246.

Multiplication in A is not necessarily the numerical multiplication, it is something more general and it is defined abstractly, that is for $x, y \in A$ we have $xy \in A$, e.g. composition or convolution, etc.

For nice examples about Banach algebras see [16], pp. 247–248, Sect. 10.3.

We also make

Remark 7.9 Next we mention about integration of A-valued functions, see [16], p. 259, Sect. 10.22:

If A is a Banach algebra and f is a continuous A-valued function on some compact Hausdorff space Q on which a complex Borel measure μ is defined, then $\int f d\mu$ exists and has all the properties that were discussed in Chap. 3 of [16], simply because A is a Banach space. However, an additional property can be added to these, namely: If $x \in A$, then

$$x \int_Q f \, d\mu = \int_Q xf(p) \, d\mu(p) \tag{7.19}$$

and

$$\left(\int_Q f \, d\mu \right) x = \int_Q f(p) \, x \, d\mu(p). \tag{7.20}$$

The Bochner integrals we will involve in our chapter follow (7.19) and (7.20). Also, let $f \in C([a, b], X)$, where $[a, b] \subset \mathbb{R}$, $(X, \|\cdot\|)$ is a Banach space. By [5], p. 3, f is Bochner integrable.

7.4 *p*-Schatten Norms Background

In this basic section all come from [11].

Let $(H, \langle\cdot,\cdot\rangle)$ be a complex Hilbert space and $\mathcal{B}(H)$ the Banach algebra of all bounded linear operators on H. If $\{e_i\}_{i\in I}$ an orthonormal basis of H, we say that $A \in \mathcal{B}(H)$ is of trace class if

$$\|A\|_1 := \sum_{i\in I} \langle |A| e_i, e_i\rangle < \infty. \tag{7.21}$$

The definition of $\|A\|_1$ does not depend on the choice of the orthornormal basis $\{e_i\}_{i\in I}$. We denote by $\mathcal{B}_1(H)$ the set of trace class operators in $\mathcal{B}(H)$.

We define the trace of a trace class operator $A \in \mathcal{B}_1(H)$ to be

$$tr(A) := \sum_{i\in I} \langle Ae_i, e_i\rangle, \tag{7.22}$$

where $\{e_i\}_{i\in I}$ an orthonormal basis of H. Note that this coincides with the usual definition of the trace if H is finite-dimensional. We observe that the series (7.22) converges absolutely and it is independent from the choice of basis.

The following result collects some properties of the trace:

Theorem 7.10 *We have:*

(i) If $A \in \mathcal{B}_1(H)$ then $A^* \in \mathcal{B}_1(H)$ and

$$tr\left(A^*\right) = \overline{tr(A)}; \tag{7.23}$$

(ii) If $A \in \mathcal{B}_1(H)$ and $T \in \mathcal{B}(H)$, then $AT, TA \in \mathcal{B}_1(H)$ and

$$tr(AT) = tr(TA) \text{ and } |tr(AT)| \le \|A\|_1 \|T\|; \tag{7.24}$$

(iii) $tr(\cdot)$ is a bounded linear functional on $\mathcal{B}_1(H)$ with $\|tr\| = 1$;
(iv) If $A, B \in \mathcal{B}_2(H)$ then $AB, BA \in \mathcal{B}_1(H)$ and $tr(AB) = tr(BA)$;
(v) $\mathcal{B}_{fin}(H)$ (finite rank operators) is a dense subspace of $\mathcal{B}_1(H)$.

An operator $A \in \mathcal{B}(H)$ is said to belong to the von Neumann-Schatten class $\mathcal{B}_p(H)$, $1 \le p < \infty$ if the p-Schatten norm is finite [18, pp. 60–64]

$$\|A\|_p := \left[tr\left(|A|^p\right)\right]^{\frac{1}{p}} < \infty,$$

$|A|^p$ is an operator notation and not a power.

For $1 < p < q < \infty$ we have that

$$\mathcal{B}_1(H) \subset \mathcal{B}_p(H) \subset \mathcal{B}_q(H) \subset \mathcal{B}(H) \tag{7.25}$$

and

$$\|A\|_1 \ge \|A\|_p \ge \|A\|_q \ge \|A\|. \tag{7.26}$$

For $p \geq 1$ the functional $\|\cdot\|_p$ is a norm on the $*$-ideal $\mathcal{B}_p(H)$, which is a Banach algebra, and $(\mathcal{B}_p(H), \|\cdot\|_p)$ is a Banach space.

Also, see for instance [18, pp. 60–64], for $p \geq 1$,

$$\|A\|_p = \|A^*\|_p, A \in \mathcal{B}_p(H) \tag{7.27}$$

$$\|AB\|_p \leq \|A\|_p \|B\|_p, \ A, B \in \mathcal{B}_p(H) \tag{7.28}$$

and

$$\|AB\|_p \leq \|A\|_p \|B\|, \ \|BA\|_p \leq \|B\| \|A\|_p, \ A \in \mathcal{B}_p(H), B \in \mathcal{B}(H). \tag{7.29}$$

This implies that

$$\|CAB\|_p \leq \|C\| \|A\|_p \|B\|, \ A \in \mathcal{B}_p(H), B, C \in \mathcal{B}(H). \tag{7.30}$$

In terms of p-Schatten norm we have the Hölder inequality for $p, q > 1$ with $\frac{1}{p} + \frac{1}{q} = 1$:

$$(|tr(AB)| \leq) \|AB\|_1 \leq \|A\|_p \|B\|_q, \ A \in \mathcal{B}_p(H), B \in \mathcal{B}_q(H). \tag{7.31}$$

For the theory of trace functionals and their applications the interested reader is referred to [17, 18].

For some classical trace inequalities see [9, 10, 14], which are continuations of the work of Bellman [7].

7.5 Main Results

We start with 1-Schatten norm weighted mixed generalized fractional Ostrowski type inequalities involving several functions taking values in the Banach algebra $\mathcal{B}_2(H) \subset \mathcal{B}(H)$:

Theorem 7.11 *Let the $*$-ideal $\mathcal{B}_2(H)$, which $(\mathcal{B}_2(H), \|\cdot\|_2)$ is a Banach algebra; $x_0 \in [a, b] \subset \mathbb{R}$, $\alpha > 0$, $n = \lceil \alpha \rceil$; $A_i \in C^n([a, b], \mathcal{B}_2(H))$, $i = 1, \ldots, r \in \mathbb{N} - \{1\}$; $g \in C^1([a, b])$, strictly increasing such that $g^{-1} \in C^n([g(a), g(b)])$, with $(A_i \circ g^{-1})^{(k)}(g(x_0)) = 0$, $k = 1, \ldots, n-1$; $i = 1, \ldots, r$.*

Then

(1) it holds,

$$\Phi(A_1, \ldots, A_r)(x_0) :=$$

$$\sum_{i=1}^{r}\left[\int_a^b\left(\prod_{\substack{j=1\\j\neq i}}^{r}A_j(x)\right)A_i(x)\,dx-\left(\int_a^b\left(\prod_{\substack{j=1\\j\neq i}}^{r}A_j(x)\right)dx\right)A_i(x_0)\right]= \quad (7.32)$$

$$\frac{1}{\Gamma(\alpha)}\sum_{i=1}^{r}\left[\left[\int_a^{x_0}\left(\prod_{\substack{j=1\\j\neq i}}^{r}A_j(x)\right)\left(\int_{g(x)}^{g(x_0)}(z-g(x))^{\alpha-1}\left(\left(D_{x_0-;g}^{\alpha}A_i\right)\circ g^{-1}\right)(z)\,dz\right)dx\right]\right.$$

$$\left.+\left[\int_{x_0}^{b}\left(\prod_{\substack{j=1\\j\neq i}}^{r}A_j(x)\right)\left(\int_{g(x_0)}^{g(x)}(g(x)-z)^{\alpha-1}\left(\left(D_{x_0+;g}^{\alpha}A_i\right)\circ g^{-1}\right)(z)\,dz\right)dx\right]\right],$$

(2) for $\gamma,\delta>1:\frac{1}{\gamma}+\frac{1}{\delta}=1$, with $\alpha>\frac{1}{\delta}$ we have that

$$\left\|\Phi(A_1,\ldots,A_r)(x_0)\right\|_1\leq\frac{1}{\Gamma(\alpha)(\gamma(\alpha-1)+1)^{\frac{1}{\gamma}}} \quad (7.33)$$

$$\sum_{i=1}^{r}\left[\left\|\left\|\left(D_{x_0-;g}^{\alpha}A_i\right)\circ g^{-1}\right\|_2\right\|_{\delta,[g(a),g(x_0)]}\int_a^{x_0}\prod_{\substack{j=1\\j\neq i}}^{r}\left\|A_j(x)\right\|_2(g(x_0)-g(x))^{\alpha-\frac{1}{\delta}}\,dx+\right.$$

$$\left.\left\|\left\|\left(D_{x_0+;g}^{\alpha}A_i\right)\circ g^{-1}\right\|_2\right\|_{\delta,[g(x_0),g(b)]}\int_{x_0}^{b}\prod_{\substack{j=1\\j\neq i}}^{r}\left\|A_j(x)\right\|_2(g(x)-g(x_0))^{\alpha-\frac{1}{\delta}}\,dx\right],$$

(3) if $\alpha\geq 1$, we obtain

$$\left\|\Phi(A_1,\ldots,A_r)(x_0)\right\|_1\leq\frac{1}{\Gamma(\alpha)} \quad (7.34)$$

$$\sum_{i=1}^{r}\left[\left\|\left\|\left(D_{x_0-;g}^{\alpha}A_i\right)\circ g^{-1}\right\|_2\right\|_{L_1([g(a),g(x_0)])}\int_a^{x_0}\prod_{\substack{j=1\\j\neq i}}^{r}\left\|A_j(x)\right\|_2(g(x_0)-g(x))^{\alpha-1}\,dx+\right.$$

$$\left.\left\|\left\|\left(D_{x_0+;g}^{\alpha}A_i\right)\circ g^{-1}\right\|_2\right\|_{L_1([g(x_0),g(b)])}\int_{x_0}^{b}\prod_{\substack{j=1\\j\neq i}}^{r}\left\|A_j(x)\right\|_2(g(x)-g(x_0))^{\alpha-1}\,dx\right],$$

and

(4)

$$\left\|\Phi\left(A_1,\ldots,A_r\right)(x_0)\right\|_1 \leq \frac{1}{\Gamma(\alpha+1)}$$

$$\sum_{i=1}^{r}\left[\left\|\left\|\left(D_{x_0-;g}^{\alpha}A_i\right)\circ g^{-1}\right\|_2\right\|_{\infty,[g(a),g(x_0)]}\int_a^{x_0}\prod_{\substack{j=1\\ j\neq i}}^{r}\left\|A_j(x)\right\|_2\left(g(x_0)-g(x)\right)^{\alpha}dx+\right.$$

$$\left.\left\|\left\|\left(D_{x_0+;g}^{\alpha}A_i\right)\circ g^{-1}\right\|_2\right\|_{\infty,[g(x_0),g(b)]}\int_{x_0}^{b}\prod_{\substack{j=1\\ j\neq i}}^{r}\left\|A_j(x)\right\|_2\left(g(x)-g(x_0)\right)^{\alpha}dx\right]. \quad (7.35)$$

Proof Since $\left(A_i\circ g^{-1}\right)^{(k)}\left(g(x_0)\right)=0, k=1,\ldots,n-1; i=1,\ldots,r$, we have by Theorem 7.5 that

$$A_i(x)-A_i(x_0)=\frac{1}{\Gamma(\alpha)}\int_{g(x_0)}^{g(x)}\left(g(x)-z\right)^{\alpha-1}\left(\left(D_{x_0+;g}^{\alpha}A_i\right)\circ g^{-1}\right)(z)\,dz, \quad (7.36)$$

$\forall\, x\in[x_0,b]$,

and by Theorem 7.6 that

$$A_i(x)-A_i(x_0)=\frac{1}{\Gamma(\alpha)}\int_{g(x)}^{g(x_0)}\left(z-g(x)\right)^{\alpha-1}\left(\left(D_{x_0-;g}^{\alpha}A_i\right)\circ g^{-1}\right)(z)\,dz, \quad (7.37)$$

$\forall\, x\in[a,x_0]$; for all $i=1,\ldots,r$.

Let multiplying (7.36) and (7.37) with $\left(\prod_{\substack{j=1\\ j\neq i}}^{r}A_j(x)\right)$ we get, respectively,

$$\left(\prod_{\substack{j=1\\ j\neq i}}^{r}A_j(x)\right)A_i(x)-\left(\prod_{\substack{j=1\\ j\neq i}}^{r}A_j(x)\right)A_i(x_0)=$$

$$\frac{\left(\prod_{\substack{j=1\\ j\neq i}}^{r}A_j(x)\right)}{\Gamma(\alpha)}\int_{g(x_0)}^{g(x)}\left(g(x)-z\right)^{\alpha-1}\left(\left(D_{x_0+;g}^{\alpha}A_i\right)\circ g^{-1}\right)(z)\,dz, \quad (7.38)$$

$\forall\, x\in[x_0,b]$,

and

$$\left(\prod_{\substack{j=1\\j\neq i}}^{r} A_j(x)\right) A_i(x) - \left(\prod_{\substack{j=1\\j\neq i}}^{r} A_j(x)\right) A_i(x_0) =$$

$$\frac{\left(\prod_{\substack{j=1\\j\neq i}}^{r} A_j(x)\right)}{\Gamma(\alpha)} \int_{g(x)}^{g(x_0)} (z - g(x))^{\alpha-1} \left(\left(D_{x_0-;g}^{\alpha} A_i\right) \circ g^{-1}\right)(z)\, dz, \tag{7.39}$$

$\forall\, x \in [a, x_0]$; for all $i = 1, \ldots, r$.

Adding (7.38) and (7.39) as separate groups, we obtain

$$\sum_{i=1}^{r} \left(\prod_{\substack{j=1\\j\neq i}}^{r} A_j(x)\right) A_i(x) - \sum_{i=1}^{r} \left(\prod_{\substack{j=1\\j\neq i}}^{r} A_j(x)\right) A_i(x_0) =$$

$$\frac{1}{\Gamma(\alpha)} \sum_{i=1}^{r} \left(\prod_{\substack{j=1\\j\neq i}}^{r} A_j(x)\right) \int_{g(x_0)}^{g(x)} (g(x) - z)^{\alpha-1} \left(\left(D_{x_0+;g}^{\alpha} A_i\right) \circ g^{-1}\right)(z)\, dz, \tag{7.40}$$

$\forall\, x \in [x_0, b]$,

and

$$\sum_{i=1}^{r} \left(\prod_{\substack{j=1\\j\neq i}}^{r} A_j(x)\right) A_i(x) - \sum_{i=1}^{r} \left(\prod_{\substack{j=1\\j\neq i}}^{r} A_j(x)\right) A_i(x_0) =$$

$$\frac{1}{\Gamma(\alpha)} \sum_{i=1}^{r} \left(\prod_{\substack{j=1\\j\neq i}}^{r} A_j(x)\right) \int_{g(x)}^{g(x_0)} (z - g(x))^{\alpha-1} \left(\left(D_{x_0-;g}^{\alpha} A_i\right) \circ g^{-1}\right)(z)\, dz, \tag{7.41}$$

$\forall\, x \in [a, x_0]$.

Next we integrate (7.40) and (7.41) with respect to $x \in [a, b]$. We have

$$\sum_{i=1}^{r} \int_{x_0}^{b} \left(\prod_{\substack{j=1\\j\neq i}}^{r} A_j(x)\right) A_i(x)\, dx - \sum_{i=1}^{r} \left(\int_{x_0}^{b} \left(\prod_{\substack{j=1\\j\neq i}}^{r} A_j(x)\right) dx\right) A_i(x_0) =$$

$$\frac{1}{\Gamma(\alpha)}\sum_{i=1}^{r}\left[\int_{x_0}^{b}\left(\prod_{\substack{j=1\\j\neq i}}^{r}A_j(x)\right)\left(\int_{g(x_0)}^{g(x)}(g(x)-z)^{\alpha-1}\left(\left(D_{x_0+;g}^{\alpha}A_i\right)\circ g^{-1}\right)(z)\,dz\right)dx\right], \tag{7.42}$$

and

$$\sum_{i=1}^{r}\int_{a}^{x_0}\left(\prod_{\substack{j=1\\j\neq i}}^{r}A_j(x)\right)A_i(x)\,dx-\sum_{i=1}^{r}\left(\int_{a}^{x_0}\left(\prod_{\substack{j=1\\j\neq i}}^{r}A_j(x)\right)dx\right)A_i(x_0)=$$

$$\frac{1}{\Gamma(\alpha)}\sum_{i=1}^{r}\left[\int_{a}^{x_0}\left(\prod_{\substack{j=1\\j\neq i}}^{r}A_j(x)\right)\left(\int_{g(x)}^{g(x_0)}(z-g(x))^{\alpha-1}\left(\left(D_{x_0-;g}^{\alpha}A_i\right)\circ g^{-1}\right)(z)\,dz\right)dx\right]. \tag{7.43}$$

Finally, adding (7.42) and (7.43) we obtain the useful identity

$$\Phi(A_1,\ldots,A_r)(x_0):=$$

$$\sum_{i=1}^{r}\left[\int_{a}^{b}\left(\prod_{\substack{j=1\\j\neq i}}^{r}A_j(x)\right)A_i(x)\,dx-\left(\int_{a}^{b}\left(\prod_{\substack{j=1\\j\neq i}}^{r}A_j(x)\right)dx\right)A_i(x_0)\right]= \tag{7.44}$$

$$\frac{1}{\Gamma(\alpha)}\sum_{i=1}^{r}\left[\left[\int_{a}^{x_0}\left(\prod_{\substack{j=1\\j\neq i}}^{r}A_j(x)\right)\left(\int_{g(x)}^{g(x_0)}(z-g(x))^{\alpha-1}\left(\left(D_{x_0-;g}^{\alpha}A_i\right)\circ g^{-1}\right)(z)\,dz\right)dx\right]\right.$$

$$\left.+\left[\int_{x_0}^{b}\left(\prod_{\substack{j=1\\j\neq i}}^{r}A_j(x)\right)\left(\int_{g(x_0)}^{g(x)}(g(x)-z)^{\alpha-1}\left(\left(D_{x_0+;g}^{\alpha}A_i\right)\circ g^{-1}\right)(z)\,dz\right)dx\right]\right].$$

Next we take the 1-Schatten norm in (7.44) and we get:

$$\|\Phi(A_1,\ldots,A_r)(x_0)\|_1\leq\frac{1}{\Gamma(\alpha)}$$

$$\sum_{i=1}^{r}\left[\left\|\int_{a}^{x_0}\left(\prod_{\substack{j=1\\j\neq i}}^{r}A_j(x)\right)\left(\int_{g(x)}^{g(x_0)}(z-g(x))^{\alpha-1}\left(\left(D_{x_0-;g}^{\alpha}A_i\right)\circ g^{-1}\right)(z)\,dz\right)dx\right\|_1\right.$$

$$+\left\|\int_{x_0}^{b}\left(\prod_{\substack{j=1\\j\neq i}}^{r}A_j(x)\right)\left(\int_{g(x_0)}^{g(x)}(g(x)-z)^{\alpha-1}\left(\left(D_{x_0+;g}^{\alpha}A_i\right)\circ g^{-1}\right)(z)\,dz\right)dx\right\|_1\Bigg]\leq$$

$$\frac{1}{\Gamma(\alpha)}\sum_{i=1}^{r}\left[\int_a^{x_0}\left\|\left(\prod_{\substack{j=1\\j\neq i}}^{r}A_j(x)\right)\left(\int_{g(x)}^{g(x_0)}(z-g(x))^{\alpha-1}\left(\left(D_{x_0-;g}^{\alpha}A_i\right)\circ g^{-1}\right)(z)\,dz\right)\right\|_1 dx\right. \tag{7.45}$$

$$\left.+\int_{x_0}^{b}\left\|\left(\prod_{\substack{j=1\\j\neq i}}^{r}A_j(x)\right)\left(\int_{g(x_0)}^{g(x)}(g(x)-z)^{\alpha-1}\left(\left(D_{x_0+;g}^{\alpha}A_i\right)\circ g^{-1}\right)(z)\,dz\right)\right\|_1 dx\right]\leq$$

(by using the p-Schatten norm and Hölder's type inequality (7.31) for $p=q=2$)

$$\frac{1}{\Gamma(\alpha)}\sum_{i=1}^{r}\left[\int_a^{x_0}\left\|\prod_{\substack{j=1\\j\neq i}}^{r}A_j(x)\right\|_2\left\|\int_{g(x)}^{g(x_0)}(z-g(x))^{\alpha-1}\left(\left(D_{x_0-;g}^{\alpha}A_i\right)\circ g^{-1}\right)(z)\,dz\right\|_2 dx\right.$$

$$\left.+\int_{x_0}^{b}\left\|\prod_{\substack{j=1\\j\neq i}}^{r}A_j(x)\right\|_2\left\|\int_{g(x_0)}^{g(x)}(g(x)-z)^{\alpha-1}\left(\left(D_{x_0+;g}^{\alpha}A_i\right)\circ g^{-1}\right)(z)\,dz\right\|_2 dx\right]\leq \tag{7.46}$$

$$\frac{1}{\Gamma(\alpha)}\sum_{i=1}^{r}\left[\int_a^{x_0}\prod_{\substack{j=1\\j\neq i}}^{r}\left\|A_j(x)\right\|_2\left(\int_{g(x)}^{g(x_0)}(z-g(x))^{\alpha-1}\left\|\left(\left(D_{x_0-;g}^{\alpha}A_i\right)\circ g^{-1}\right)(z)\right\|_2 dz\right)dx\right.$$

$$\left.+\int_{x_0}^{b}\prod_{\substack{j=1\\j\neq i}}^{r}\left\|A_j(x)\right\|_2\left(\int_{g(x_0)}^{g(x)}(g(x)-z)^{\alpha-1}\left\|\left(\left(D_{x_0+;g}^{\alpha}A_i\right)\circ g^{-1}\right)(z)\right\|_2 dz\right)dx\right].$$

We have proved, so far, that

$$\left\|\Phi\left(A_1,\ldots,A_r\right)(x_0)\right\|_1\leq\frac{1}{\Gamma(\alpha)}$$

$$\sum_{i=1}^{r}\left[\int_{a}^{x_0}\prod_{\substack{j=1\\ j\neq i}}^{r}\left\|A_j(x)\right\|_2\left(\int_{g(x)}^{g(x_0)}(z-g(x))^{\alpha-1}\left\|\left(\left(D_{x_0-;g}^{\alpha}A_i\right)\circ g^{-1}\right)(z)\right\|_2 dz\right)dx+\right.$$

$$\left.\int_{x_0}^{b}\prod_{\substack{j=1\\ j\neq i}}^{r}\left\|A_j(x)\right\|_2\left(\int_{g(x_0)}^{g(x)}(g(x)-z)^{\alpha-1}\left\|\left(\left(D_{x_0+;g}^{\alpha}A_i\right)\circ g^{-1}\right)(z)\right\|_2 dz\right)dx\right]=:(\xi). \tag{7.47}$$

Let now $\gamma, \delta > 1$ such that $\frac{1}{\gamma}+\frac{1}{\delta}=1$, and we apply the usual Hölder's inequality in (7.47); $\alpha > \frac{1}{\delta}$. Then we have

$$\left\|\Phi\left(A_1,\ldots,A_r\right)(x_0)\right\|_1 \leq (\xi) \leq \frac{1}{\Gamma(\alpha)\left(\gamma(\alpha-1)+1\right)^{\frac{1}{\gamma}}}$$

$$\sum_{i=1}^{r}\left[\int_{a}^{x_0}\prod_{\substack{j=1\\ j\neq i}}^{r}\left\|A_j(x)\right\|_2\left(g(x_0)-g(x)\right)^{\frac{\gamma(\alpha-1)+1}{\gamma}}\left(\int_{g(x)}^{g(x_0)}\left\|\left(\left(D_{x_0-;g}^{\alpha}A_i\right)\circ g^{-1}\right)(z)\right\|_2^{\delta}dz\right)^{\frac{1}{\delta}}dx+\right.$$

$$\left.\int_{x_0}^{b}\prod_{\substack{j=1\\ j\neq i}}^{r}\left\|A_j(x)\right\|_2\left(g(x)-g(x_0)\right)^{\frac{\gamma(\alpha-1)+1}{\gamma}}\left(\int_{g(x_0)}^{g(x)}\left\|\left(\left(D_{x_0+;g}^{\alpha}A_i\right)\circ g^{-1}\right)(z)\right\|_2^{\delta}dz\right)^{\frac{1}{\delta}}dx\right] \tag{7.48}$$

$$\leq \frac{1}{\Gamma(\alpha)\left(\gamma(\alpha-1)+1\right)^{\frac{1}{\gamma}}}$$

$$\sum_{i=1}^{r}\left[\left\|\left\|\left(D_{x_0-;g}^{\alpha}A_i\right)\circ g^{-1}\right\|_2\right\|_{\delta,[g(a),g(x_0)]}\int_{a}^{x_0}\prod_{\substack{j=1\\ j\neq i}}^{r}\left\|A_j(x)\right\|_2\left(g(x_0)-g(x)\right)^{\alpha-\frac{1}{\delta}}dx+\right.$$

$$\left.\left\|\left\|\left(D_{x_0+;g}^{\alpha}A_i\right)\circ g^{-1}\right\|_2\right\|_{\delta,[g(x_0),g(b)]}\int_{x_0}^{b}\prod_{\substack{j=1\\ j\neq i}}^{r}\left\|A_j(x)\right\|_2\left(g(x)-g(x_0)\right)^{\alpha-\frac{1}{\delta}}dx\right], \tag{7.49}$$

proving (7.33).

If $\alpha \geq 1$ we obtain

$$\left\|\Phi\left(A_1,\ldots,A_r\right)(x_0)\right\|_1 \leq (\xi) \leq \frac{1}{\Gamma(\alpha)}$$

$$\sum_{i=1}^{r}\left[\left\|\left\|\left(D_{x_0-;g}^{\alpha}A_i\right)\circ g^{-1}\right\|_2\right\|_{L_1([g(a),g(x_0)])}\int_a^{x_0}\prod_{\substack{j=1\\j\neq i}}^{r}\left\|A_j(x)\right\|_2(g(x_0)-g(x))^{\alpha-1}dx+\right.$$

$$\left.\left\|\left\|\left(D_{x_0+;g}^{\alpha}A_i\right)\circ g^{-1}\right\|_2\right\|_{L_1([g(x_0),g(b)])}\int_{x_0}^{b}\prod_{\substack{j=1\\j\neq i}}^{r}\left\|A_j(x)\right\|_2(g(x)-g(x_0))^{\alpha-1}dx\right], \tag{7.50}$$

proving (7.34).

At last we derive

$$\|\Phi(A_1,\ldots,A_r)(x_0)\|_1\le(\xi)\le\frac{1}{\Gamma(\alpha+1)}$$

$$\sum_{i=1}^{r}\left[\left\|\left\|\left(D_{x_0-;g}^{\alpha}A_i\right)\circ g^{-1}\right\|_2\right\|_{\infty,[g(a),g(x_0)]}\int_a^{x_0}\prod_{\substack{j=1\\j\neq i}}^{r}\left\|A_j(x)\right\|_2(g(x_0)-g(x))^{\alpha}dx+\right.$$

$$\left.\left\|\left\|\left(D_{x_0+;g}^{\alpha}A_i\right)\circ g^{-1}\right\|_2\right\|_{\infty,[g(x_0),g(b)]}\int_{x_0}^{b}\prod_{\substack{j=1\\j\neq i}}^{r}\left\|A_j(x)\right\|_2(g(x)-g(x_0))^{\alpha}dx\right], \tag{7.51}$$

proving (7.35).

The theorem is proved. ∎

When $r=2$ we obtain the following operator related Ostrowski type fractional inequalities.

Theorem 7.12 *Let $p,q>1:\frac{1}{p}+\frac{1}{q}=1$, and let the $*$-ideals $\mathcal{B}_p(H)$, $\mathcal{B}_q(H)$, for which $\left(\mathcal{B}_p(H),\|\cdot\|_p\right)$, $\left(\mathcal{B}_q(H),\|\cdot\|_q\right)$ are Banach algebras; $x_0\in[a,b]\subset\mathbb{R}$, $\alpha>0$, $n=\lceil\alpha\rceil$; $A_1\in C^n\left([a,b],\mathcal{B}_p(H)\right)$, $A_2\in C^n\left([a,b],\mathcal{B}_q(H)\right)$; $g\in C^1([a,b])$, strictly increasing, such that $g^{-1}\in C^n([g(a),g(b)])$, with $\left(A_i\circ g^{-1}\right)^{(k)}(g(x_0))=0$, $k=1,\ldots,n-1$; $i=1,2$. Then*

(1) it holds

$$\Phi(A_1,A_2)(x_0):=\int_a^b A_2(x)A_1(x)dx+\int_a^b A_1(x)A_2(x)dx-$$

$$\left(\int_a^b A_2(x)dx\right)A_1(x_0)-\left(\int_a^b A_1(x)dx\right)A_2(x_0)=$$

$$\frac{1}{\Gamma(\alpha)}\left\{\left[\int_a^{x_0} A_2(x)\left(\int_{g(x)}^{g(x_0)}(z-g(x))^{\alpha-1}\left(\left(D_{x_0-;g}^{\alpha}A_1\right)\circ g^{-1}\right)(z)\,dz\right)dx\right]+\right.$$

$$\left[\int_{x_0}^{b} A_2(x)\left(\int_{g(x_0)}^{g(x)}(g(x)-z)^{\alpha-1}\left(\left(D_{x_0+;g}^{\alpha}A_1\right)\circ g^{-1}\right)(z)\,dz\right)dx\right]+$$

$$\left[\int_a^{x_0} A_1(x)\left(\int_{g(x)}^{g(x_0)}(z-g(x))^{\alpha-1}\left(\left(D_{x_0-;g}^{\alpha}A_2\right)\circ g^{-1}\right)(z)\,dz\right)dx\right]+$$

$$\left.\left[\int_{x_0}^{b} A_1(x)\left(\int_{g(x_0)}^{g(x)}(g(x)-z)^{\alpha-1}\left(\left(D_{x_0+;g}^{\alpha}A_2\right)\circ g^{-1}\right)(z)\,dz\right)dx\right]\right\}, \quad (7.52)$$

(2) for $\gamma, \delta > 1 : \frac{1}{\gamma}+\frac{1}{\delta}=1$, with $\alpha > \frac{1}{\delta}$, we have that

$$\|\Phi(A_1,A_2)(x_0)\|_1 \leq \frac{1}{\Gamma(\alpha)(\gamma(\alpha-1)+1)^{\frac{1}{\gamma}}}$$

$$\left\{\left[\left\|\left\|\left(D_{x_0-;g}^{\alpha}A_1\right)\circ g^{-1}\right\|_p\right\|_{\delta,[g(a),g(x_0)]}\int_a^{x_0}\|A_2(x)\|_q\,(g(x_0)-g(x))^{\alpha-\frac{1}{\delta}}\,dx\right]+\right.$$

$$\left[\left\|\left\|\left(D_{x_0+;g}^{\alpha}A_1\right)\circ g^{-1}\right\|_p\right\|_{\delta,[g(x_0),g(b)]}\int_{x_0}^{b}\|A_2(x)\|_q\,(g(x)-g(x_0))^{\alpha-\frac{1}{\delta}}\,dx\right]+ \quad (7.53)$$

$$\left[\left\|\left\|\left(D_{x_0-;g}^{\alpha}A_2\right)\circ g^{-1}\right\|_q\right\|_{\delta,[g(a),g(x_0)]}\int_a^{x_0}\|A_1(x)\|_p\,(g(x_0)-g(x))^{\alpha-\frac{1}{\delta}}\,dx\right]+$$

$$\left.\left[\left\|\left\|\left(D_{x_0+;g}^{\alpha}A_2\right)\circ g^{-1}\right\|_q\right\|_{\delta,[g(x_0),g(b)]}\int_{x_0}^{b}\|A_1(x)\|_p\,(g(x)-g(x_0))^{\alpha-\frac{1}{\delta}}\,dx\right]\right\},$$

(3) if $\alpha \geq 1$, we obtain

$$\|\Phi(A_1,A_2)(x_0)\|_1 \leq \frac{1}{\Gamma(\alpha)}$$

$$\left\{\left[\left\|\left\|\left(D_{x_0-;g}^{\alpha}A_1\right)\circ g^{-1}\right\|_p\right\|_{L_1([g(a),g(x_0)])}\int_a^{x_0}\|A_2(x)\|_q\,(g(x_0)-g(x))^{\alpha-1}\,dx\right]+\right.$$

$$\left[\left\|\left\|\left(D_{x_0+;g}^{\alpha}A_1\right)\circ g^{-1}\right\|_p\right\|_{L_1([g(x_0),g(b)])}\int_{x_0}^{b}\|A_2(x)\|_q\,(g(x)-g(x_0))^{\alpha-1}\,dx\right]+$$

$$\left[\left\|\left\|\left(D_{x_0-;g}^{\alpha}A_2\right)\circ g^{-1}\right\|_q\right\|_{L_1([g(a),g(x_0)])}\int_a^{x_0}\|A_1(x)\|_p\,(g(x_0)-g(x))^{\alpha-1}\,dx\right]+ \quad (7.54)$$

$$\left[\left\|\left\|\left(D^{\alpha}_{x_0+;g}A_2\right)\circ g^{-1}\right\|_q\right\|_{L_1([g(x_0),g(b)])}\int_{x_0}^{b}\|A_1(x)\|_p\,(g(x)-g(x_0))^{\alpha-1}\,dx\right]\Bigg\},$$

and

(4)

$$\|\Phi(A_1,A_2)(x_0)\|_1\le\frac{1}{\Gamma(\alpha+1)}$$

$$\Bigg\{\left[\left\|\left\|\left(D^{\alpha}_{x_0-;g}A_1\right)\circ g^{-1}\right\|_p\right\|_{\infty,[g(a),g(x_0)]}\int_{a}^{x_0}\|A_2(x)\|_q\,(g(x_0)-g(x))^{\alpha}\,dx\right]+$$

$$\left[\left\|\left\|\left(D^{\alpha}_{x_0+;g}A_1\right)\circ g^{-1}\right\|_p\right\|_{\infty,[g(x_0),g(b)]}\int_{x_0}^{b}\|A_2(x)\|_q\,(g(x)-g(x_0))^{\alpha}\,dx\right]+$$

$$\left[\left\|\left\|\left(D^{\alpha}_{x_0-;g}A_2\right)\circ g^{-1}\right\|_q\right\|_{\infty,[g(a),g(x_0)]}\int_{a}^{x_0}\|A_1(x)\|_p\,(g(x_0)-g(x))^{\alpha}\,dx\right]+$$

$$\left[\left\|\left\|\left(D^{\alpha}_{x_0+;g}A_2\right)\circ g^{-1}\right\|_q\right\|_{\infty,[g(x_0),g(b)]}\int_{x_0}^{b}\|A_1(x)\|_p\,(g(x)-g(x_0))^{\alpha}\,dx\right]\Bigg\}. \tag{7.55}$$

Proof Here we have that (acting as in the proof of Theorem 7.11 for $r=2$)

$$\Phi(A_1,A_2)(x_0):=\int_a^b A_2(x)A_1(x)\,dx+\int_a^b A_1(x)A_2(x)\,dx-$$

$$\left(\int_a^b A_2(x)\,dx\right)A_1(x_0)-\left(\int_a^b A_1(x)\,dx\right)A_2(x_0)\overset{(7.32)}{=}$$

$$\frac{1}{\Gamma(\alpha)}\Bigg\{\left[\int_a^{x_0}A_2(x)\left(\int_{g(x)}^{g(x_0)}(z-g(x))^{\alpha-1}\left(\left(D^{\alpha}_{x_0-;g}A_1\right)\circ g^{-1}\right)(z)\,dz\right)dx\right]+$$

$$\left[\int_{x_0}^{b}A_2(x)\left(\int_{g(x_0)}^{g(x)}(g(x)-z)^{\alpha-1}\left(\left(D^{\alpha}_{x_0+;g}A_1\right)\circ g^{-1}\right)(z)\,dz\right)dx\right]+$$

$$\left[\int_a^{x_0}A_1(x)\left(\int_{g(x)}^{g(x_0)}(z-g(x))^{\alpha-1}\left(\left(D^{\alpha}_{x_0-;g}A_2\right)\circ g^{-1}\right)(z)\,dz\right)dx\right]+$$

$$\left[\int_{x_0}^{b}A_1(x)\left(\int_{g(x_0)}^{g(x)}(g(x)-z)^{\alpha-1}\left(\left(D^{\alpha}_{x_0+;g}A_2\right)\circ g^{-1}\right)(z)\,dz\right)dx\right]\Bigg\}. \tag{7.56}$$

Therefore it holds by taking the 1-Schatten norm that

$$\|\Phi(A_1, A_2)(x_0)\|_1 = \left\| \int_a^b A_2(x) A_1(x) + \int_a^b A_1(x) A_2(x) dx - \right.$$

$$\left. \left(\int_a^b A_2(x) dx \right) A_1(x_0) - \left(\int_a^b A_1(x) dx \right) A_2(x_0) \right\|_1 \leq$$

$$\frac{1}{\Gamma(\alpha)} \left\{ \left[\left\| \int_a^{x_0} A_2(x) \left(\int_{g(x)}^{g(x_0)} (z - g(x))^{\alpha-1} \left(\left(D_{x_0-;g}^{\alpha} A_1 \right) \circ g^{-1} \right)(z) dz \right) dx \right\|_1 \right] + \right.$$

$$\left[\left\| \int_{x_0}^b A_2(x) \left(\int_{g(x_0)}^{g(x)} (g(x) - z)^{\alpha-1} \left(\left(D_{x_0+;g}^{\alpha} A_1 \right) \circ g^{-1} \right)(z) dz \right) dx \right\|_1 \right] +$$

$$\left[\left\| \int_a^{x_0} A_1(x) \left(\int_{g(x)}^{g(x_0)} (z - g(x))^{\alpha-1} \left(\left(D_{x_0-;g}^{\alpha} A_2 \right) \circ g^{-1} \right)(z) dz \right) dx \right\|_1 \right] +$$

$$\left. \left[\left\| \int_{x_0}^b A_1(x) \left(\int_{g(x_0)}^{g(x)} (g(x) - z)^{\alpha-1} \left(\left(D_{x_0+;g}^{\alpha} A_2 \right) \circ g^{-1} \right)(z) dz \right) dx \right\|_1 \right] \right\} \leq \tag{7.57}$$

$$\frac{1}{\Gamma(\alpha)} \left\{ \left[\int_a^{x_0} \left\| A_2(x) \left(\int_{g(x)}^{g(x_0)} (z - g(x))^{\alpha-1} \left(\left(D_{x_0-;g}^{\alpha} A_1 \right) \circ g^{-1} \right)(z) dz \right) \right\|_1 dx \right] + \right.$$

$$\left[\int_{x_0}^b \left\| A_2(x) \left(\int_{g(x_0)}^{g(x)} (g(x) - z)^{\alpha-1} \left(\left(D_{x_0+;g}^{\alpha} A_1 \right) \circ g^{-1} \right)(z) dz \right) \right\|_1 dx \right] +$$

$$\left[\int_a^{x_0} \left\| A_1(x) \left(\int_{g(x)}^{g(x_0)} (z - g(x))^{\alpha-1} \left(\left(D_{x_0-;g}^{\alpha} A_2 \right) \circ g^{-1} \right)(z) dz \right) \right\|_1 dx \right] +$$

$$\left. \left[\int_{x_0}^b \left\| A_1(x) \left(\int_{g(x_0)}^{g(x)} (g(x) - z)^{\alpha-1} \left(\left(D_{x_0+;g}^{\alpha} A_2 \right) \circ g^{-1} \right)(z) dz \right) \right\|_1 dx \right] \right\} \leq \tag{7.58}$$

(by using the p-Schatten norm and Hölder's type inequality (7.31) for $p, q > 1$: $\frac{1}{p} + \frac{1}{q} = 1$)

$$\frac{1}{\Gamma(\alpha)} \left\{ \left[\int_a^{x_0} \|A_2(x)\|_q \left\| \left(\int_{g(x)}^{g(x_0)} (z - g(x))^{\alpha-1} \left(\left(D_{x_0-;g}^{\alpha} A_1 \right) \circ g^{-1} \right)(z) dz \right) \right\|_p dx \right] + \right.$$

$$\left[\int_{x_0}^b \|A_2(x)\|_q \left\| \left(\int_{g(x_0)}^{g(x)} (g(x) - z)^{\alpha-1} \left(\left(D_{x_0+;g}^{\alpha} A_1 \right) \circ g^{-1} \right)(z) dz \right) \right\|_p dx \right] +$$

$$\left[\int_a^{x_0} \|A_1(x)\|_p \left\| \left(\int_{g(x)}^{g(x_0)} (z - g(x))^{\alpha-1} \left(\left(D_{x_0-;g}^{\alpha} A_2 \right) \circ g^{-1} \right)(z) dz \right) \right\|_q dx \right] +$$

$$\left[\int_{x_0}^{b} \|A_1(x)\|_p \left\| \left(\int_{g(x_0)}^{g(x)} (g(x) - z)^{\alpha-1} \left(\left(D_{x_0+;g}^{\alpha} A_2 \right) \circ g^{-1} \right)(z)\, dz \right) \right\|_q dx \right] \Bigg\} \leq \tag{7.59}$$

$$\frac{1}{\Gamma(\alpha)} \Bigg\{ \left[\int_a^{x_0} \|A_2(x)\|_q \left(\int_{g(x)}^{g(x_0)} (z - g(x))^{\alpha-1} \left\| \left(\left(D_{x_0-;g}^{\alpha} A_1 \right) \circ g^{-1} \right)(z) \right\|_p dz \right) dx \right] +$$

$$\left[\int_{x_0}^{b} \|A_2(x)\|_q \left(\int_{g(x_0)}^{g(x)} (g(x) - z)^{\alpha-1} \left\| \left(\left(D_{x_0+;g}^{\alpha} A_1 \right) \circ g^{-1} \right)(z) \right\|_p dz \right) dx \right] +$$

$$\left[\int_a^{x_0} \|A_1(x)\|_p \left(\int_{g(x)}^{g(x_0)} (z - g(x))^{\alpha-1} \left\| \left(\left(D_{x_0-;g}^{\alpha} A_2 \right) \circ g^{-1} \right)(z) \right\|_q dz \right) dx \right] +$$

$$\left[\int_{x_0}^{b} \|A_1(x)\|_p \left(\int_{g(x_0)}^{g(x)} (g(x) - z)^{\alpha-1} \left\| \left(\left(D_{x_0+;g}^{\alpha} A_2 \right) \circ g^{-1} \right)(z) \right\|_q dz \right) dx \right] \Bigg\}. \tag{7.60}$$

We have proved, so far, that

$$\|\Phi(A_1, A_2)(x_0)\|_1 \leq$$

$$\frac{1}{\Gamma(\alpha)} \Bigg\{ \left[\int_a^{x_0} \|A_2(x)\|_q \left(\int_{g(x)}^{g(x_0)} (z - g(x))^{\alpha-1} \left\| \left(\left(D_{x_0-;g}^{\alpha} A_1 \right) \circ g^{-1} \right)(z) \right\|_p dz \right) dx \right] +$$

$$\left[\int_{x_0}^{b} \|A_2(x)\|_q \left(\int_{g(x_0)}^{g(x)} (g(x) - z)^{\alpha-1} \left\| \left(\left(D_{x_0+;g}^{\alpha} A_1 \right) \circ g^{-1} \right)(z) \right\|_p dz \right) dx \right] +$$

$$\left[\int_a^{x_0} \|A_1(x)\|_p \left(\int_{g(x)}^{g(x_0)} (z - g(x))^{\alpha-1} \left\| \left(\left(D_{x_0-;g}^{\alpha} A_2 \right) \circ g^{-1} \right)(z) \right\|_q dz \right) dx \right] +$$

$$\left[\int_{x_0}^{b} \|A_1(x)\|_p \left(\int_{g(x_0)}^{g(x)} (g(x) - z)^{\alpha-1} \left\| \left(\left(D_{x_0+;g}^{\alpha} A_2 \right) \circ g^{-1} \right)(z) \right\|_q dz \right) dx \right] \Bigg\} =: (\lambda). \tag{7.61}$$

Let now $\gamma, \delta > 1$ such that $\frac{1}{\gamma} + \frac{1}{\delta} = 1$, and we apply the usual Hölder's inequality in (7.61); $\alpha > \frac{1}{\delta}$. Then we have that

$$\|\Phi(A_1, A_2)(x_0)\|_1 \leq (\lambda) \leq \frac{1}{\Gamma(\alpha)(\gamma(\alpha-1)+1)^{\frac{1}{\gamma}}}$$

$$\Bigg\{ \left[\int_a^{x_0} \|A_2(x)\|_q (g(x_0) - g(x))^{\frac{\gamma(\alpha-1)+1}{\gamma}} \left(\int_{g(x)}^{g(x_0)} \left\| \left(\left(D_{x_0-;g}^{\alpha} A_1 \right) \circ g^{-1} \right)(z) \right\|_p^{\delta} dz \right)^{\frac{1}{\delta}} dx \right] +$$

$$\left[\int_{x_0}^{b} \|A_2(x)\|_q (g(x) - g(x_0))^{\frac{\gamma(\alpha-1)+1}{\gamma}} \left(\int_{g(x_0)}^{g(x)} \left\|\left(\left(D_{x_0+;g}^{\alpha} A_1\right) \circ g^{-1}\right)(z)\right\|_p^{\delta} dz\right)^{\frac{1}{\delta}} dx\right] +$$

$$\left[\int_{a}^{x_0} \|A_1(x)\|_p (g(x_0) - g(x))^{\frac{\gamma(\alpha-1)+1}{\gamma}} \left(\int_{g(x)}^{g(x_0)} \left\|\left(\left(D_{x_0-;g}^{\alpha} A_2\right) \circ g^{-1}\right)(z)\right\|_q^{\delta} dz\right)^{\frac{1}{\delta}} dx\right] + \tag{7.62}$$

$$\left[\int_{x_0}^{b} \|A_1(x)\|_p (g(x) - g(x_0))^{\frac{\gamma(\alpha-1)+1}{\gamma}} \left(\int_{g(x_0)}^{g(x)} \left\|\left(\left(D_{x_0+;g}^{\alpha} A_2\right) \circ g^{-1}\right)(z)\right\|_q^{\delta} dz\right)^{\frac{1}{\delta}} dx\right]\right\}$$

$$\leq \frac{1}{\Gamma(\alpha)(\gamma(\alpha-1)+1)^{\frac{1}{\gamma}}}$$

$$\left\{\left[\left\|\left\|\left(D_{x_0-;g}^{\alpha} A_1\right) \circ g^{-1}\right\|_p\right\|_{\delta,[g(a),g(x_0)]} \int_a^{x_0} \|A_2(x)\|_q (g(x_0) - g(x))^{\alpha-\frac{1}{\delta}} dx\right] +\right.$$

$$\left[\left\|\left\|\left(D_{x_0+;g}^{\alpha} A_1\right) \circ g^{-1}\right\|_p\right\|_{\delta,[g(x_0),g(b)]} \int_{x_0}^{b} \|A_2(x)\|_q (g(x) - g(x_0))^{\alpha-\frac{1}{\delta}} dx\right] + \tag{7.63}$$

$$\left[\left\|\left\|\left(D_{x_0-;g}^{\alpha} A_2\right) \circ g^{-1}\right\|_q\right\|_{\delta,[g(a),g(x_0)]} \int_a^{x_0} \|A_1(x)\|_p (g(x_0) - g(x))^{\alpha-\frac{1}{\delta}} dx\right] +$$

$$\left.\left[\left\|\left\|\left(D_{x_0+;g}^{\alpha} A_2\right) \circ g^{-1}\right\|_q\right\|_{\delta,[g(x_0),g(b)]} \int_{x_0}^{b} \|A_1(x)\|_p (g(x) - g(x_0))^{\alpha-\frac{1}{\delta}} dx\right]\right\},$$

proving (7.53).

If $\alpha \geq 1$, we obtain

$$\|\Phi(A_1, A_2)(x_0)\|_1 \leq (\lambda) \leq \frac{1}{\Gamma(\alpha)}$$

$$\left\{\left[\left\|\left\|\left(D_{x_0-;g}^{\alpha} A_1\right) \circ g^{-1}\right\|_p\right\|_{L_1([g(a),g(x_0)])} \int_a^{x_0} \|A_2(x)\|_q (g(x_0) - g(x))^{\alpha-1} dx\right] +\right.$$

$$\left[\left\|\left\|\left(D_{x_0+;g}^{\alpha} A_1\right) \circ g^{-1}\right\|_p\right\|_{L_1([g(x_0),g(b)])} \int_{x_0}^{b} \|A_2(x)\|_q (g(x) - g(x_0))^{\alpha-1} dx\right] +$$

$$\left[\left\|\left\|\left(D_{x_0-;g}^{\alpha} A_2\right) \circ g^{-1}\right\|_q\right\|_{L_1([g(a),g(x_0)])} \int_a^{x_0} \|A_1(x)\|_p (g(x_0) - g(x))^{\alpha-1} dx\right] + \tag{7.64}$$

$$\left.\left[\left\|\left\|\left(D_{x_0+;g}^{\alpha} A_2\right) \circ g^{-1}\right\|_q\right\|_{L_1([g(x_0),g(b)])} \int_{x_0}^{b} \|A_1(x)\|_p (g(x) - g(x_0))^{\alpha-1} dx\right]\right\},$$

proving (7.54).

At last we derive

$$\|\Phi(A_1, A_2)(x_0)\|_1 \leq (\lambda) \leq \frac{1}{\Gamma(\alpha+1)}$$

$$\left\{\left[\left\|\left\|\left(D^{\alpha}_{x_0-;g}A_1\right)\circ g^{-1}\right\|_p\right\|_{\infty,[g(a),g(x_0)]}\int_a^{x_0}\|A_2(x)\|_q (g(x_0)-g(x))^{\alpha}dx\right]+\right.$$

$$\left[\left\|\left\|\left(D^{\alpha}_{x_0+;g}A_1\right)\circ g^{-1}\right\|_p\right\|_{\infty,[g(x_0),g(b)]}\int_{x_0}^{b}\|A_2(x)\|_q (g(x)-g(x_0))^{\alpha}dx\right]+$$

$$\left[\left\|\left\|\left(D^{\alpha}_{x_0-;g}A_2\right)\circ g^{-1}\right\|_q\right\|_{\infty,[g(a),g(x_0)]}\int_a^{x_0}\|A_1(x)\|_p (g(x_0)-g(x))^{\alpha}dx\right]+$$

$$\left.\left[\left\|\left\|\left(D^{\alpha}_{x_0+;g}A_2\right)\circ g^{-1}\right\|_q\right\|_{\infty,[g(x_0),g(b)]}\int_{x_0}^{b}\|A_1(x)\|_p (g(x)-g(x_0))^{\alpha}dx\right]\right\}, \tag{7.65}$$

proving (7.55).

The theorem is proved. ■

We continue with p-Schatten norm weighted mixed generalized fractional Ostrowski type inequalities involving several functions taking values in the Banach algebra $\mathcal{B}_p(H) \subset \mathcal{B}(H)$, $p > 1$.

Theorem 7.13 *Let the $*$-ideal $\mathcal{B}_p(H)$, which $\left(\mathcal{B}_p(H), \|\cdot\|_p\right)$, $p > 1$, is a Banach algebra; $x_0 \in [a,b] \subset \mathbb{R}$, $\alpha > 0$, $n = \lceil \alpha \rceil$; $A_i \in C^n\left([a,b], \mathcal{B}_p(H)\right)$, $i = 1, \ldots, r \in \mathbb{N} - \{1\}$; $g \in C^1([a,b])$, strictly increasing such that $g^{-1} \in C^n([g(a), g(b)])$, with $\left(A_i \circ g^{-1}\right)^{(k)}(g(x_0)) = 0$, $k = 1, \ldots, n-1$; $i = 1, \ldots, r$. Let $\Phi(A_1, \ldots, A_r)(x_0)$ be as in (7.32).*

Then

(1) for $\gamma, \delta > 1 : \frac{1}{\gamma} + \frac{1}{\delta} = 1$, with $\alpha > \frac{1}{\delta}$ we have that

$$\|\Phi(A_1, \ldots, A_r)(x_0)\|_p \leq \frac{1}{\Gamma(\alpha)(\gamma(\alpha-1)+1)^{\frac{1}{\gamma}}}$$

$$\sum_{i=1}^{r}\left[\left\|\left\|\left(D^{\alpha}_{x_0-;g}A_i\right)\circ g^{-1}\right\|_p\right\|_{\delta,[g(a),g(x_0)]}\int_a^{x_0}\prod_{\substack{j=1\\ j\neq i}}^{r}\|A_j(x)\|_p (g(x_0)-g(x))^{\alpha-\frac{1}{\delta}}dx+\right.$$

$$\left\| \left\| \left(D^{\alpha}_{x_0+;g} A_i \right) \circ g^{-1} \right\|_p \right\|_{\delta,[g(x_0),g(b)]} \int_{x_0}^{b} \prod_{\substack{j=1 \\ j \neq i}}^{r} \left\| A_j(x) \right\|_p (g(x) - g(x_0))^{\alpha - \frac{1}{\delta}} dx \Bigg], \tag{7.66}$$

(2) if $\alpha \geq 1$, we obtain

$$\left\| \Phi(A_1, \ldots, A_r)(x_0) \right\|_p \leq \frac{1}{\Gamma(\alpha)}$$

$$\sum_{i=1}^{r} \Bigg[\left\| \left\| \left(D^{\alpha}_{x_0-;g} A_i \right) \circ g^{-1} \right\|_p \right\|_{L_1([g(a),g(x_0)])} \int_{a}^{x_0} \prod_{\substack{j=1 \\ j \neq i}}^{r} \left\| A_j(x) \right\|_p (g(x_0) - g(x))^{\alpha - 1} dx +$$

$$\left\| \left\| \left(D^{\alpha}_{x_0+;g} A_i \right) \circ g^{-1} \right\|_p \right\|_{L_1([g(x_0),g(b)])} \int_{x_0}^{b} \prod_{\substack{j=1 \\ j \neq i}}^{r} \left\| A_j(x) \right\|_p (g(x) - g(x_0))^{\alpha - 1} dx \Bigg], \tag{7.67}$$

and

(3)

$$\left\| \Phi(A_1, \ldots, A_r)(x_0) \right\|_p \leq \frac{1}{\Gamma(\alpha + 1)}$$

$$\sum_{i=1}^{r} \Bigg[\left\| \left\| \left(D^{\alpha}_{x_0-;g} A_i \right) \circ g^{-1} \right\|_p \right\|_{\infty,[g(a),g(x_0)]} \int_{a}^{x_0} \prod_{\substack{j=1 \\ j \neq i}}^{r} \left\| A_j(x) \right\|_p (g(x_0) - g(x))^{\alpha} dx +$$

$$\left\| \left\| \left(D^{\alpha}_{x_0+;g} A_i \right) \circ g^{-1} \right\|_p \right\|_{\infty,[g(x_0),g(b)]} \int_{x_0}^{b} \prod_{\substack{j=1 \\ j \neq i}}^{r} \left\| A_j(x) \right\|_p (g(x) - g(x_0))^{\alpha} dx \Bigg]. \tag{7.68}$$

Proof As similar to the proof of Theorem 7.11 is omitted. Use of (7.28). ■

We make

Remark 7.14 (*to Theorem* 7.11)

(i) for $\gamma, \delta > 1 : \frac{1}{\gamma} + \frac{1}{\delta} = 1$, with $\alpha > \frac{1}{\delta}$; case of inequality (7.33):

Call and assume

$$M_1(f_1, \ldots, f_r) := \tag{7.69}$$

$$\max_{i=1,\ldots,r} \left\{ \sup_{x_0 \in [a,b]} \left\| \left\| \left(D^{\alpha}_{x_0-;g} A_i \right) \circ g^{-1} \right\|_2 \right\|_{\delta,[g(a),g(x_0)]}, \right.$$

$$\left. \sup_{x_0 \in [a,b]} \left\| \left\| \left(D^{\alpha}_{x_0+;g} A_i \right) \circ g^{-1} \right\|_2 \right\|_{\delta,[g(x_0),g(b)]} \right\} < +\infty.$$

Then

$$\|\Phi(A_1, \ldots, A_r)(x_0)\|_1 \leq \text{Right hand side (7.33)} \leq$$

$$\frac{M_1(f_1, \ldots, f_r)(g(b) - g(a))^{\alpha - \frac{1}{\delta}}}{\Gamma(\alpha)(\gamma(\alpha - 1) + 1)^{\frac{1}{\gamma}}} \sum_{i=1}^{r} \left(\int_a^b \prod_{\substack{j=1 \\ j \neq i}}^{r} \|A_j(x)\|_2 \, dx \right). \tag{7.70}$$

(ii) Case of inequality (7.35):
Call and assume

$$M_2(f_1, \ldots, f_r) := \tag{7.71}$$

$$\max_{i=1,\ldots,r} \left\{ \sup_{x_0 \in [a,b]} \left\| \left\| \left(D^{\alpha}_{x_0-;g} A_i \right) \circ g^{-1} \right\|_2 \right\|_{\infty,[g(a),g(x_0)]}, \right.$$

$$\left. \sup_{x_0 \in [a,b]} \left\| \left\| \left(D^{\alpha}_{x_0+;g} A_i \right) \circ g^{-1} \right\|_2 \right\|_{\infty,[g(x_0),g(b)]} \right\} < +\infty.$$

Then

$$\|\Phi(A_1, \ldots, A_r)(x_0)\|_1 \leq \text{Right hand side (7.35)} \leq$$

$$\frac{M_2(f_1, \ldots, f_r)(g(b) - g(a))^{\alpha}}{\Gamma(\alpha + 1)} \sum_{i=1}^{r} \left(\int_a^b \prod_{\substack{j=1 \\ j \neq i}}^{r} \|A_j(x)\|_2 \, dx \right). \tag{7.72}$$

We make

Remark 7.15 (*to Theorem* 7.12)
(i) for $\gamma, \delta > 1 : \frac{1}{\gamma} + \frac{1}{\delta} = 1$, with $\alpha > \frac{1}{\delta}$; case of inequality (7.53):
Call and assume

$$N_1(f_1, f_2) :=$$

$$\max \left\{ \sup_{x_0 \in [a,b]} \left\| \left\| \left(D^{\alpha}_{x_0-;g} A_1 \right) \circ g^{-1} \right\|_p \right\|_{\delta,[g(a),g(x_0)]}, \right.$$

$$\sup_{x_0 \in [a,b]} \left\| \left\| \left(D^{\alpha}_{x_0+;g} A_1 \right) \circ g^{-1} \right\|_p \right\|_{\delta,[g(x_0),g(b)]},$$

$$\sup_{x_0 \in [a,b]} \left\| \left\| \left(D^{\alpha}_{x_0-;g} A_2 \right) \circ g^{-1} \right\|_q \right\|_{\delta,[g(a),g(x_0)]},$$

$$\left.\sup_{x_0\in[a,b]}\left\|\left\|\left(D^{\alpha}_{x_0+;g}A_2\right)\circ g^{-1}\right\|_q\right\|_{\delta,[g(x_0),g(b)]}\right\}<+\infty. \quad (7.73)$$

Then

$$\|\Phi(A_1,A_2)(x_0)\|_1\leq \text{right hand side (7.53)} \leq$$

$$\frac{N_1(f_1,f_2)(g(b)-g(a))^{\alpha-\frac{1}{\delta}}}{\Gamma(\alpha)(\gamma(\alpha-1)+1)^{\frac{1}{\gamma}}}\left[\int_a^b\|A_1(x)\|_p\,dx+\int_a^b\|A_2(x)\|_q\,dx\right]. \quad (7.74)$$

(ii) Case of inequality (7.55):
Call and assume

$$N_2(f_1,f_2):=$$

$$\max\left\{\sup_{x_0\in[a,b]}\left\|\left\|\left(D^{\alpha}_{x_0-;g}A_1\right)\circ g^{-1}\right\|_p\right\|_{\infty,[g(a),g(x_0)]},\right.$$

$$\sup_{x_0\in[a,b]}\left\|\left\|\left(D^{\alpha}_{x_0+;g}A_1\right)\circ g^{-1}\right\|_p\right\|_{\infty,[g(x_0),g(b)]},$$

$$\sup_{x_0\in[a,b]}\left\|\left\|\left(D^{\alpha}_{x_0-;g}A_2\right)\circ g^{-1}\right\|_q\right\|_{\infty,[g(a),g(x_0)]},$$

$$\left.\sup_{x_0\in[a,b]}\left\|\left\|\left(D^{\alpha}_{x_0+;g}A_2\right)\circ g^{-1}\right\|_q\right\|_{\infty,[g(x_0),g(b)]}\right\}<+\infty. \quad (7.75)$$

Then

$$\|\Phi(A_1,A_2)(x_0)\|_1\leq \text{right hand side (7.55)} \leq$$

$$\frac{N_2(f_1,f_2)(g(b)-g(a))^{\alpha}}{\Gamma(\alpha+1)}\left[\int_a^b\|A_1(x)\|_p\,dx+\int_a^b\|A_2(x)\|_q\,dx\right]. \quad (7.76)$$

We also make

Remark 7.16 (*to Theorem* 7.13)
(i) for $\gamma,\delta>1:\frac{1}{\gamma}+\frac{1}{\delta}=1$, with $\alpha>\frac{1}{\delta}$; case of inequality (7.66):
Call and assume

$$R_1(f_1,\ldots,f_r):= \quad (7.77)$$

$$\max_{i=1,\ldots,r}\left\{\sup_{x_0\in[a,b]}\left\|\left\|\left(D^{\alpha}_{x_0-;g}A_i\right)\circ g^{-1}\right\|_p\right\|_{\delta,[g(a),g(x_0)]},\right.$$

$$\left.\sup_{x_0\in[a,b]}\left\|\left\|\left(D^{\alpha}_{x_0+;g}A_i\right)\circ g^{-1}\right\|_p\right\|_{\delta,[g(x_0),g(b)]}\right\}<+\infty.$$

Then $(p > 1)$

$$\left\| \Phi\left(A_1, \ldots, A_r\right)(x_0) \right\|_p \leq \text{right hand side (7.66)} \leq$$

$$\frac{R_1\left(f_1, \ldots, f_r\right)\left(g\left(b\right) - g\left(a\right)\right)^{\alpha - \frac{1}{\delta}}}{\Gamma\left(\alpha\right)\left(\gamma\left(\alpha - 1\right) + 1\right)^{\frac{1}{\gamma}}} \sum_{i=1}^{r} \left(\int_a^b \prod_{\substack{j=1 \\ j \neq i}}^{r} \left\| A_j(x) \right\|_p dx \right). \tag{7.78}$$

(ii) Case of inequality (7.68):
Call and assume

$$R_2\left(f_1, \ldots, f_r\right) := \tag{7.79}$$

$$\max_{i=1,\ldots,r} \left\{ \sup_{x_0 \in [a,b]} \left\| \left\| \left(D_{x_0-;g}^{\alpha} A_i \right) \circ g^{-1} \right\|_p \right\|_{\infty,[g(a),g(x_0)]}, \right.$$

$$\left. \sup_{x_0 \in [a,b]} \left\| \left\| \left(D_{x_0+;g}^{\alpha} A_i \right) \circ g^{-1} \right\|_p \right\|_{\infty,[g(x_0),g(b)]} \right\} < +\infty.$$

Then $(p > 1)$

$$\left\| \Phi\left(A_1, \ldots, A_r\right)(x_0) \right\|_p \leq \text{Right hand side (7.68)} \leq$$

$$\frac{R_2\left(f_1, \ldots, f_r\right)\left(g\left(b\right) - g\left(a\right)\right)^{\alpha}}{\Gamma\left(\alpha + 1\right)} \sum_{i=1}^{r} \left(\int_a^b \prod_{\substack{j=1 \\ j \neq i}}^{r} \left\| A_j(x) \right\|_p dx \right). \tag{7.80}$$

Next come 1-Schatten norm generalized fractional Grüss type inequalities involving several functions taking values in the Banach algebra $\mathcal{B}_2(H) \subset \mathcal{B}(H)$:

Theorem 7.17 *Let the* $*$*-ideal* $\mathcal{B}_2(H)$, *which* $\left(\mathcal{B}_2(H), \|\cdot\|_2\right)$ *is a Banach algebra;* $0 < \alpha < 1$, *and* $A_i \in C^1\left([a,b], \mathcal{B}_2(H)\right)$, $i = 1, \ldots, r \in \mathbb{N} - \{1\}$; $g \in C^1([a,b])$, *strictly increasing such that* $g^{-1} \in C^1\left([g(a), g(b)]\right)$. *Here* $M_1\left(f_1, \ldots, f_r\right)$ *is as in* (7.69), *and* $M_2\left(f_1, \ldots, f_r\right)$ *is as in* (7.71). *Denote by*

$$\Delta\left(A_1, \ldots, A_r\right) := \int_a^b \Phi\left(A_1, \ldots, A_r\right)(x_0)\, dx_0 =$$

$$\sum_{i=1}^{r} \left[(b-a) \int_a^b \left(\prod_{\substack{j=1 \\ j \neq i}}^{r} A_j(x) \right) A_i(x)\, dx - \left(\int_a^b \left(\prod_{\substack{j=1 \\ j \neq i}}^{r} A_j(x) \right) dx \right) \left(\int_a^b A_i(x)\, dx \right) \right], \tag{7.81}$$

where $\Phi\left(A_1, \ldots, A_r\right)(x_0)$ *as in* (7.32).

Then
(i) for $\gamma, \delta > 1 : \frac{1}{\gamma} + \frac{1}{\delta} = 1$, *with* $\alpha > \frac{1}{\delta}$ *we have that*

$$\|\Delta (A_1, \ldots, A_r)\|_1 \leq \frac{M_1 (f_1, \ldots, f_r) (g (b) - g (a))^{\alpha - \frac{1}{\delta}} (b - a)}{\Gamma (\alpha) (\gamma (\alpha - 1) + 1)^{\frac{1}{\gamma}}}$$

$$\sum_{i=1}^{r} \left(\int_a^b \prod_{\substack{j=1 \\ j \neq i}}^{r} \|A_j (x)\|_2 \, dx \right), \tag{7.82}$$

and
(ii)

$$\|\Delta (A_1, \ldots, A_r)\|_1 \leq \frac{M_2 (f_1, \ldots, f_r) (g (b) - g (a))^{\alpha} (b - a)}{\Gamma (\alpha + 1)}$$

$$\sum_{i=1}^{r} \left(\int_a^b \prod_{\substack{j=1 \\ j \neq i}}^{r} \|A_j (x)\|_2 \, dx \right). \tag{7.83}$$

Proof (i) By (7.81) we have that

$$\|\Delta (A_1, \ldots, A_r)\|_1 \leq \int_a^b \|\Phi (A_1, \ldots, A_r) (x)\|_1 \, dx \overset{(7.70)}{\leq}$$

$$\frac{M_1 (f_1, \ldots, f_r) (g (b) - g (a))^{\alpha - \frac{1}{\delta}} (b - a)}{\Gamma (\alpha) (\gamma (\alpha - 1) + 1)^{\frac{1}{\gamma}}} \sum_{i=1}^{r} \left(\int_a^b \prod_{\substack{j=1 \\ j \neq i}}^{r} \|A_j (x)\|_2 \, dx \right), \tag{7.84}$$

proving (7.82).
(ii) Also it holds

$$\|\Delta (A_1, \ldots, A_r)\|_1 \leq \int_a^b \|\Phi (A_1, \ldots, A_r) (x)\|_1 \, dx \overset{(7.72)}{\leq}$$

$$\frac{M_2 (f_1, \ldots, f_r) (g (b) - g (a))^{\alpha} (b - a)}{\Gamma (\alpha + 1)} \sum_{i=1}^{r} \left(\int_a^b \prod_{\substack{j=1 \\ j \neq i}}^{r} \|A_j (x)\|_2 \, dx \right), \tag{7.85}$$

proving (7.83).
The theorem is proved. ■

When $r = 2$ we obtain the following operator related Grüss type fractional inequalities.

Theorem 7.18 *Let $p, q > 1 : \frac{1}{p} + \frac{1}{q} = 1$, and let the $*$-ideals $\mathcal{B}_p(H)$, $\mathcal{B}_q(H)$, for which $(\mathcal{B}_p(H), \|\cdot\|_p)$, $(\mathcal{B}_q(H), \|\cdot\|_q)$ are Banach algebras; $0 < \alpha < 1$, $A_1 \in C^1([a,b], \mathcal{B}_p(H))$, $A_2 \in C^1([a,b], \mathcal{B}_q(H))$; $g \in C^1([a,b])$, strictly increasing, such that $g^{-1} \in C^1([g(a), g(b)])$. Here $N_1(f_1, f_2)$ is as in (7.73), and $N_2(f_1, f_2)$ is as in (7.75).*

Denote by

$$\Delta(A_1, A_2) := \int_a^b \Phi(A_1, A_2)(x_0)\, dx_0 =$$

$$(b-a)\left(\int_a^b A_2(x) A_1(x)\, dx + \int_a^b A_1(x) A_2(x)\, dx\right) - \tag{7.86}$$

$$\left(\int_a^b A_2(x)\, dx\right)\left(\int_a^b A_1(x)\, dx\right) - \left(\int_a^b A_1(x)\, dx\right)\left(\int_a^b A_2(x)\, dx\right),$$

where $\Phi(A_1, A_2)(x_0)$ as in (7.52)

Then

(i) for $\gamma, \delta > 1 : \frac{1}{\gamma} + \frac{1}{\delta} = 1$, with $\alpha > \frac{1}{\delta}$, we have that

$$\|\Delta(A_1, A_2)\|_1 \leq \frac{N_1(f_1, f_2)(g(b) - g(a))^{\alpha - \frac{1}{\delta}}(b-a)}{\Gamma(\alpha)(\gamma(\alpha - 1) + 1)^{\frac{1}{\gamma}}}$$

$$\left[\int_a^b \|A_1(x)\|_p\, dx + \int_a^b \|A_2(x)\|_q\, dx\right], \tag{7.87}$$

and

(ii)

$$\|\Delta(A_1, A_2)\|_1 \leq \frac{N_2(f_1, f_2)(g(b) - g(a))^{\alpha}(b-a)}{\Gamma(\alpha + 1)}$$

$$\left[\int_a^b \|A_1(x)\|_p\, dx + \int_a^b \|A_2(x)\|_q\, dx\right]. \tag{7.88}$$

Proof (i) By (7.86) we have that

$$\|\Delta(A_1, A_2)\|_1 \leq \int_a^b \|\Phi(A_1, A_2)(x)\|_1\, dx \overset{(7.74)}{\leq}$$

$$\frac{N_1 (f_1, f_2) (g (b) - g (a))^{\alpha - \frac{1}{\delta}} (b - a)}{\Gamma (\alpha) (\gamma (\alpha - 1) + 1)^{\frac{1}{\gamma}}} \left[\int_a^b \| A_1 (x) \|_p \, dx + \int_a^b \| A_2 (x) \|_q \, dx \right], \tag{7.89}$$

proving (7.87).

(ii) Also we get

$$\| \Delta (A_1, A_2) \|_1 \leq \int_a^b \| \Phi (A_1, A_2) (x) \|_1 \, dx \overset{(7.76)}{\leq}$$

$$\frac{N_2 (f_1, f_2) (g (b) - g (a))^{\alpha} (b - a)}{\Gamma (\alpha + 1)} \left[\int_a^b \| A_1 (x) \|_p \, dx + \int_a^b \| A_2 (x) \|_q \, dx \right], \tag{7.90}$$

proving (7.88). The theorem is proved. ■

We continue with p-Schatten norm generalized fractional Grüss type inequalities involving several functions takin values in the Banach algebra $\mathcal{B}_p (H) \subset \mathcal{B} (H)$, $p > 1$.

Theorem 7.19 *Let the $*$-ideal $\mathcal{B}_p (H)$, which $\left(\mathcal{B}_p (H), \| \cdot \|_p \right)$, $p > 1$, is a Banach algebra, $0 < \alpha < 1$. Here $A_i \in C^1 \left([a, b], \mathcal{B}_p (H) \right)$, $i = 1, \ldots, r \in \mathbb{N} - \{1\}$; $g \in C^1 ([a, b])$, strictly increasing such that $g^{-1} \in C^1 ([g (a), g (b)])$. Here $R_1 (f_1, \ldots, f_r)$ is as in (7.77), and $R_2 (f_1, \ldots, f_r)$ is as in (7.79). Furthermore $\Delta (A_1, \ldots, A_r)$ is as in (7.81) with $\Phi (A_1, \ldots, A_r) (x_0)$ as in (7.32). Then*

(i) for $\gamma, \delta > 1 : \frac{1}{\gamma} + \frac{1}{\delta} = 1$, with $\alpha > \frac{1}{\delta}$ we have that

$$\| \Delta (A_1, \ldots, A_r) \|_p \leq \frac{R_1 (f_1, \ldots, f_r) (g (b) - g (a))^{\alpha - \frac{1}{\delta}} (b - a)}{\Gamma (\alpha) (\gamma (\alpha - 1) + 1)^{\frac{1}{\gamma}}}$$

$$\sum_{i=1}^{r} \left(\int_a^b \prod_{\substack{j=1 \\ j \neq i}}^{r} \| A_j (x) \|_p \, dx \right), \tag{7.91}$$

and

(ii)

$$\| \Delta (A_1, \ldots, A_r) \|_p \leq \frac{R_2 (f_1, \ldots, f_r) (g (b) - g (a))^{\alpha} (b - a)}{\Gamma (\alpha + 1)}$$

$$\sum_{i=1}^{r} \left(\int_a^b \prod_{\substack{j=1 \\ j \neq i}}^{r} \| A_j (x) \|_p \, dx \right). \tag{7.92}$$

Proof (i) By (7.81) we have

$$\|\Delta(A_1,\ldots,A_r)\|_p \le \int_a^b \|\Phi(A_1,\ldots,A_r)(x)\|_p \, dx \overset{(7.78)}{\le}$$

$$\frac{R_1(f_1,\ldots,f_r)(g(b)-g(a))^{\alpha-\frac{1}{\delta}}(b-a)}{\Gamma(\alpha)(\gamma(\alpha-1)+1)^{\frac{1}{\gamma}}} \sum_{i=1}^{r} \left(\int_a^b \prod_{\substack{j=1 \\ j \ne i}}^{r} \left\| A_j(x) \right\|_p dx \right), \tag{7.93}$$

proving (7.91).

(ii) Also we have

$$\|\Delta(A_1,\ldots,A_r)\|_p \le \int_a^b \|\Phi(A_1,\ldots,A_r)(x)\|_p \, dx \overset{(7.80)}{\le}$$

$$\frac{R_2(f_1,\ldots,f_r)(g(b)-g(a))^{\alpha}(b-a)}{\Gamma(\alpha+1)} \sum_{i=1}^{r} \left(\int_a^b \prod_{\substack{j=1 \\ j \ne i}}^{r} \left\| A_j(x) \right\|_p dx \right), \tag{7.94}$$

proving (7.92). The theorem is proved. ■

7.6 Applications

We start with special Ostrowski type inequalities. We give

Corollary 7.20 (to Theorem 7.11) *All as in Theorem 7.11, with $g(t)=t$, $\forall\, t \in [a,b]$. Then*

(1) for $\gamma, \delta > 1 : \frac{1}{\gamma} + \frac{1}{\delta} = 1$, $\alpha > \frac{1}{\delta}$ we have that

$$\|\Phi(A_1,\ldots,A_r)(x_0)\|_1 \le \frac{1}{\Gamma(\alpha)(\gamma(\alpha-1)+1)^{\frac{1}{\gamma}}}$$

$$\sum_{i=1}^{r} \left[\left\| \left\| \left(D_{x_0-}^{\alpha} A_i \right) \right\|_2 \right\|_{\delta,[a,x_0]} \int_a^{x_0} \prod_{\substack{j=1 \\ j \ne i}}^{r} \left\| A_j(x) \right\|_2 (x_0-x)^{\alpha-\frac{1}{\delta}} dx + \right. \tag{7.95}$$

$$\left. \left\| \left\| \left(D_{*x_0}^{\alpha} A_i \right) \right\|_2 \right\|_{\delta,[x_0,b]} \int_{x_0}^{b} \prod_{\substack{j=1 \\ j \ne i}}^{r} \left\| A_j(x) \right\|_2 (x-x_0)^{\alpha-\frac{1}{\delta}} dx \right],$$

(2) if $\alpha \ge 1$, we obtain

$$\|\Phi(A_1, \ldots, A_r)(x_0)\|_1 \leq \frac{1}{\Gamma(\alpha)}$$

$$\sum_{i=1}^{r} \left[\left\| \left\| \left(D_{x_0-}^{\alpha} A_i \right) \right\|_2 \right\|_{L_1([a,x_0])} \int_a^{x_0} \prod_{\substack{j=1 \\ j \neq i}}^{r} \left\| A_j(x) \right\|_2 (x_0 - x)^{\alpha-1} dx + \right.$$

$$\left. \left\| \left\| \left(D_{*x_0}^{\alpha} A_i \right) \right\|_2 \right\|_{L_1([x_0,b])} \int_{x_0}^{b} \prod_{\substack{j=1 \\ j \neq i}}^{r} \left\| A_j(x) \right\|_2 (x - x_0)^{\alpha-1} dx \right], \quad (7.96)$$

and

(3)

$$\|\Phi(A_1, \ldots, A_r)(x_0)\|_1 \leq \frac{1}{\Gamma(\alpha+1)}$$

$$\sum_{i=1}^{r} \left[\left\| \left\| \left(D_{x_0-}^{\alpha} A_i \right) \right\|_2 \right\|_{\infty,[a,x_0]} \int_a^{x_0} \prod_{\substack{j=1 \\ j \neq i}}^{r} \left\| A_j(x) \right\|_2 (x_0 - x)^{\alpha} dx + \right.$$

$$\left. \left\| \left\| \left(D_{*x_0}^{\alpha} A_i \right) \right\|_2 \right\|_{\infty,[x_0,b]} \int_{x_0}^{b} \prod_{\substack{j=1 \\ j \neq i}}^{r} \left\| A_j(x) \right\|_2 (x - x_0)^{\alpha} dx \right]. \quad (7.97)$$

We present

Corollary 7.21 *(to Theorem 7.12) All as in Theorem 7.12, with $g(t) = e^t$, $\forall\, t \in [a, b]$. Then*

(1) for $\gamma, \delta > 1 : \frac{1}{\gamma} + \frac{1}{\delta} = 1$, with $\alpha > \frac{1}{\delta}$, we have that

$$\|\Phi(A_1, A_2)(x_0)\|_1 \leq \frac{1}{\Gamma(\alpha)(\gamma(\alpha-1)+1)^{\frac{1}{\gamma}}}$$

$$\left\{ \left[\left\| \left\| \left(D_{x_0-;e^t}^{\alpha} A_1 \right) \circ \log \right\|_p \right\|_{\delta,[e^a,e^{x_0}]} \int_a^{x_0} \|A_2(x)\|_q \left(e^{x_0} - e^x \right)^{\alpha-\frac{1}{\delta}} dx \right] + \right.$$

$$\left[\left\| \left\| \left(D_{x_0+;e^t}^{\alpha} A_1 \right) \circ \log \right\|_p \right\|_{\delta,[e^{x_0},e^b]} \int_{x_0}^{b} \|A_2(x)\|_q \left(e^x - e^{x_0} \right)^{\alpha-\frac{1}{\delta}} dx \right] + \quad (7.98)$$

$$\left[\left\| \left\| \left(D_{x_0-;e^t}^{\alpha} A_2 \right) \circ \log \right\|_q \right\|_{\delta,[e^a,e^{x_0}]} \int_a^{x_0} \|A_1(x)\|_p \left(e^{x_0} - e^x \right)^{\alpha-\frac{1}{\delta}} dx \right] +$$

$$\left[\left\|\left\|\left(D^{\alpha}_{x_0+;e^t}A_2\right)\circ\log\right\|_q\right\|_{\delta,[e^{x_0},e^b]}\int_{x_0}^{b}\|A_1(x)\|_p\left(e^x-e^{x_0}\right)^{\alpha-\frac{1}{\delta}}dx\right]\Bigg\},$$

(2) if $\alpha \geq 1$, we obtain

$$\|\Phi(A_1,A_2)(x_0)\|_1 \leq \frac{1}{\Gamma(\alpha)}$$

$$\Bigg\{\left[\left\|\left\|\left(D^{\alpha}_{x_0-;e^t}A_1\right)\circ\log\right\|_p\right\|_{L_1([e^a,e^{x_0}])}\int_{a}^{x_0}\|A_2(x)\|_q\left(e^{x_0}-e^x\right)^{\alpha-1}dx\right]+$$

$$\left[\left\|\left\|\left(D^{\alpha}_{x_0+;e^t}A_1\right)\circ\log\right\|_p\right\|_{L_1([e^{x_0},e^b])}\int_{x_0}^{b}\|A_2(x)\|_q\left(e^x-e^{x_0}\right)^{\alpha-1}dx\right]+$$

$$\left[\left\|\left\|\left(D^{\alpha}_{x_0-;e^t}A_2\right)\circ\log\right\|_q\right\|_{L_1([e^a,e^{x_0}])}\int_{a}^{x_0}\|A_1(x)\|_p\left(e^{x_0}-e^x\right)^{\alpha-1}dx\right]+ \quad (7.99)$$

$$\left[\left\|\left\|\left(D^{\alpha}_{x_0+;e^t}A_2\right)\circ\log\right\|_q\right\|_{L_1([e^{x_0},e^b])}\int_{x_0}^{b}\|A_1(x)\|_p\left(e^x-e^{x_0}\right)^{\alpha-1}dx\right]\Bigg\},$$

and

(3)

$$\|\Phi(A_1,A_2)(x_0)\|_1 \leq \frac{1}{\Gamma(\alpha+1)}$$

$$\Bigg\{\left[\left\|\left\|\left(D^{\alpha}_{x_0-;e^t}A_1\right)\circ\log\right\|_p\right\|_{\infty,[e^a,e^{x_0}]}\int_{a}^{x_0}\|A_2(x)\|_q\left(e^{x_0}-e^x\right)^{\alpha}dx\right]+$$

$$\left[\left\|\left\|\left(D^{\alpha}_{x_0+;e^t}A_1\right)\circ\log\right\|_p\right\|_{\infty,[e^{x_0},e^b]}\int_{x_0}^{b}\|A_2(x)\|_q\left(e^x-e^{x_0}\right)^{\alpha}dx\right]+$$

$$\left[\left\|\left\|\left(D^{\alpha}_{x_0-;e^t}A_2\right)\circ\log\right\|_q\right\|_{\infty,[e^a,e^{x_0}]}\int_{a}^{x_0}\|A_1(x)\|_p\left(e^{x_0}-e^x\right)^{\alpha}dx\right]+$$

$$\left[\left\|\left\|\left(D^{\alpha}_{x_0+;e^t}A_2\right)\circ\log\right\|_q\right\|_{\infty,[e^{x_0},e^b]}\int_{x_0}^{b}\|A_1(x)\|_p\left(e^x-e^{x_0}\right)^{\alpha}dx\right]\Bigg\}. \quad (7.100)$$

We continue with

Corollary 7.22 (to Theorem 7.13) *All as in Theorem 7.13, eith $g(t)=\log t$, $\forall$ $t\in[a,b]\subset\mathbb{R}_+-\{0\}$. Then*

(1) for $\gamma,\delta>1:\frac{1}{\gamma}+\frac{1}{\delta}=1$, with $\alpha>\frac{1}{\delta}$ we have that

$$\|\Phi(A_1,\dots,A_r)(x_0)\|_p \le \frac{1}{\Gamma(\alpha)(\gamma(\alpha-1)+1)^{\frac{1}{\gamma}}}$$

$$\sum_{i=1}^{r}\Bigg[\left\|\left\|\left(D^{\alpha}_{x_0-;\log}A_i\right)\circ e^t\right\|_p\right\|_{\delta,[\log a,\log x_0]}\int_a^{x_0}\prod_{\substack{j=1\\j\neq i}}^{r}\|A_j(x)\|_p\left(\log\frac{x_0}{x}\right)^{\alpha-\frac{1}{\delta}}dx+$$

$$\left\|\left\|\left(D^{\alpha}_{x_0+;\log}A_i\right)\circ e^t\right\|_p\right\|_{\delta,[\log x_0,\log b]}\int_{x_0}^{b}\prod_{\substack{j=1\\j\neq i}}^{r}\|A_j(x)\|_p\left(\log\frac{x}{x_0}\right)^{\alpha-\frac{1}{\delta}}dx\Bigg], \tag{7.101}$$

(2) if $\alpha \ge 1$*, we obtain*

$$\|\Phi(A_1,\dots,A_r)(x_0)\|_p \le \frac{1}{\Gamma(\alpha)}$$

$$\sum_{i=1}^{r}\Bigg[\left\|\left\|\left(D^{\alpha}_{x_0-;\log}A_i\right)\circ e^t\right\|_p\right\|_{L_1([\log a,\log x_0])}\int_a^{x_0}\prod_{\substack{j=1\\j\neq i}}^{r}\|A_j(x)\|_p\left(\log\frac{x_0}{x}\right)^{\alpha-1}dx+$$

$$\left\|\left\|\left(D^{\alpha}_{x_0+;\log}A_i\right)\circ e^t\right\|_p\right\|_{L_1([\log x_0,\log b])}\int_{x_0}^{b}\prod_{\substack{j=1\\j\neq i}}^{r}\|A_j(x)\|_p\left(\log\frac{x}{x_0}\right)^{\alpha-1}dx\Bigg], \tag{7.102}$$

and

(3)

$$\|\Phi(A_1,\dots,A_r)(x_0)\|_p \le \frac{1}{\Gamma(\alpha+1)}$$

$$\sum_{i=1}^{r}\Bigg[\left\|\left\|\left(D^{\alpha}_{x_0-;\log}A_i\right)\circ e^t\right\|_p\right\|_{\infty,[\log a,\log x_0]}\int_a^{x_0}\prod_{\substack{j=1\\j\neq i}}^{r}\|A_j(x)\|_p\left(\log\frac{x_0}{x}\right)^{\alpha}dx+$$

$$\left\|\left\|\left(D^{\alpha}_{x_0+;\log}A_i\right)\circ e^t\right\|_p\right\|_{\infty,[\log x_0,\log b]}\int_{x_0}^{b}\prod_{\substack{j=1\\j\neq i}}^{r}\|A_j(x)\|_p\left(\log\frac{x}{x_0}\right)^{\alpha}dx\Bigg]. \tag{7.103}$$

We continue with special Grüss type inequalities.

Corollary 7.23 (to Theorem 7.17) *All as in Theorem 7.17, with* $g(t) = t$, $\forall\, t \in [a, b]$. *Then*

(i) for $\gamma, \delta > 1 : \frac{1}{\gamma} + \frac{1}{\delta} = 1$, $\alpha > \frac{1}{\delta}$ *we have that*

$$\|\Delta(A_1, \ldots, A_r)\|_1 \leq \frac{M_1(f_1, \ldots, f_r)(b-a)^{\alpha + \frac{1}{\gamma}}}{\Gamma(\alpha)(\gamma(\alpha - 1) + 1)^{\frac{1}{\gamma}}} \sum_{i=1}^{r} \left(\int_a^b \prod_{\substack{j=1 \\ j \neq i}}^{r} \|A_j(x)\|_2 \, dx \right), \tag{7.104}$$

and

(ii)

$$\|\Delta(A_1, \ldots, A_r)\|_1 \leq \frac{M_2(f_1, \ldots, f_r)(b-a)^{\alpha + 1}}{\Gamma(\alpha + 1)} \sum_{i=1}^{r} \left(\int_a^b \prod_{\substack{j=1 \\ j \neq i}}^{r} \|A_j(x)\|_2 \, dx \right). \tag{7.105}$$

Next comes

Corollary 7.24 (to Theorem 7.18) *All as in Theorem 7.18, with* $g(t) = e^t$, $\forall\, t \in [a, b]$. *Then*

(i) for $\gamma, \delta > 1 : \frac{1}{\gamma} + \frac{1}{\delta} = 1$, $\alpha > \frac{1}{\delta}$, *we have that*

$$\|\Delta(A_1, A_2)\|_1 \leq \frac{N_1(f_1, f_2)\left(e^b - e^a\right)^{\alpha - \frac{1}{\delta}}(b-a)}{\Gamma(\alpha)(\gamma(\alpha - 1) + 1)^{\frac{1}{\gamma}}}$$
$$\left[\int_a^b \|A_1(x)\|_p \, dx + \int_a^b \|A_2(x)\|_q \, dx \right], \tag{7.106}$$

and

(ii)

$$\|\Delta(A_1, A_2)\|_1 \leq \frac{N_2(f_1, f_2)\left(e^b - e^a\right)^{\alpha}(b-a)}{\Gamma(\alpha + 1)}$$
$$\left[\int_a^b \|A_1(x)\|_p \, dx + \int_a^b \|A_2(x)\|_q \, dx \right]. \tag{7.107}$$

We finish with

Corollary 7.25 (to Theorem 7.19) *All as in Theorem 7.19, with* $g(t) = \log t$, $\forall\, t \in [a, b] \subset \mathbb{R}_+ - \{0\}$. *Then*

(i) for $\gamma, \delta > 1 : \frac{1}{\gamma} + \frac{1}{\delta} = 1$, $\alpha > \frac{1}{\delta}$ *we have that*

$$\left\| \Delta \left(A_1, \ldots, A_r \right) \right\|_p \leq \frac{R_1 \left(f_1, \ldots, f_r \right) \left(\log \frac{b}{a} \right)^{\alpha - \frac{1}{\delta}} (b - a)}{\Gamma (\alpha) \left(\gamma (\alpha - 1) + 1 \right)^{\frac{1}{\gamma}}}$$

$$\sum_{i=1}^{r} \left(\int_a^b \prod_{\substack{j=1 \\ j \neq i}}^{r} \left\| A_j (x) \right\|_p dx \right), \tag{7.108}$$

and

(ii)

$$\left\| \Delta \left(A_1, \ldots, A_r \right) \right\|_p \leq \frac{R_2 \left(f_1, \ldots, f_r \right) \left(\log \frac{b}{a} \right)^{\alpha} (b - a)}{\Gamma (\alpha + 1)}$$

$$\sum_{i=1}^{r} \left(\int_a^b \prod_{\substack{j=1 \\ j \neq i}}^{r} \left\| A_j (x) \right\|_p dx \right). \tag{7.109}$$

References

1. Aliprantis, C., Border, K.: Infinite Dimensional Analysis, 3rd edn. Springer, New York (2006)
2. Anastassiou, G.A.: Fractional Differentiation Inequalities. Research Monograph, Springer, New York (2009)
3. Anastassiou, G.A.: Advances on Fractional Inequalities. Research Monograph, Springer, New York (2011)
4. Anastassiou, G.A.: Intelligent Comparisons: Analytic Inequalities. Springer, Heidelberg, New York (2016)
5. Anastassiou, G.A.: Intelligent Computations: Abstract Fractional Calculus. Inequalities, Approximations, Springer, Heidelberg, New York (2018)
6. Anastassiou, G.A.: *p*-Schatten norm generalized fractional Ostrowski and Grüss type inequalities involving several functions. In: Progress in Fractional Differentiation and Applications, accepted for publication (2021)
7. Bellman, R.: Some inequalities for positive definite matrices. In: Beckenbach, E.F. (ed.) General Inequalities 2, Proceedings of the 2nd International Conference on General Inequalities, pp. 89–90. Birkhauser, Basel (1980)
8. Čebyšev, Sur les expressions approximatives des intégrales définies par les aures prises entre les mêmes limites. Proc. Math. Soc. Charkov **2**(1882), 93–98
9. Chang, D.: A matrix trace inequality for products of Hermitian matrices. J. Math. Anal. Appl. **237**, 721–725 (1999)
10. Coop, I.D.: On matrix trace inequalities and related topics for products of Hermitian matrix. J. Math. Anal. Appl. **188**, 999–1001 (1994)
11. Dragomir, S.S.: *p*-Schatten norm inequalities of Ostrowski's type. RGMIA Res. Rep. Coll. **24**(108), 19 (2021)
12. Dragomir, S.S.: *p*-Schatten norm inequalities of Grüss type. RGMIA Res. Rep. Coll. **24**(115), 16 (2021)
13. Mikusinski, J.: The Bochner Integral. Academic Press, New York (1978)

14. Neudecker, H.: A matrix trace inequality. J. Math. Anal. Appl. **166**, 302–303 (1992)
15. Ostrowski, A.: Über die Absolutabweichung einer differentiabaren Funcktion von ihrem Integralmittelwert. Comment. Math. Helv. **10**, 226–227 (1938)
16. Rudin, W.: Functional Analysis, 2nd edn. McGraw-Hill Inc, New York (1991)
17. Simon, B.: Trace Ideals and Their Applications. Cambridge University Press, Cambridge (1979)
18. Zagrebvov, V.A.: Gibbs Semigroups. Operator Theory: Advances and Applications, vol. 273. Birkhauser (2019)

Chapter 8
p-Schatten Norm Iterated Generalized Fractional Ostrowski and Grüss Inequalities for Multiple Functions

Using iterated generalized Caputo fractional left and right vectorial Taylor formulae we establish sequential generalized fractional Ostrowski and Grüss inequalities for several functions that take values in the von Neumann–Schatten class $\mathcal{B}_p(H)$, $1 \leq p < \infty$. The estimates are given for all p-Schatten norms, $1 \leq p < \infty$. We finish with applications. It follows [6].

8.1 Introduction

The following results inspire this chapter.

Theorem 8.1 (1938, Ostrowski [15]) *Let $f : [a, b] \to \mathbb{R}$ be continuous on $[a, b]$ and differentiable on (a, b) whose derivative $f' : (a, b) \to \mathbb{R}$ is bounded on (a, b), i.e., $\|f'\|_{\infty}^{\sup} := \sup_{t \in (a,b)} |f'(t)| < +\infty$. Then*

$$\left| \frac{1}{b-a} \int_a^b f(t)\, dt - f(x) \right| \leq \left[\frac{1}{4} + \frac{\left(x - \frac{a+b}{2}\right)^2}{(b-a)^2} \right] (b-a) \|f'\|_{\infty}^{\sup}, \tag{8.1}$$

for any $x \in [a, b]$. The constant $\frac{1}{4}$ is the best possible.

Ostrowski type inequalities have great applications to integral approximations in Numerical Analysis.

Theorem 8.2 (1882, Čebyšev [8]) *Let $f, g : [a, b] \to \mathbb{R}$ be absolutely continuous functions with $f', g' \in L_{\infty}([a, b])$. Then*

G. A. Anastassiou, *Fractional Inequalities In Banach Algebras*,
Studies in Systems, Decision and Control 441,
https://doi.org/10.1007/978-3-031-05148-7_8

$$\left|\frac{1}{b-a}\int_a^b f(x)\,g(x)\,dx-\left(\frac{1}{b-a}\int_a^b f(x)\,dx\right)\left(\frac{1}{b-a}\int_a^b g(x)\,dx\right)\right|$$

$$\leq \frac{1}{12}(b-a)^2\left\|f'\right\|_\infty\left\|g'\right\|_\infty. \tag{8.2}$$

The above integrals are assumed to exist.

The related Grüss type inequalities have many applications to Probability Theory. We presented also ([4], Chaps. 8, 9) mixed fractional Ostrowski and Grüss–Cebysev type inequalities for several functions, acting to all possible directions. The estimates involve the left and right Caputo fractional derivatives. See also the monographs written by the author [2], Chaps. 24–26 and [3], Chaps. 2–6.

We are motivated also by Dragomir [11] recent work:

An operator $A\in\mathcal{B}(H)$ is said to belong to the von Neumann–Schatten class $\mathcal{B}_p(H)$, $1\leq p<\infty$ if the p-Schatten norm is finite

$$\|A\|_p := \left[tr\left(|A|^p\right)\right]^{\frac{1}{p}} < \infty.$$

Assume that $A:[a,b]\to\mathcal{B}_p(H)$, $B:[a,b]\to\mathcal{B}_q(H)$, $p,q>1$ with $\frac{1}{p}+\frac{1}{q}=1$, are continuous and B is strongly differentiable on (a,b), then

$$\left\|\int_a^b A(t)\,B(t)\,dt-\left(\int_a^b A(s)\,ds\right)B(u)\right\|_1\leq$$

$$\sup_{t\in[a,b]}\left\|B'(t)\right\|_q\times\begin{cases}\left[\frac{1}{2}(b-a)+\left|u-\frac{a+b}{2}\right|\right]\int_a^b\|A(t)\|_p\,dt,\\[2ex] \left[\frac{(u-a)^{\beta+1}+(b-u)^{\beta+1}}{\beta+1}\right]^{\frac{1}{\beta}}\left(\int_a^b\|A(t)\|_p^\alpha\right)^{\frac{1}{\alpha}},\\ \text{for } \alpha,\beta>1 \text{ with } \frac{1}{\alpha}+\frac{1}{\beta}=1\\[2ex] \left[\frac{1}{4}(b-a)^2+\left(u-\frac{a+b}{2}\right)^2\right]\sup_{t\in[a,b]}\|A(t)\|_p,\end{cases} \tag{8.3}$$

for all $u\in[a,b]$, an Ostrowski type inequality.

Further inspiration comes from Dragomir [12] recent work on Grüss inequalities:

For two continuous functions $A,B:[a,b]\to\mathcal{B}(H)$ we define the noncommutative Cebysev fractional

$$D(A,B):=(b-a)\int_a^b A(t)\,B(t)\,dt-\int_a^b A(t)\,dt\int_a^b B(t)\,dt. \tag{8.4}$$

If $p,q>1$ with $\frac{1}{p}+\frac{1}{q}=1$, let $A:[a,b]\to\mathcal{B}_p(H)$, $B:[a,b]\to\mathcal{B}_q(H)$ be strongly differentiable functions on the interval (a,b), then

$$\|D(A,B)\|_1 \leq D\left(\int_a^b \|A'(u)\|_p du, \int_a^b \|B'(u)\|_q du\right) \leq$$

$$\frac{1}{4}(b-a)^2 \int_a^b \|A'(u)\|_p du \int_a^b \|B'(u)\|_q du.$$

In this chapter we generalize [4], Chaps. 8, 9 for several Banach algebra $\mathcal{B}_p(H)$ valued functions, in the sense of developing sequential fractional Ostrowski and Grüss type inequalities. Now our left and right sequential generalized Caputo fractional derivatives are for Banach space valued functions and our integrals are of Bochner type [1, 13]. Applications finish the chapter.

8.2 Vectorial Sequential Generalized Fractional Calculus Background

We need

Definition 8.3 ([5], *p. 106*) Let $0 < \alpha \leq 1$, $f \in C^1([a,b], X)$, where $[a,b] \subset \mathbb{R}$, and $(X, \|\cdot\|)$ is a Banach space. Let $g \in C^1([a,b])$, strictly increasing, such that $g^{-1} \in C^1([g(a), g(b)])$.

We define the left generalized g-fractional derivative X-valued of f of order α as follows:

$$\left(D_{a+;g}^{\alpha} f\right)(x) := \frac{1}{\Gamma(1-\alpha)} \int_a^x (g(x) - g(t))^{-\alpha} g'(t) \left(f \circ g^{-1}\right)'(g(t))\, dt, \tag{8.5}$$

$\forall\, x \in [a,b]$, where Γ is the gamma function. The last integral is of Bochner type [13].

If $0 < \alpha < 1$, by Theorem 4.10, p. 98, [5], we have that $\left(D_{a+;g}^{\alpha} f\right) \in C([a,b], X)$.

We set

$$D_{a+;g}^{1} f(x) := \left(\left(f \circ g^{-1}\right)' \circ g\right)(x) \in C([a,b], X), \tag{8.6}$$

$$D_{a+;g}^{0} f(x) = f(x), \ \forall\, x \in [a,b].$$

When $g = id$, then

$$D_{a+;g}^{\alpha} f = D_{a+;id}^{\alpha} f = D_{*a}^{\alpha} f, \tag{8.7}$$

the usual left X-valued Caputo fractional derivative, see [5], Chap. 1.

We make

Remark 8.4 By (8.5) we have

$$\left\| \left(D_{a+;g}^{\alpha} f \right) (x) \right\| \leq \frac{1}{\Gamma (1-\alpha)} \int_a^x (g(x) - g(t))^{-\alpha} g'(t) \left\| \left(f \circ g^{-1} \right)' (g(t)) \right\| dt \leq$$

$$\frac{\left\| \left\| \left(f \circ g^{-1} \right)' \circ g \right\| \right\|_{\infty,[a,b]}}{\Gamma (1-\alpha)} \int_a^x (g(x) - g(t))^{-\alpha} g'(t) \, dt =$$

$$\frac{\left\| \left\| \left(f \circ g^{-1} \right)' \circ g \right\| \right\|_{\infty,[a,b]}}{\Gamma (1-\alpha)} \frac{(g(x) - g(t))^{1-\alpha}}{1-\alpha} = \tag{8.8}$$

$$\frac{\left\| \left\| \left(f \circ g^{-1} \right)' \circ g \right\| \right\|_{\infty,[a,b]}}{\Gamma (2-\alpha)} (g(x) - g(a))^{1-\alpha}, \ \forall \, x \in [a,b].$$

Hence

$$\left(D_{a+;g}^{\alpha} f \right) (a) = 0. \tag{8.9}$$

We need

Definition 8.5 ([5], *p. 107*) Let $0 < \alpha \leq 1$, $f \in C^1([a,b], X)$, where $[a,b] \subset \mathbb{R}$, and $(X, \|\cdot\|)$ is a Banach space. Let $g \in C^1([a,b])$, strictly increasing, such that $g^{-1} \in C^1([g(a), g(b)])$.

We define the right generalized g-fractional derivative X-valued of f of order α as follows:

$$\left(D_{b-;g}^{\alpha} f \right) (x) := \frac{-1}{\Gamma (1-\alpha)} \int_x^b (g(t) - g(x))^{-\alpha} g'(t) \left(f \circ g^{-1} \right)' (g(t)) \, dt, \tag{8.10}$$

$\forall \, x \in [a,b]$. The last integral is of Bochner type.

If $0 < \alpha < 1$, by Theorem 4.11, p. 101 [5], we have that $\left(D_{b-;g}^{\alpha} f \right) \in C([a,b], X)$.

We set

$$D_{b-;g}^{1} f(x) := -\left(\left(f \circ g^{-1} \right)' \circ g \right) (x) \in C([a,b], X), \tag{8.11}$$

$$D_{b-;g}^{0} f(x) := f(x), \ \forall \, x \in [a,b].$$

When $g = id$, then

$$D_{b-;g}^{\alpha} f(x) = D_{b-;id}^{\alpha} f(x) = D_{b-}^{\alpha} f, \tag{8.12}$$

the usual right X-valued Caputo fractional derivative, see [5], Chap. 2.

We make

Remark 8.6 By (8.10) we have

$$\left\|\left(D_{b-;g}^{\alpha}f\right)(x)\right\| \leq \frac{1}{\Gamma(1-\alpha)}\int_{x}^{b}(g(t)-g(x))^{-\alpha}g'(t)\left\|\left(f\circ g^{-1}\right)'(g(t))\right\|dt \leq$$

$$\frac{\left\|\left(f\circ g^{-1}\right)'\circ g\right\|_{\infty,[a,b]}}{\Gamma(1-\alpha)}\int_{x}^{b}(g(t)-g(x))^{-\alpha}g'(t)\,dt = \tag{8.13}$$

$$\frac{\left\|\left(f\circ g^{-1}\right)'\circ g\right\|_{\infty,[a,b]}}{\Gamma(1-\alpha)}\frac{(g(b)-g(x))^{1-\alpha}}{1-\alpha} =$$

$$\frac{\left\|\left(f\circ g^{-1}\right)'\circ g\right\|_{\infty,[a,b]}}{\Gamma(2-\alpha)}(g(b)-g(x))^{1-\alpha}, \ \forall\, x\in[a,b].$$

Hence

$$\left(D_{b-;g}^{\alpha}f\right)(b) = 0. \tag{8.14}$$

We need

Definition 8.7 ([5], *p. 115*) Denote by $(0<\alpha\leq 1)$

$$D_{a+;g}^{n\alpha} := D_{a+;g}^{\alpha}D_{a+;g}^{\alpha}\ldots D_{a+;g}^{\alpha}\ (n\text{ times}),\ n\in\mathbb{N} \tag{8.15}$$

and $D_{a+;g}^{0} = I$ (identity operator).

We also need

Definition 8.8 ([5], *p. 118*)

$$D_{b-;g}^{n\alpha} := D_{b-;g}^{\alpha}D_{b-;g}^{\alpha}\ldots D_{b-;g}^{\alpha}\ (n\text{ times}),\ n\in\mathbb{N} \tag{8.16}$$

and $D_{b-;g}^{0} = I$ (identity operator).

Based on (8.9) and Theorem 4.30, p. 117, [5], we have the following g-left generalized modified X-valued Taylor's formula:

Theorem 8.9 *Let* $0<\alpha\leq 1$, $n\in\mathbb{N}$, $f\in C^{1}([a,b],X)$, $(X,\|\cdot\|)$ *a Banach space,* $g\in C^{1}([a,b])$, *strictly increasing, such that* $g^{-1}\in C^{1}([g(a),g(b)])$. *Let* $F_k := D_{a+;g}^{k\alpha}f$, $k=1,\ldots,n$, *that fulfill* $F_k\in C^{1}([a,b],X)$, *and* $F_{n+1}\in C([a,b],X)$. *Then*

$$f(x) - f(a) = \sum_{i=2}^{n} \frac{(g(x) - g(a))^{i\alpha}}{\Gamma(i\alpha + 1)} \left(D_{a+;g}^{i\alpha} f\right)(a) +$$

$$\frac{1}{\Gamma((n+1)\alpha)} \int_a^x (g(x) - g(t))^{(n+1)\alpha - 1} g'(t) \left(D_{a+;g}^{(n+1)\alpha} f\right)(t)\, dt, \quad (8.17)$$

$\forall\, x \in [a, b]$.

When $n = 1$ we obtain

Corollary 8.10 *Let* $0 < \alpha \leq 1$, $f \in C^1([a, b], X)$, $(X, \|\cdot\|)$ *is a Banach space,* $g \in C^1([a, b])$, *strictly increasing, such that* $g^{-1} \in C^1([g(a), g(b)])$. *Assume that* $D_{a+;g}^{\alpha} f \in C^1([a, b], X)$, *and* $D_{a+;g}^{2\alpha} f \in C([a, b], X)$. *Then*

$$f(x) - f(a) = \frac{1}{\Gamma(2\alpha)} \int_a^x (g(x) - g(t))^{2\alpha - 1} g'(t) \left(D_{a+;g}^{2\alpha} f\right)(t)\, dt, \quad (8.18)$$

$\forall\, x \in [a, b]$.

Based on (8.14) and Theorem 4.33, p. 120, [5], we have the following g-right generalized modified X-valued Taylor's formula:

Theorem 8.11 *Let* $f \in C^1([a, b], X)$, $(X, \|\cdot\|)$ *a Banach space,* $g \in C^1([a, b])$, *strictly increasing, such that* $g^{-1} \in C^1([g(a), g(b)])$. *Suppose* $F_k := D_{b-;g}^{k\alpha} f$, $k = 1, ..., n$, *fulfill* $F_k \in C^1([a, b], X)$, *and* $F_{n+1} \in C([a, b], X)$, *where* $0 < \alpha \leq 1$, $n \in \mathbb{N}$. *Then*

$$f(x) - f(b) = \sum_{i=2}^{n} \frac{(g(b) - g(x))^{i\alpha}}{\Gamma(i\alpha + 1)} \left(D_{b-;g}^{i\alpha} f\right)(b) +$$

$$\frac{1}{\Gamma((n+1)\alpha)} \int_x^b (g(t) - g(x))^{(n+1)\alpha - 1} g'(t) \left(D_{b-;g}^{(n+1)\alpha} f\right)(t)\, dt, \quad (8.19)$$

$\forall\, x \in [a, b]$.

When $n = 1$ we obtain

Corollary 8.12 *Let* $0 < \alpha \leq 1$, $f \in C^1([a, b], X)$, $(X, \|\cdot\|)$ *is a Banach space,* $g \in C^1([a, b])$, *strictly increasing, such that* $g^{-1} \in C^1([g(a), g(b)])$. *Assume that* $D_{b-;g}^{\alpha} f \in C^1([a, b], X)$, *and* $D_{b-;g}^{2\alpha} f \in C([a, b], X)$. *Then*

$$f(x) - f(b) = \frac{1}{\Gamma(2\alpha)} \int_x^b (g(t) - g(x))^{2\alpha - 1} g'(t) \left(D_{b-;g}^{2\alpha} f\right)(t)\, dt, \quad (8.20)$$

$\forall\, x \in [a, b]$.

We are greatly motivated by the following sequential generalized fractional Ostrowski type inequality:

Theorem 8.13 (p. 140, [5]) *Let* $g \in C^1([a,b])$ *and strictly increasing, such that* $g^{-1} \in C^1([g(a), g(b)])$, *and* $0 < \alpha < 1$, $n \in \mathbb{N}$, $f \in C^1([a,b], X)$, *where* $(X, \|\cdot\|)$ *is a Banach space. Let* $x_0 \in [a,b]$ *be fixed. Assume that* $F_k^{x_0} := D_{x_0-;g}^{k\alpha} f$, *for* $k = 1, \ldots, n$, *fulfill* $F_k^{x_0} \in C^1([a,b], X)$ *and* $F_{n+1}^{x_0} \in C([a, x_0], X)$ *and* $\left(D_{x_0-;g}^{i\alpha} f\right)(x_0) = 0$, $i = 1, \ldots, n$.

Similarly, we assume that $G_k^{x_0} := D_{x_0+;g}^{k\alpha} f$, *for* $k = 1, \ldots, n$, *fulfill* $G_k^{x_0} \in C^1([x_0, b], X)$ *and* $G_{n+1}^{x_0} \in C([x_0, b], X)$ *and* $\left(D_{x_0+;g}^{i\alpha} f\right)(x_0) = 0$, $i = 1, \ldots, n$.

Then

$$\left\| \frac{1}{b-a} \int_a^b f(x)\, dx - f(x_0) \right\| \leq \frac{1}{(b-a)\,\Gamma((n+1)\alpha + 1)} \cdot$$

$$\left\{ (g(b) - g(x_0))^{(n+1)\alpha} (b - x_0) \left\| D_{x_0+;g}^{(n+1)\alpha} f \right\|_{\infty,[x_0,b]} + \right. \tag{8.21}$$

$$\left. (g(x_0) - g(a))^{(n+1)\alpha} (x_0 - a) \left\| D_{x_0-;g}^{(n+1)\alpha} f \right\|_{\infty,[a,x_0]} \right\}.$$

8.3 Banach Algebras Basic Background

All here come from [16].

We need

Definition 8.14 ([16], *p. 245*) A complex algebra is a vector space A over the complex filed $\mathbb{C}$ in which a multiplication is defined that satisfies

$$x(yz) = (xy)z, \tag{8.22}$$

$$(x+y)z = xz + yz, \; x(y+z) = xy + xz, \tag{8.23}$$

and

$$\alpha(xy) = (\alpha x)y = x(\alpha y), \tag{8.24}$$

for all x, y and z in A and for all scalars α.

Additionally if A is a Banach space with respect to a norm that satisfies the multiplicative inequality

$$\|xy\| \leq \|x\|\,\|y\| \quad (x \in A, y \in A) \tag{8.25}$$

and if A contains a unit element e such that

$$xe = ex = x \ (x \in A) \tag{8.26}$$

and

$$\|e\| = 1, \tag{8.27}$$

then A is called a Banach algebra.

A is commutative iff $xy = yx$ for all $x, y \in A$.

We make

Remark 8.15 There exists at most one $e \in A$ that satisfies (8.26).

Inequality (8.25) makes multiplication to be continuous, more precisely left and right continuous, see [16], p. 246.

Multiplication in A is not necessarily the numerical multiplication, it is something more general and it is defined abstractly, that is for $x, y \in A$ we have $xy \in A$, e.g. composition or convolution, etc.

For nice examples about Banach algebras see [16], pp. 247–248, Sect. 10.3.

We also make

Remark 8.16 Next we mention about integration of A-valued functions, see [16], p. 259, Sect. 10.22:

If A is a Banach algebra and f is a continuous A-valued function on some compact Hausdorff space Q on which a complex Borel measure μ is defined, then $\int f d\mu$ exists and has all the properties that were discussed in Chap. 3 of [16], simply because A is a Banach space. However, an additional property can be added to these, namely: If $x \in A$, then

$$x \int_Q f \, d\mu = \int_Q x f(p) \, d\mu(p) \tag{8.28}$$

and

$$\left(\int_Q f \, d\mu \right) x = \int_Q f(p) \, x \, d\mu(p). \tag{8.29}$$

The Bochner integrals we will involve in our chapter follow (8.28) and (8.29). Also, let $f \in C([a, b], X)$, where $[a, b] \subset \mathbb{R}$, $(X, \|\cdot\|)$ is a Banach space. By [5], p. 3, f is Bochner integrable.

8.4 p-Schatten Norms Background

In this advanced section all come from [11].

Let $(H, \langle \cdot, \cdot \rangle)$ be a complex Hilbert space and $\mathcal{B}(H)$ the Banach algebra of all bounded linear operators on H. If $\{e_i\}_{i \in I}$ an orthonormal basis of H, we say that $A \in \mathcal{B}(H)$ is of trace class if

$$\|A\|_1 := \sum_{i \in I} \langle |A| e_i, e_i \rangle < \infty. \tag{8.30}$$

The definition of $\|A\|_1$ does not depend on the choice of the orthornormal basis $\{e_i\}_{i \in I}$. We denote by $\mathcal{B}_1(H)$ the set of trace class operators in $\mathcal{B}(H)$.

We define the trace of a trace class operator $A \in \mathcal{B}_1(H)$ to be

$$tr(A) := \sum_{i \in I} \langle Ae_i, e_i \rangle, \tag{8.31}$$

where $\{e_i\}_{i \in I}$ an orthonormal basis of H. Note that this coincides with the usual definition of the trace if H is finite-dimensional. We observe that the series (8.31) converges absolutely and it is independent from the choice of basis.

The following result collects some properties of the trace:

Theorem 8.17 *We have:*

(i) If $A \in \mathcal{B}_1(H)$ then $A^* \in \mathcal{B}_1(H)$ and

$$tr(A^*) = \overline{tr(A)}; \tag{8.32}$$

(ii) If $A \in \mathcal{B}_1(H)$ and $T \in \mathcal{B}(H)$, then $AT, TA \in \mathcal{B}_1(H)$ and

$$tr(AT) = tr(TA) \text{ and } |tr(AT)| \leq \|A\|_1 \|T\|; \tag{8.33}$$

(iii) $tr(\cdot)$ is a bounded linear functional on $\mathcal{B}_1(H)$ with $\|tr\| = 1$;

(iv) If $A, B \in \mathcal{B}_2(H)$ then $AB, BA \in \mathcal{B}_1(H)$ and $tr(AB) = tr(BA)$;

(v) $\mathcal{B}_{fin}(H)$ (finite rank operators) is a dense subspace of $\mathcal{B}_1(H)$.

An operator $A \in \mathcal{B}(H)$ is said to belong to the von Neumann–Schatten class $\mathcal{B}_p(H)$, $1 \leq p < \infty$ if the p-Schatten norm is finite [18, pp. 60–64]

$$\|A\|_p := \left[tr\left(|A|^p\right)\right]^{\frac{1}{p}} < \infty,$$

$|A|^p$ is an operator notation and not a power.

For $1 < p < q < \infty$ we have that

$$\mathcal{B}_1(H) \subset \mathcal{B}_p(H) \subset \mathcal{B}_q(H) \subset \mathcal{B}(H) \tag{8.34}$$

and

$$\|A\|_1 \geq \|A\|_p \geq \|A\|_q \geq \|A\|. \tag{8.35}$$

For $p \geq 1$ the functional $\|\cdot\|_p$ is a norm on the $*$-ideal $\mathcal{B}_p(H)$, which is a Banach algebra, and $\left(\mathcal{B}_p(H), \|\cdot\|_p\right)$ is a Banach space.

Also, see for instance [18, pp. 60–64], for $p \geq 1$,

$$\|A\|_p = \left\|A^*\right\|_p, \quad A \in \mathcal{B}_p(H) \tag{8.36}$$

$$\|AB\|_p \leq \|A\|_p \|B\|_p, \quad A, B \in \mathcal{B}_p(H) \tag{8.37}$$

and

$$\|AB\|_p \leq \|A\|_p \|B\|, \quad \|BA\|_p \leq \|B\| \|A\|_p, \quad A \in \mathcal{B}_p(H), B \in \mathcal{B}(H). \tag{8.38}$$

This implies that

$$\|CAB\|_p \leq \|C\| \|A\|_p \|B\|, \quad A \in \mathcal{B}_p(H), B, C \in \mathcal{B}(H). \tag{8.39}$$

In terms of p-Schatten norm we have the Hölder inequality for $p, q > 1$ with $\frac{1}{p} + \frac{1}{q} = 1$:

$$(|tr(AB)| \leq) \|AB\|_1 \leq \|A\|_p \|B\|_q, \quad A \in \mathcal{B}_p(H), B \in \mathcal{B}_q(H). \tag{8.40}$$

For the theory of trace functionals and their applications the interested reader is referred to [17, 18].

For some classical trace inequalities see [9, 10, 14], which are continuations of the work of Bellman [7].

8.5 Main Results

We start with 1-Schatten norm weighted mixed sequential generalized fractional Ostrowski type inequalities for several functions taking values in the Banach algebra $\mathcal{B}_2(H) \subset \mathcal{B}(H)$:

Theorem 8.18 *Let the $*$-ideal $\mathcal{B}_2(H)$, which $(\mathcal{B}_2(H), \|\cdot\|_2)$ is a Banach algebra; $x_0 \in [a, b] \subset \mathbb{R}$, $0 < \alpha < 1$; $A_i \in C^1([a, b], \mathcal{B}_2(H))$, $i = 1, ..., r \in \mathbb{N} - \{1\}$; $g \in C^1([a, b])$, strictly increasing, such that $g^{-1} \in C^1([g(a), g(b)])$. Assume that $F_{ki}^{x_0} := D_{x_0-;g}^{k\alpha} A_i$, for $k = 1, ..., n \in \mathbb{N}$, fulfill $F_{ki}^{x_0} \in C^1([a, x_0], \mathcal{B}_2(H))$ and $F_{(n+1)i}^{x_0} \in C([a, x_0], \mathcal{B}_2(H))$, and $\left(D_{x_0-;g}^{j\alpha} A_i\right)(x_0) = 0$, $j = 2, ..., n$; $i = 1, ..., r$. Similarly, we assume that $G_{ki}^{x_0} := D_{x_0+;g}^{k\alpha} A_i$, $k = 1, ..., n$, fulfill $G_{ki}^{x_0} \in C^1([x_0, b], \mathcal{B}_2(H))$ and $G_{(n+1)i}^{x_0} \in C([x_0, b], \mathcal{B}_2(H))$, and $\left(D_{x_0+;g}^{j\alpha} A_i\right)(x_0) = 0$, $j = 2, ..., n$; $i = 1, ..., r$.*

Denote by

$$\Omega\left(A_1,\dots,A_r\right)\left(x_0\right) :=$$

$$\sum_{i=1}^{r}\left[\int_a^b\left(\prod_{\substack{j=1\\ j\neq i}}^{r}A_j\left(x\right)\right)A_i\left(x\right)dx-\left(\int_a^b\left(\prod_{\substack{j=1\\ j\neq i}}^{r}A_j\left(x\right)\right)dx\right)A_i\left(x_0\right)\right]. \quad (8.41)$$

Then

$$\left\|\Omega\left(A_1,\dots,A_r\right)\left(x_0\right)\right\|_1 \leq \frac{1}{\Gamma\left(\left(n+1\right)\alpha+1\right)}\sum_{i=1}^{r}\left[\left[\left\|\left\|\left(D_{x_0-;g}^{(n+1)\alpha}A_i\right)\right\|_2\right\|_{\infty,[a,x_0]}\right.\right.$$

$$\left.\left(g\left(x_0\right)-g\left(a\right)\right)^{(n+1)\alpha}\left(\int_a^{x_0}\left(\prod_{\substack{j=1\\ j\neq i}}^{r}\left\|A_j\left(x\right)\right\|_2\right)dx\right)\right]+ \quad (8.42)$$

$$\left.\left[\left\|\left\|\left(D_{x_0+;g}^{(n+1)\alpha}A_i\right)\right\|_2\right\|_{\infty,[x_0,b]}\left(g\left(b\right)-g\left(x_0\right)\right)^{(n+1)\alpha}\left(\int_{x_0}^{b}\left(\prod_{\substack{j=1\\ j\neq i}}^{r}\left\|A_j\left(x\right)\right\|_2\right)dx\right)\right]\right].$$

Proof By Theorem 8.11 we obtain

$$A_i\left(x\right)-A_i\left(x_0\right)=\frac{1}{\Gamma\left(\left(n+1\right)\alpha\right)}\int_x^{x_0}\left(g\left(t\right)-g\left(x\right)\right)^{(n+1)\alpha-1}g'\left(t\right)\left(D_{x_0-;g}^{(n+1)\alpha}A_i\right)\left(t\right)dt, \quad (8.43)$$

$\forall\, x\in\left[a,x_0\right],\ i=1,\dots,r.$

Also, by Theorem 8.9, we get

$$A_i\left(x\right)-A_i\left(x_0\right)=\frac{1}{\Gamma\left(\left(n+1\right)\alpha\right)}\int_{x_0}^{x}\left(g\left(x\right)-g\left(t\right)\right)^{(n+1)\alpha-1}g'\left(t\right)\left(D_{x_0+;g}^{(n+1)\alpha}A_i\right)\left(t\right)dt, \quad (8.44)$$

$\forall\, x\in\left[x_0,b\right],\ i=1,\dots,r.$

Left multiplying (8.43) and (8.44) with $\left(\prod_{\substack{j=1\\ j\neq i}}^{r}A_j\left(x\right)\right)$ we get, respectively,

$$\left(\prod_{\substack{j=1\\ j\neq i}}^{r}A_j\left(x\right)\right)A_i\left(x\right)-\left(\prod_{\substack{j=1\\ j\neq i}}^{r}A_j\left(x\right)\right)A_i\left(x_0\right)= \quad (8.45)$$

$$\frac{\left(\prod_{\substack{j=1\\ j\neq i}}^{r}A_j\left(x\right)\right)}{\Gamma\left(\left(n+1\right)\alpha\right)}\int_x^{x_0}\left(g\left(t\right)-g\left(x\right)\right)^{(n+1)\alpha-1}g'\left(t\right)\left(D_{x_0-;g}^{(n+1)\alpha}A_i\right)\left(t\right)dt,$$

$\forall\, x \in [a, x_0]\,, i = 1, ..., r,$

and

$$\left(\prod_{\substack{j=1\\ j\neq i}}^{r} A_j\,(x)\right) A_i\,(x) - \left(\prod_{\substack{j=1\\ j\neq i}}^{r} A_j\,(x)\right) A_i\,(x_0) =$$

$$\frac{\left(\prod_{\substack{j=1\\ j\neq i}}^{r} A_j\,(x)\right)}{\Gamma\left((n+1)\,\alpha\right)} \int_{x_0}^{x} \left(g\,(x) - g\,(t)\right)^{(n+1)\alpha-1} g'\,(t) \left(D_{x_0+;g}^{(n+1)\alpha} A_i\right)(t)\,dt, \tag{8.46}$$

$\forall\, x \in [x_0, b]\,, i = 1, ..., r.$

Adding (8.45) and (8.46) as separate groups, we obtain

$$\sum_{i=1}^{r}\left(\prod_{\substack{j=1\\ j\neq i}}^{r} A_j\,(x)\right) A_i\,(x) - \sum_{i=1}^{r}\left(\prod_{\substack{j=1\\ j\neq i}}^{r} A_j\,(x)\right) A_i\,(x_0) =$$

$$\frac{\sum_{i=1}^{r}\left(\prod_{\substack{j=1\\ j\neq i}}^{r} A_j\,(x)\right)}{\Gamma\left((n+1)\,\alpha\right)} \int_{x}^{x_0} \left(g\,(t) - g\,(x)\right)^{(n+1)\alpha-1} g'\,(t) \left(D_{x_0-;g}^{(n+1)\alpha} A_i\right)(t)\,dt, \tag{8.47}$$

$\forall\, x \in [a, x_0]\,,$

and

$$\sum_{i=1}^{r}\left(\prod_{\substack{j=1\\ j\neq i}}^{r} A_j\,(x)\right) A_i\,(x) - \sum_{i=1}^{r}\left(\prod_{\substack{j=1\\ j\neq i}}^{r} A_j\,(x)\right) A_i\,(x_0) =$$

$$\frac{\sum_{i=1}^{r}\left(\prod_{\substack{j=1\\ j\neq i}}^{r} A_j\,(x)\right)}{\Gamma\left((n+1)\,\alpha\right)} \int_{x_0}^{x} \left(g\,(x) - g\,(t)\right)^{(n+1)\alpha-1} g'\,(t) \left(D_{x_0+;g}^{(n+1)\alpha} A_i\right)(t)\,dt, \tag{8.48}$$

$\forall\, x \in [x_0, b]\,.$

Next, we integrate (8.47) and (8.48) with respect to $x \in [a, b]$. We have

$$\sum_{i=1}^{r}\int_{a}^{x_0}\left(\prod_{\substack{j=1\\ j\neq i}}^{r} A_j\,(x)\right) A_i\,(x)\,dx - \sum_{i=1}^{r}\left(\int_{a}^{x_0}\left(\prod_{\substack{j=1\\ j\neq i}}^{r} A_j\,(x)\right) dx\right) A_i\,(x_0) =$$

$$\frac{1}{\Gamma((n+1)\alpha)}\sum_{i=1}^{r}\left[\int_a^{x_0}\left(\prod_{\substack{j=1\\ j\neq i}}^{r}A_j(x)\right)\right.$$

$$\left.\left(\int_x^{x_0}(g(t)-g(x))^{(n+1)\alpha-1}g'(t)\left(D_{x_0-;g}^{(n+1)\alpha}A_i\right)(t)\,dt\right)dx\right], \tag{8.49}$$

and

$$\sum_{i=1}^{r}\int_{x_0}^{b}\left(\prod_{\substack{j=1\\ j\neq i}}^{r}A_j(x)\right)A_i(x)\,dx-\sum_{i=1}^{r}\left(\int_{x_0}^{b}\left(\prod_{\substack{j=1\\ j\neq i}}^{r}A_j(x)\right)dx\right)A_i(x_0)=$$

$$\frac{1}{\Gamma((n+1)\alpha)}\sum_{i=1}^{r}\left[\int_{x_0}^{b}\left(\prod_{\substack{j=1\\ j\neq i}}^{r}A_j(x)\right)\right.$$

$$\left.\left(\int_{x_0}^{x}(g(x)-g(t))^{(n+1)\alpha-1}g'(t)\left(D_{x_0+;g}^{(n+1)\alpha}A_i\right)(t)\,dt\right)dx\right]. \tag{8.50}$$

Finally, adding (8.49) and (8.50) we obtain the useful identity

$$\Omega(A_1,\ldots,A_r)(x_0):=$$

$$\sum_{i=1}^{r}\left[\int_a^{b}\left(\prod_{\substack{j=1\\ j\neq i}}^{r}A_j(x)\right)A_i(x)\,dx-\left(\int_a^{b}\left(\prod_{\substack{j=1\\ j\neq i}}^{r}A_j(x)\right)dx\right)A_i(x_0)\right]=$$

$$\frac{1}{\Gamma((n+1)\alpha)}\sum_{i=1}^{r}\left[\left[\int_a^{x_0}\left(\prod_{\substack{j=1\\ j\neq i}}^{r}A_j(x)\right)\right.\right.$$

$$\left.\left(\int_x^{x_0}(g(t)-g(x))^{(n+1)\alpha-1}g'(t)\left(D_{x_0-;g}^{(n+1)\alpha}A_i\right)(t)\,dt\right)dx\right]$$

$$\left.+\left[\int_{x_0}^{b}\left(\prod_{\substack{j=1\\ j\neq i}}^{r}A_j(x)\right)\left(\int_{x_0}^{x}(g(x)-g(t))^{(n+1)\alpha-1}g'(t)\left(D_{x_0+;g}^{(n+1)\alpha}A_i\right)(t)\,dt\right)dx\right]\right]. \tag{8.51}$$

Therefore, we get (by using the p-Schatten norm and Hölder's type inequality (8.40) for $p = q = 2$)

$$\|\Omega(A_1, ..., A_r)(x_0)\|_1 \le \frac{1}{\Gamma((n+1)\alpha)} \sum_{i=1}^{r} \Bigg[\Bigg\|\Bigg[\int_a^{x_0} \left(\prod_{\substack{j=1\\ j\ne i}}^{r} A_j(x)\right)$$

$$\left(\int_x^{x_0} (g(t)-g(x))^{(n+1)\alpha-1} g'(t) \left(D_{x_0-;g}^{(n+1)\alpha} A_i\right)(t)\,dt\right) dx\Bigg]\Bigg\|_1 +$$

$$\Bigg\|\Bigg[\int_{x_0}^{b} \left(\prod_{\substack{j=1\\ j\ne i}}^{r} A_j(x)\right) \left(\int_{x_0}^{x} (g(x)-g(t))^{(n+1)\alpha-1} g'(t) \left(D_{x_0+;g}^{(n+1)\alpha} A_i\right)(t)\,dt\right) dx\Bigg]\Bigg\|_1\Bigg] \tag{8.52}$$

$$\le \frac{1}{\Gamma((n+1)\alpha)} \sum_{i=1}^{r} \Bigg[\Bigg[\int_a^{x_0} \Bigg\|\left(\prod_{\substack{j=1\\ j\ne i}}^{r} A_j(x)\right)$$

$$\left(\int_x^{x_0} (g(t)-g(x))^{(n+1)\alpha-1} g'(t) \left(D_{x_0-;g}^{(n+1)\alpha} A_i\right)(t)\,dt\right)\Bigg\|_1 dx\Bigg] +$$

$$\Bigg[\int_{x_0}^{b} \Bigg\|\left(\prod_{\substack{j=1\\ j\ne i}}^{r} A_j(x)\right) \left(\int_{x_0}^{x} (g(x)-g(t))^{(n+1)\alpha-1} g'(t) \left(D_{x_0+;g}^{(n+1)\alpha} A_i\right)(t)\,dt\right)\Bigg\|_1 dx\Bigg]\Bigg]$$

(by (8.37), (8.40))

$$\le \frac{1}{\Gamma((n+1)\alpha)} \sum_{i=1}^{r} \Bigg[\Bigg[\int_a^{x_0} \left(\prod_{\substack{j=1\\ j\ne i}}^{r} \|A_j(x)\|_2\right)$$

$$\left(\int_x^{x_0} (g(t)-g(x))^{(n+1)\alpha-1} g'(t) \left\|\left(D_{x_0-;g}^{(n+1)\alpha} A_i\right)(t)\right\|_2 dt\right) dx\Bigg] +$$

$$\Bigg[\int_{x_0}^{b} \left(\prod_{\substack{j=1\\ j\ne i}}^{r} \|A_j(x)\|_2\right) \left(\int_{x_0}^{x} (g(x)-g(t))^{(n+1)\alpha-1} g'(t) \left\|\left(D_{x_0+;g}^{(n+1)\alpha} A_i\right)(t)\right\|_2 dt\right) dx\Bigg]\Bigg] \tag{8.53}$$

$$=: (*).$$

Hence it holds

$$\|\Omega(A_1, ..., A_r)(x_0)\|_1 \leq (*) . \tag{8.54}$$

We have that

$$(*) \leq \frac{1}{\Gamma((n+1)\alpha+1)}$$

$$\sum_{i=1}^{r}\left[\left[\left\|\left\|\left(D_{x_0-;g}^{(n+1)\alpha}A_i\right)\right\|_2\right\|_{\infty,[a,x_0]}\int_a^{x_0}\left(\prod_{\substack{j=1\\j\neq i}}^{r}\|A_j(x)\|_2\right)(g(x_0)-g(x))^{(n+1)\alpha}dx\right]\right.$$

$$\left.+\left[\left\|\left\|\left(D_{x_0+;g}^{(n+1)\alpha}A_i\right)\right\|_2\right\|_{\infty,[x_0,b]}\int_{x_0}^{b}\left(\prod_{\substack{j=1\\j\neq i}}^{r}\|A_j(x)\|_2\right)(g(x)-g(x_0))^{(n+1)\alpha}dx\right]\right] \leq$$

$$\frac{1}{\Gamma((n+1)\alpha+1)}\sum_{i=1}^{r}\left[\left[\left\|\left\|\left(D_{x_0-;g}^{(n+1)\alpha}A_i\right)\right\|_2\right\|_{\infty,[a,x_0]}\right.\right.$$

$$\left.(g(x_0)-g(a))^{(n+1)\alpha}\left(\int_a^{x_0}\left(\prod_{\substack{j=1\\j\neq i}}^{r}\|A_j(x)\|_2\right)dx\right)\right]+ \tag{8.55}$$

$$\left.\left[\left\|\left\|\left(D_{x_0+;g}^{(n+1)\alpha}A_i\right)\right\|_2\right\|_{\infty,[x_0,b]}(g(b)-g(x_0))^{(n+1)\alpha}\left(\int_{x_0}^{b}\left(\prod_{\substack{j=1\\j\neq i}}^{r}\|A_j(x)\|_2\right)dx\right)\right]\right],$$

proving (8.42). ■

Next comes an L_1 estimate.

Theorem 8.19 *All as in Theorem 8.18, plus* $\frac{1}{(n+1)} \leq \alpha < 1$, $n \in \mathbb{N}$. *Then*

$$\|\Omega(A_1, ..., A_r)(x_0)\|_1 \leq \frac{\|g'\|_{\infty,[a,b]}}{\Gamma((n+1)\alpha)}$$

$$\sum_{i=1}^{r}\left[\left[\left\|\left\|\left(D_{x_0-;g}^{(n+1)\alpha}A_i\right)\right\|_2\right\|_{L_1([a,x_0])}(g(x_0)-g(a))^{(n+1)\alpha-1}\left(\int_a^{x_0}\left(\prod_{\substack{j=1\\j\neq i}}^{r}\|A_j(x)\|_2\right)dx\right)\right]\right.$$

$$+\left[\left\|\left\|\left(D_{x_0+;g}^{(n+1)\alpha}A_i\right)\right\|_2\right\|_{L_1([x_0,b])}(g(b)-g(x_0))^{(n+1)\alpha-1}\int_{x_0}^{b}\left(\prod_{\substack{j=1\\j\neq i}}^{r}\|A_j(x)\|_2\right)dx\right]\Bigg]. \tag{8.56}$$

Proof We have that (by (8.53), (8.54))

$$(*)\leq\frac{\|g'\|_{\infty,[a,b]}}{\Gamma((n+1)\alpha)}$$

$$\sum_{i=1}^{r}\Bigg[\Bigg[\left\|\left\|\left(D_{x_0-;g}^{(n+1)\alpha}A_i\right)\right\|_2\right\|_{L_1([a,x_0])}\int_a^{x_0}\left(\prod_{\substack{j=1\\j\neq i}}^{r}\|A_j(x)\|_2\right)(g(x_0)-g(x))^{(n+1)\alpha-1}dx\Bigg]$$

$$+\left[\left\|\left\|\left(D_{x_0+;g}^{(n+1)\alpha}A_i\right)\right\|_2\right\|_{L_1([x_0,b])}\int_{x_0}^{b}\left(\prod_{\substack{j=1\\j\neq i}}^{r}\|A_j(x)\|_2\right)(g(x)-g(x_0))^{(n+1)\alpha-1}dx\right]\Bigg]$$

$$\leq\frac{\|g'\|_{\infty,[a,b]}}{\Gamma((n+1)\alpha)}$$

$$\sum_{i=1}^{r}\Bigg[\Bigg[\left\|\left\|\left(D_{x_0-;g}^{(n+1)\alpha}A_i\right)\right\|_2\right\|_{L_1([a,x_0])}(g(x_0)-g(a))^{(n+1)\alpha-1}\left(\int_a^{x_0}\left(\prod_{\substack{j=1\\j\neq i}}^{r}\|A_j(x)\|_2\right)dx\right)\Bigg] \tag{8.57}$$

$$+\left[\left\|\left\|\left(D_{x_0+;g}^{(n+1)\alpha}A_i\right)\right\|_2\right\|_{L_1([x_0,b])}(g(b)-g(x_0))^{(n+1)\alpha-1}\left(\int_{x_0}^{b}\left(\prod_{\substack{j=1\\j\neq i}}^{r}\|A_j(x)\|_2\right)dx\right)\right]\Bigg],$$

proving the claim. ■

An L_γ estimate follows.

Theorem 8.20 *All as in Theorem 8.18, plus* $\gamma,\delta>1:\frac{1}{\gamma}+\frac{1}{\delta}=1$, *and* $\frac{1}{\delta(n+1)}<\alpha<1$. *Then*

$$\|\Omega(A_1,\ldots,A_r)(x_0)\|_1\leq\frac{1}{\Gamma((n+1)\alpha)(\gamma((n+1)\alpha-1)+1)^{\frac{1}{\gamma}}}$$

$$\sum_{i=1}^{r}\Bigg[\Bigg[\left\|\left\|\left(D_{x_0-;g}^{(n+1)\alpha}A_i\right)\circ g^{-1}\right\|_2\right\|_{\delta,[g(a),g(x_0)]}$$

$$(g(x_0) - g(a))^{(n+1)\alpha - \frac{1}{\delta}} \left(\int_a^{x_0} \left(\prod_{\substack{j=1 \\ j \neq i}}^{r} \|A_j(x)\|_2 \right) dx \right) \Bigg] +$$

$$\left[\left\| \left\| \left(D_{x_0+;g}^{(n+1)\alpha} A_i \right) \circ g^{-1} \right\|_2 \right\|_{\delta, [g(x_0), g(b)]} (g(b) - g(x_0))^{(n+1)\alpha - \frac{1}{\delta}} \right.$$

$$\left. \left(\int_{x_0}^{b} \left(\prod_{\substack{j=1 \\ j \neq i}}^{r} \|A_j(x)\|_2 \right) dx \right) \right] \Bigg]. \tag{8.58}$$

Proof By (8.53) and (8.54) we obtain

$$(*) = \frac{1}{\Gamma((n+1)\alpha)} \sum_{i=1}^{r} \Bigg[\Bigg[\int_a^{x_0} \left(\prod_{\substack{j=1 \\ j \neq i}}^{r} \|A_j(x)\|_2 \right)$$

$$\left(\int_x^{x_0} (g(t) - g(x))^{(n+1)\alpha - 1} \left\| \left(\left(D_{x_0-;g}^{(n+1)\alpha} A_i \right) \circ g^{-1} \right) (g(t)) \right\|_2 dg(t) \right) dx \Bigg] +$$

$$\Bigg[\int_{x_0}^{b} \left(\prod_{\substack{j=1 \\ j \neq i}}^{r} \|A_j(x)\|_2 \right)$$

$$\left(\int_{x_0}^{x} (g(x) - g(t))^{(n+1)\alpha - 1} \left\| \left(\left(D_{x_0+;g}^{(n+1)\alpha} A_i \right) \circ g^{-1} \right) (g(t)) \right\|_2 dg(t) \right) dx \Bigg] \Bigg] \tag{8.59}$$

$$= \frac{1}{\Gamma((n+1)\alpha)} \sum_{i=1}^{r} \Bigg[\Bigg[\int_a^{x_0} \left(\prod_{\substack{j=1 \\ j \neq i}}^{r} \|A_j(x)\|_2 \right)$$

$$\left(\int_{g(x)}^{g(x_0)} (z - g(x))^{(n+1)\alpha - 1} \left\| \left(\left(D_{x_0-;g}^{(n+1)\alpha} A_i \right) \circ g^{-1} \right) (z) \right\|_2 dz \right) dx \Bigg] +$$

$$\Bigg[\int_{x_0}^{b} \left(\prod_{\substack{j=1 \\ j \neq i}}^{r} \|A_j(x)\|_2 \right) \left(\int_{g(x_0)}^{g(x)} (g(x) - z)^{(n+1)\alpha - 1} \left\| \left(\left(D_{x_0+;g}^{(n+1)\alpha} A_i \right) \circ g^{-1} \right) (z) \right\|_2 dz \right) dx \Bigg] \Bigg]$$

$$\leq \frac{1}{\Gamma((n+1)\alpha)} \sum_{i=1}^{r} \Bigg[\Bigg[\int_a^{x_0} \left(\prod_{\substack{j=1 \\ j \neq i}}^{r} \| A_j(x) \|_2 \right) \left(\int_{g(x)}^{g(x_0)} (z - g(x))^{\gamma((n+1)\alpha - 1)} dz \right)^{\frac{1}{\gamma}}$$

$$\left(\int_{g(x)}^{g(x_0)} \left\| \left(\left(D_{x_0-;g}^{(n+1)\alpha} A_i \right) \circ g^{-1} \right)(z) \right\|_2^{\delta} dz \right)^{\frac{1}{\delta}} dx \Bigg] +$$

$$\Bigg[\int_{x_0}^{b} \left(\prod_{\substack{j=1 \\ j \neq i}}^{r} \| A_j(x) \|_2 \right) \left(\int_{g(x_0)}^{g(x)} (g(x) - z)^{\gamma((n+1)\alpha - 1)} dz \right)^{\frac{1}{\gamma}}$$

$$\left(\int_{g(x_0)}^{g(x)} \left\| \left(\left(D_{x_0+;g}^{(n+1)\alpha} A_i \right) \circ g^{-1} \right)(z) \right\|_2^{\delta} dz \right)^{\frac{1}{\delta}} dx \Bigg] \Bigg] \leq \tag{8.60}$$

$$\frac{1}{\Gamma((n+1)\alpha)(\gamma((n+1)\alpha - 1) + 1)^{\frac{1}{\gamma}}}$$

$$\sum_{i=1}^{r} \Bigg[\Bigg[\left\| \left\| \left(\left(D_{x_0-;g}^{(n+1)\alpha} A_i \right) \circ g^{-1} \right) \right\|_2 \right\|_{\delta, [g(a), g(x_0)]}$$

$$\left(\int_a^{x_0} \left(\prod_{\substack{j=1 \\ j \neq i}}^{r} \| A_j(x) \|_2 \right) (g(x_0) - g(x))^{\frac{\gamma((n+1)\alpha - 1) + 1}{\gamma}} dx \right) \Bigg] +$$

$$\Bigg[\left\| \left\| \left(\left(D_{x_0+;g}^{(n+1)\alpha} A_i \right) \circ g^{-1} \right) \right\|_2 \right\|_{\delta, [g(x_0), g(b)]}$$

$$\left(\int_{x_0}^{b} \left(\prod_{\substack{j=1 \\ j \neq i}}^{r} \| A_j(x) \|_2 \right) (g(x) - g(x_0))^{\frac{\gamma((n+1)\alpha - 1) + 1}{\gamma}} dx \right) \Bigg] \Bigg]$$

$$\leq \frac{1}{\Gamma((n+1)\alpha)(\gamma((n+1)\alpha - 1) + 1)^{\frac{1}{\gamma}}}$$

$$\sum_{i=1}^{r} \Bigg[\left\| \left\| \left(D_{x_0-;g}^{(n+1)\alpha} A_i \right) \circ g^{-1} \right\|_2 \right\|_{\delta, [g(a), g(x_0)]}$$

$$(g(x_0) - g(a))^{(n+1)\alpha - \frac{1}{\delta}} \left(\int_a^{x_0} \left(\prod_{\substack{j=1 \\ j \neq i}}^{r} \|A_j(x)\|_2 \right) dx \right) \tag{8.61}$$

$$+ \left\| \left\| \left(D_{x_0+;g}^{(n+1)\alpha} A_i \right) \circ g^{-1} \right\|_2 \right\|_{\delta,[g(x_0),g(b)]} (g(b) - g(x_0))^{(n+1)\alpha - \frac{1}{\delta}}$$

$$\left. \left(\int_{x_0}^{b} \left(\prod_{\substack{j=1 \\ j \neq i}}^{r} \|A_j(x)\|_2 \right) dx \right) \right],$$

proving the claim. ■

When $n = 1$ we get the following results:

Corollary 8.21 (to Theorem 8.18) *Let the* $*$*-ideal* $\mathcal{B}_2(H)$*;* $x_0 \in [a, b] \subset \mathbb{R}$, $0 < \alpha < 1$*;* $A_i \in C^1([a, b], \mathcal{B}_2(H))$, $i = 1, ..., r \in \mathbb{N} - \{1\}$*;* $g \in C^1([a, b])$*, strictly increasing, such that* $g^{-1} \in C^1([g(a), g(b)])$. *Assume that* $D_{x_0-;g}^{\alpha} A_i \in C^1([a, x_0], \mathcal{B}_2(H))$ *and* $D_{x_0-;g}^{2\alpha} A_i \in C([a, x_0], \mathcal{B}_2(H))$, $i = 1, ..., r$. *Similarly, assume that* $D_{x_0+;g}^{\alpha} A_i \in C^1([x_0, b], \mathcal{B}_2(H))$ *and* $D_{x_0+;g}^{2\alpha} A_i \in C([x_0, b], \mathcal{B}_2(H))$*;* $i = 1, ..., r$.

Denote by

$$\Omega(A_1, ..., A_r)(x_0) :=$$

$$\sum_{i=1}^{r} \left[\int_a^b \left(\prod_{\substack{j=1 \\ j \neq i}}^{r} A_j(x) \right) A_i(x) \, dx - \left(\int_a^b \left(\prod_{\substack{j=1 \\ j \neq i}}^{r} A_j(x) \right) dx \right) A_i(x_0) \right]. \tag{8.62}$$

Then

$$\|\Omega(A_1, ..., A_r)(x_0)\|_1 \leq \frac{1}{\Gamma(2\alpha + 1)} \sum_{i=1}^{r} \left[\left[\left\| \left\| \left(D_{x_0-;g}^{2\alpha} A_i \right) \right\|_2 \right\|_{\infty,[a,x_0]} \right. \right.$$

$$\left. (g(x_0) - g(a))^{2\alpha} \left(\int_a^{x_0} \left(\prod_{\substack{j=1 \\ j \neq i}}^{r} \|A_j(x)\|_2 \right) dx \right) \right] + \tag{8.63}$$

$$\left. \left[\left\| \left\| \left(D_{x_0+;g}^{2\alpha} A_i \right) \right\|_2 \right\|_{\infty,[x_0,b]} (g(b) - g(x_0))^{2\alpha} \left(\int_{x_0}^{b} \left(\prod_{\substack{j=1 \\ j \neq i}}^{r} \|A_j(x)\|_2 \right) dx \right) \right] \right]$$

$$=: \Lambda_1(x_0).$$

Proof By Theorem 8.18 and Corollaries 8.10, 8.12. ■

We continue with

Corollary 8.22 (to Theorem 8.19) *All as in Corollary 8.21, plus $\frac{1}{2} \leq \alpha < 1$. Then*

$$\|\Omega(A_1, ..., A_r)(x_0)\|_1 \leq \frac{\|g'\|_{\infty,[a,b]}}{\Gamma(2\alpha)} \sum_{i=1}^{r} \Bigg[\Bigg[\Big\| \big\| \big(D_{x_0-;g}^{2\alpha} A_i\big) \big\|_2 \Big\|_{L_1([a,x_0])}$$

$$(g(x_0) - g(a))^{2\alpha-1} \left(\int_a^{x_0} \left(\prod_{\substack{j=1 \\ j \neq i}}^{r} \|A_j(x)\|_2 \right) dx \right) \Bigg] +$$

$$\Bigg[\Big\| \big\| \big(D_{x_0+;g}^{2\alpha} A_i\big) \big\|_2 \Big\|_{L_1([x_0,b])} (g(b) - g(x_0))^{2\alpha-1} \left(\int_{x_0}^{b} \left(\prod_{\substack{j=1 \\ j \neq i}}^{r} \|A_j(x)\|_2 \right) dx \right) \Bigg] \Bigg] \tag{8.64}$$

$$=: \Lambda_2(x_0).$$

Proof By Theorem 8.19. ■

We also give

Corollary 8.23 (to Theorem 8.20) *All as in Corollary 8.21, plus $\gamma, \delta > 1 : \frac{1}{\gamma} + \frac{1}{\delta} = 1$, and $\frac{1}{2\delta} < \alpha < 1$. Then*

$$\|\Omega(A_1, ..., A_r)(x_0)\|_1 \leq \frac{1}{\Gamma(2\alpha)(\gamma(2\alpha-1)+1)^{\frac{1}{\gamma}}}$$

$$\sum_{i=1}^{r} \Bigg[\Bigg[\Big\| \big\| \big(D_{x_0-;g}^{2\alpha} A_i\big) \circ g^{-1} \big\|_2 \Big\|_{\delta,[g(a),g(x_0)]}$$

$$(g(x_0) - g(a))^{2\alpha-\frac{1}{\delta}} \left(\int_a^{x_0} \left(\prod_{\substack{j=1 \\ j \neq i}}^{r} \|A_j(x)\|_2 \right) dx \right) \Bigg] +$$

$$\Bigg[\Big\| \big\| \big(D_{x_0+;g}^{2\alpha} A_i\big) \circ g^{-1} \big\|_2 \Big\|_{\delta,[g(x_0),g(b)]} (g(b) - g(x_0))^{2\alpha-\frac{1}{\delta}}$$

$$\left(\int_{x_0}^{b} \left(\prod_{\substack{j=1 \\ j \neq i}}^{r} \|A_j(x)\|_2 \right) dx \right) \Bigg] \Bigg] = \Lambda_3(x_0). \quad (8.65)$$

Proof By Theorem 8.20. ■

Next come p-Schatten norm, $p > 1$, fractional Ostrowski type inequalities for $\mathcal{B}_p(H)$ valued functions, $\mathcal{B}_p(H) \subset \mathcal{B}(H)$:

Theorem 8.24 *Let $p > 1$, the $*$-ideal $\mathcal{B}_p(H)$, which $\left(\mathcal{B}_p(H), \|\cdot\|_p\right)$ is a Banach algebra; $x_0 \in [a,b] \subset \mathbb{R}$, $0 < \alpha < 1$; $A_i \in C^1\left([a,b], \mathcal{B}_p(H)\right)$, $i = 1, ..., r \in \mathbb{N} - \{1\}$; $g \in C^1([a,b])$, strictly increasing, such that $g^{-1} \in C^1([g(a), g(b)])$. Assume that $F_{ki}^{x_0} := D_{x_0-;g}^{k\alpha} A_i$, for $k = 1, ..., n \in \mathbb{N}$, fulfill $F_{ki}^{x_0} \in C^1\left([a, x_0], \mathcal{B}_p(H)\right)$ and $F_{(n+1)i}^{x_0} \in C\left([a, x_0], \mathcal{B}_p(H)\right)$, and $\left(D_{x_0-;g}^{j\alpha} A_i\right)(x_0) = 0$, $j = 2, ..., n$; $i = 1, ..., r$. Similarly, we assume that $G_{ki}^{x_0} := D_{x_0+;g}^{k\alpha} A_i$, $k = 1, ..., n$, fulfill $G_{ki}^{x_0} \in C^1\left([x_0, b], \mathcal{B}_p(H)\right)$ and $G_{(n+1)i}^{x_0} \in C\left([x_0, b], \mathcal{B}_p(H)\right)$, and $\left(D_{x_0+;g}^{j\alpha} A_i\right)(x_0) = 0$, $j = 2, ..., n$; $i = 1, ..., r$.*

Denote by

$$\Omega(A_1, ..., A_r)(x_0) :=$$

$$\sum_{i=1}^{r} \left[\int_a^b \left(\prod_{\substack{j=1 \\ j \neq i}}^{r} A_j(x) \right) A_i(x)\, dx - \left(\int_a^b \left(\prod_{\substack{j=1 \\ j \neq i}}^{r} A_j(x) \right) dx \right) A_i(x_0) \right]. \quad (8.66)$$

Then

$$\|\Omega(A_1, ..., A_r)(x_0)\|_p \leq \frac{1}{\Gamma((n+1)\alpha + 1)} \sum_{i=1}^{r} \left[\left[\left\| \left\| \left(D_{x_0-;g}^{(n+1)\alpha} A_i \right) \right\|_p \right\|_{\infty,[a,x_0]} \right.\right.$$

$$\left. (g(x_0) - g(a))^{(n+1)\alpha} \left(\int_a^{x_0} \left(\prod_{\substack{j=1 \\ j \neq i}}^{r} \|A_j(x)\|_p \right) dx \right) \right] + \quad (8.67)$$

$$\left[\left\| \left\| \left(D_{x_0+;g}^{(n+1)\alpha} A_i \right) \right\|_p \right\|_{\infty,[x_0,b]} (g(b) - g(x_0))^{(n+1)\alpha} \left(\int_{x_0}^{b} \left(\prod_{\substack{j=1 \\ j \neq i}}^{r} \|A_j(x)\|_p \right) dx \right) \right] \Bigg].$$

Proof As similar to Theorem 8.18 is omitted. Use of (8.37). ■

Next comes an L_1 estimate.

Theorem 8.25 *All as in Theorem 8.24, plus* $\frac{1}{(n+1)} \leq \alpha < 1$, $n \in \mathbb{N}$. *Then*

$$\|\Omega(A_1, ..., A_r)(x_0)\|_p \leq \frac{\|g'\|_{\infty,[a,b]}}{\Gamma((n+1)\alpha)}$$

$$\sum_{i=1}^{r}\left[\left[\left\|\left\|\left(D_{x_0-;g}^{(n+1)\alpha}A_i\right)\right\|_p\right\|_{L_1([a,x_0])}(g(x_0)-g(a))^{(n+1)\alpha-1}\left(\int_a^{x_0}\left(\prod_{\substack{j=1\\j\neq i}}^{r}\|A_j(x)\|_p\right)dx\right)\right]\right.$$
$$\left.+\left[\left\|\left\|\left(D_{x_0+;g}^{(n+1)\alpha}A_i\right)\right\|_p\right\|_{L_1([x_0,b])}(g(b)-g(x_0))^{(n+1)\alpha-1}\int_{x_0}^{b}\left(\prod_{\substack{j=1\\j\neq i}}^{r}\|A_j(x)\|_p\right)dx\right]\right]. \tag{8.68}$$

Proof As similar to the proof of Theorem 8.19 is omitted. ■

An L_γ estimate follows.

Theorem 8.26 *All as in Theorem 8.24, plus* $\gamma, \delta > 1 : \frac{1}{\gamma} + \frac{1}{\delta} = 1$, *and* $\frac{1}{\delta(n+1)} < \alpha < 1$. *Then*

$$\|\Omega(A_1, ..., A_r)(x_0)\|_p \leq \frac{1}{\Gamma((n+1)\alpha)(\gamma((n+1)\alpha-1)+1)^{\frac{1}{\gamma}}}$$

$$\sum_{i=1}^{r}\left[\left[\left\|\left\|\left(D_{x_0-;g}^{(n+1)\alpha}A_i\right)\circ g^{-1}\right\|_p\right\|_{\delta,[g(a),g(x_0)]}\right.\right.$$

$$\left.(g(x_0)-g(a))^{(n+1)\alpha-\frac{1}{\delta}}\left(\int_a^{x_0}\left(\prod_{\substack{j=1\\j\neq i}}^{r}\|A_j(x)\|_p\right)dx\right)\right]+$$

$$\left[\left\|\left\|\left(D_{x_0+;g}^{(n+1)\alpha}A_i\right)\circ g^{-1}\right\|_p\right\|_{\delta,[g(x_0),g(b)]}(g(b)-g(x_0))^{(n+1)\alpha-\frac{1}{\delta}} \tag{8.69}$$

$$\left.\left.\left(\int_{x_0}^{b}\left(\prod_{\substack{j=1\\j\neq i}}^{r}\|A_j(x)\|_p\right)dx\right)\right]\right].$$

Proof As similar to Theorem 8.20 is omitted. ■

When $n = 1$ we get the following results:

Corollary 8.27 (to Theorem 8.24) *Let $p > 1$, the $*$-ideal $\mathcal{B}_p(H)$; $x_0 \in [a,b] \subset \mathbb{R}$, $0 < \alpha < 1$; $A_i \in C^1([a,b], \mathcal{B}_p(H))$, $i = 1, ..., r \in \mathbb{N} - \{1\}$; $g \in C^1([a,b])$, strictly increasing, such that $g^{-1} \in C^1([g(a), g(b)])$. Assume that $D^{\alpha}_{x_0-;g} A_i \in C^1([a,x_0], \mathcal{B}_p(H))$ and $D^{2\alpha}_{x_0-;g} A_i \in C([a,x_0], \mathcal{B}_p(H))$, $i = 1, ..., r$. Similarly, assume that $D^{\alpha}_{x_0+;g} A_i \in C^1([x_0,b], \mathcal{B}_p(H))$ and $D^{2\alpha}_{x_0+;g} A_i \in C([x_0,b], \mathcal{B}_p(H))$; $i = 1, ..., r$; $\Omega(A_1, ..., A_r)(x_0)$ is as in (8.66).*

Then

$$\|\Omega(A_1, ..., A_r)(x_0)\|_p \leq \frac{1}{\Gamma(2\alpha+1)} \sum_{i=1}^{r} \left[\left[\left\| \left\| \left(D^{2\alpha}_{x_0-;g} A_i\right)\right\|_p \right\|_{\infty,[a,x_0]} \right.\right.$$

$$\left.(g(x_0) - g(a))^{2\alpha} \left(\int_a^{x_0} \left(\prod_{\substack{j=1 \\ j \neq i}}^{r} \|A_j(x)\|_p \right) dx \right)\right] + \tag{8.70}$$

$$\left.\left[\left\| \left\| \left(D^{2\alpha}_{x_0+;g} A_i\right)\right\|_p \right\|_{\infty,[x_0,b]} (g(b) - g(x_0))^{2\alpha} \left(\int_{x_0}^{b} \left(\prod_{\substack{j=1 \\ j \neq i}}^{r} \|A_j(x)\|_p \right) dx \right)\right]\right]$$

$$=: \Lambda_4(x_0).$$

Proof By Theorem 8.24 and Corollaries 8.10, 8.12. ■

We continue with

Corollary 8.28 (to Theorem 8.25) *All as in Corollary 8.27, plus $\frac{1}{2} \leq \alpha < 1$. Then*

$$\|\Omega(A_1, ..., A_r)(x_0)\|_p \leq \frac{\|g'\|_{\infty,[a,b]}}{\Gamma(2\alpha)} \sum_{i=1}^{r} \left[\left[\left\| \left\| \left(D^{2\alpha}_{x_0-;g} A_i\right)\right\|_p \right\|_{L_1([a,x_0])} \right.\right.$$

$$\left.(g(x_0) - g(a))^{2\alpha-1} \left(\int_a^{x_0} \left(\prod_{\substack{j=1 \\ j \neq i}}^{r} \|A_j(x)\|_p \right) dx \right)\right] + \tag{8.71}$$

$$\left.\left[\left\| \left\| \left(D^{2\alpha}_{x_0+;g} A_i\right)\right\|_p \right\|_{L_1([x_0,b])} (g(b) - g(x_0))^{2\alpha-1} \left(\int_{x_0}^{b} \left(\prod_{\substack{j=1 \\ j \neq i}}^{r} \|A_j(x)\|_p \right) dx \right)\right]\right]$$

$$= \Lambda_5(x_0).$$

Proof By Theorem 8.25. ■

We also give

Corollary 8.29 (to Theorem 8.26) *All as in Corollary 8.27, plus* $\gamma, \delta > 1 : \frac{1}{\gamma} + \frac{1}{\delta} = 1$, *and* $\frac{1}{2\delta} < \alpha < 1$. *Then*

$$\|\Omega(A_1, ..., A_r)(x_0)\|_p \le \frac{1}{\Gamma(2\alpha)(\gamma(2\alpha-1)+1)^{\frac{1}{\gamma}}}$$

$$\sum_{i=1}^{r} \Bigg[\Bigg[\left\| \left\| \left(D^{2\alpha}_{x_0-;g} A_i\right) \circ g^{-1} \right\|_p \right\|_{\delta,[g(a),g(x_0)]}$$

$$(g(x_0) - g(a))^{2\alpha - \frac{1}{\delta}} \left(\int_a^{x_0} \left(\prod_{\substack{j=1 \\ j \ne i}}^{r} \|A_j(x)\|_p \right) dx \right) \Bigg] + \tag{8.72}$$

$$\Bigg[\left\| \left\| \left(D^{2\alpha}_{x_0+;g} A_i\right) \circ g^{-1} \right\|_p \right\|_{\delta,[g(x_0),g(b)]} (g(b) - g(x_0))^{2\alpha - \frac{1}{\delta}}$$

$$\left(\int_{x_0}^{b} \left(\prod_{\substack{j=1 \\ j \ne i}}^{r} \|A_j(x)\|_p \right) dx \right) \Bigg]\Bigg] = \Lambda_6(x_0).$$

Proof By Theorem 8.26. ■

When $r = 2$ we obtain the following operator related sequential fractional Ostrowski type inequalities.

Theorem 8.30 *Let* $p, q > 1 : \frac{1}{p} + \frac{1}{q} = 1$, *and let the* $*$*-ideals* $\mathcal{B}_p(H)$, $\mathcal{B}_q(H)$, *for which* $(\mathcal{B}_p(H), \|\cdot\|_p)$, $(\mathcal{B}_q(H), \|\cdot\|_q)$ *are Banach algebras;* $x_0 \in [a, b] \subset \mathbb{R}$, $0 < \alpha < 1$, $A_1 \in C^1([a, b], \mathcal{B}_p(H))$, $A_2 \in C^1([a, b], \mathcal{B}_q(H))$; $g \in C^1([a, b])$, *strictly increasing, such that* $g^{-1} \in C^1([g(a), g(b)])$. *Assume that* $F^{x_0}_{ki} := D^{k\alpha}_{x_0-;g} A_i$, *for* $k = 1, ..., n \in \mathbb{N}$, $i = 1, 2$, *fulfill* $F^{x_0}_{k1} \in C^1([a, x_0], \mathcal{B}_p(H))$, $F^{x_0}_{(n+1)1} \in C([a, x_0], \mathcal{B}_p(H))$; $F^{x_0}_{k2} \in C^1([a, x_0], \mathcal{B}_q(H))$, $F^{x_0}_{(n+1)2} \in C([a, x_0], \mathcal{B}_q(H))$, *and* $\left(D^{j\alpha}_{x_0-;g} A_i\right)(x_0) = 0$, $j = 2, ..., n$; $i = 1, 2$. *Similarly, we assume that* $G^{x_0}_{ki} := D^{k\alpha}_{x_0+;g} A_i$, $k = 1, ..., n$, $i = 1, 2$, *fulfill* $G^{x_0}_{k1} \in C^1([x_0, b], \mathcal{B}_p(H))$, $G^{x_0}_{(n+1)1} \in C([x_0, b], \mathcal{B}_p(H))$, $G^{x_0}_{k2} \in C^1([x_0, b], \mathcal{B}_q(H))$, $G^{x_0}_{(n+1)2} \in C([x_0, b], \mathcal{B}_q(H))$, *and* $\left(D^{j\alpha}_{x_0+;g} A_i\right)(x_0) = 0$, $j = 2, ..., n$; $i = 1, 2$.

Then

(1) it holds

$$\Omega(A_1, A_2)(x_0) := \int_a^b A_2(x) A_1(x)\,dx + \int_a^b A_1(x) A_2(x)\,dx -$$

$$\left(\int_a^b A_2(x)\,dx\right) A_1(x_0) - \left(\int_a^b A_1(x)\,dx\right) A_2(x_0) = \frac{1}{\Gamma((n+1)\alpha)}$$

$$\left\{\left[\int_a^{x_0} A_2(x)\left(\int_{g(x)}^{g(x_0)} (z - g(x))^{(n+1)\alpha-1}\left(\left(D_{x_0-;g}^{(n+1)\alpha} A_1\right)\circ g^{-1}\right)(z)\,dz\right)dx\right] + \right.$$

$$\left[\int_{x_0}^b A_2(x)\left(\int_{g(x_0)}^{g(x)} (g(x) - z)^{(n+1)\alpha-1}\left(\left(D_{x_0+;g}^{(n+1)\alpha} A_1\right)\circ g^{-1}\right)(z)\,dz\right)dx\right] + \tag{8.73}$$

$$\left[\int_a^{x_0} A_1(x)\left(\int_{g(x)}^{g(x_0)} (z - g(x))^{(n+1)\alpha-1}\left(\left(D_{x_0-;g}^{(n+1)\alpha} A_2\right)\circ g^{-1}\right)(z)\,dz\right)dx\right] +$$

$$\left.\left[\int_{x_0}^b A_1(x)\left(\int_{g(x_0)}^{g(x)} (g(x) - z)^{(n+1)\alpha-1}\left(\left(D_{x_0+;g}^{(n+1)\alpha} A_2\right)\circ g^{-1}\right)(z)\,dz\right)dx\right]\right\},$$

(2) for $\gamma, \delta > 1 : \frac{1}{\gamma} + \frac{1}{\delta} = 1$, with $\frac{1}{\delta(n+1)} < \alpha < 1$, we have that

$$\|\Omega(A_1, A_2)(x_0)\|_1 \leq \frac{1}{\Gamma((n+1)\alpha)(\gamma((n+1)\alpha - 1) + 1)^{\frac{1}{\gamma}}}$$

$$\left\{\left[\left\|\left\|\left(D_{x_0-;g}^{(n+1)\alpha} A_1\right)\circ g^{-1}\right\|_p\right\|_{\delta,[g(a),g(x_0)]} \int_a^{x_0} \|A_2(x)\|_q (g(x_0) - g(x))^{(n+1)\alpha-\frac{1}{\delta}}\,dx\right] + \right.$$

$$\left[\left\|\left\|\left(D_{x_0+;g}^{(n+1)\alpha} A_1\right)\circ g^{-1}\right\|_p\right\|_{\delta,[g(x_0),g(b)]} \int_{x_0}^b \|A_2(x)\|_q (g(x) - g(x_0))^{(n+1)\alpha-\frac{1}{\delta}}\,dx\right] + \tag{8.74}$$

$$\left[\left\|\left\|\left(D_{x_0-;g}^{(n+1)\alpha} A_2\right)\circ g^{-1}\right\|_q\right\|_{\delta,[g(a),g(x_0)]} \int_a^{x_0} \|A_1(x)\|_p (g(x_0) - g(x))^{(n+1)\alpha-\frac{1}{\delta}}\,dx\right] +$$

$$\left.\left[\left\|\left\|\left(D_{x_0+;g}^{(n+1)\alpha} A_2\right)\circ g^{-1}\right\|_q\right\|_{\delta,[g(x_0),g(b)]} \int_{x_0}^b \|A_1(x)\|_p (g(x) - g(x_0))^{(n+1)\alpha-\frac{1}{\delta}}\,dx\right]\right\},$$

(3) if $(n+1)\alpha \geq 1$, we obtain

$$\|\Omega(A_1, A_2)(x_0)\|_1 \leq \frac{1}{\Gamma((n+1)\alpha)}$$

$$\left\{\left[\left\|\left\|\left(D_{x_0-;g}^{(n+1)\alpha} A_1\right)\circ g^{-1}\right\|_p\right\|_{L_1([g(a),g(x_0)])} \int_a^{x_0} \|A_2(x)\|_q (g(x_0) - g(x))^{(n+1)\alpha-1}\,dx\right] + \right.$$

$$\left[\left\|\left\|\left(D_{x_0+;g}^{(n+1)\alpha}A_1\right)\circ g^{-1}\right\|_p\right\|_{L_1([g(x_0),g(b)])}\int_{x_0}^{b}\|A_2(x)\|_q(g(x)-g(x_0))^{(n+1)\alpha-1}dx\right]+$$

$$\left[\left\|\left\|\left(D_{x_0-;g}^{(n+1)\alpha}A_2\right)\circ g^{-1}\right\|_q\right\|_{L_1([g(a),g(x_0)])}\int_{a}^{x_0}\|A_1(x)\|_p(g(x_0)-g(x))^{(n+1)\alpha-1}dx\right]+$$

$$\left.\left[\left\|\left\|\left(D_{x_0+;g}^{(n+1)\alpha}A_2\right)\circ g^{-1}\right\|_q\right\|_{L_1([g(x_0),g(b)])}\int_{x_0}^{b}\|A_1(x)\|_p(g(x)-g(x_0))^{(n+1)\alpha-1}dx\right]\right\}, \tag{8.75}$$

and

(4)

$$\|\Omega(A_1,A_2)(x_0)\|_1\leq\frac{1}{\Gamma((n+1)\alpha+1)}$$

$$\left\{\left[\left\|\left\|\left(D_{x_0-;g}^{(n+1)\alpha}A_1\right)\circ g^{-1}\right\|_p\right\|_{\infty,[g(a),g(x_0)]}\int_{a}^{x_0}\|A_2(x)\|_q(g(x_0)-g(x))^{(n+1)\alpha}dx\right]+\right.$$

$$\left[\left\|\left\|\left(D_{x_0+;g}^{(n+1)\alpha}A_1\right)\circ g^{-1}\right\|_p\right\|_{\infty,[g(x_0),g(b)]}\int_{x_0}^{b}\|A_2(x)\|_q(g(x)-g(x_0))^{(n+1)\alpha}dx\right]+ \tag{8.76}$$

$$\left[\left\|\left\|\left(D_{x_0-;g}^{(n+1)\alpha}A_2\right)\circ g^{-1}\right\|_q\right\|_{\infty,[g(a),g(x_0)]}\int_{a}^{x_0}\|A_1(x)\|_p(g(x_0)-g(x))^{(n+1)\alpha}dx\right]+$$

$$\left.\left[\left\|\left\|\left(D_{x_0+;g}^{(n+1)\alpha}A_2\right)\circ g^{-1}\right\|_q\right\|_{\infty,[g(x_0),g(b)]}\int_{x_0}^{b}\|A_1(x)\|_p(g(x)-g(x_0))^{(n+1)\alpha}dx\right]\right\}.$$

Proof Here we have that (acting as in the proof of Theorem 8.18 for $r=2$)

$$\Omega(A_1,A_2)(x_0):=\int_a^b A_2(x)A_1(x)dx+\int_a^b A_1(x)A_2(x)dx-$$

$$\left(\int_a^b A_2(x)dx\right)A_1(x_0)-\left(\int_a^b A_1(x)dx\right)A_2(x_0)\overset{(8.51)}{=}\frac{1}{\Gamma((n+1)\alpha)}$$

$$\left\{\left[\int_a^{x_0}A_2(x)\left(\int_{g(x)}^{g(x_0)}(z-g(x))^{(n+1)\alpha-1}\left(\left(D_{x_0-;g}^{(n+1)\alpha}A_1\right)\circ g^{-1}\right)(z)dz\right)dx\right]+\right.$$

$$\left[\int_{x_0}^{b}A_2(x)\left(\int_{g(x_0)}^{g(x)}(g(x)-z)^{(n+1)\alpha-1}\left(\left(D_{x_0+;g}^{(n+1)\alpha}A_1\right)\circ g^{-1}\right)(z)dz\right)dx\right]+ \tag{8.77}$$

$$\left[\int_a^{x_0}A_1(x)\left(\int_{g(x)}^{g(x_0)}(z-g(x))^{(n+1)\alpha-1}\left(\left(D_{x_0-;g}^{(n+1)\alpha}A_2\right)\circ g^{-1}\right)(z)dz\right)dx\right]+$$

$$\left[\int_{x_0}^{b} A_1(x)\left(\int_{g(x_0)}^{g(x)} (g(x)-z)^{(n+1)\alpha-1}\left(\left(D_{x_0+;g}^{(n+1)\alpha} A_2\right)\circ g^{-1}\right)(z)\,dz\right)dx\right]\Bigg\}.$$

Therefore it holds by taking the 1-Schatten norm that

$$\left\|\Omega(A_1,A_2)(x_0)\right\|_1 = \Bigg\| \int_a^b A_2(x)A_1(x)\,dx + \int_a^b A_1(x)A_2(x)\,dx -$$

$$\left(\int_a^b A_2(x)\,dx\right)A_1(x_0) - \left(\int_a^b A_1(x)\,dx\right)A_2(x_0)\Bigg\|_1 \le \frac{1}{\Gamma((n+1)\alpha)}$$

$$\Bigg\{\left[\left\|\int_a^{x_0} A_2(x)\left(\int_{g(x)}^{g(x_0)} (z-g(x))^{(n+1)\alpha-1}\left(\left(D_{x_0-;g}^{(n+1)\alpha} A_1\right)\circ g^{-1}\right)(z)\,dz\right)dx\right\|_1\right]+$$

$$\left[\left\|\int_{x_0}^{b} A_2(x)\left(\int_{g(x_0)}^{g(x)} (g(x)-z)^{(n+1)\alpha-1}\left(\left(D_{x_0+;g}^{(n+1)\alpha} A_1\right)\circ g^{-1}\right)(z)\,dz\right)dx\right\|_1\right]+$$

$$\left[\left\|\int_a^{x_0} A_1(x)\left(\int_{g(x)}^{g(x_0)} (z-g(x))^{(n+1)\alpha-1}\left(\left(D_{x_0-;g}^{(n+1)\alpha} A_2\right)\circ g^{-1}\right)(z)\,dz\right)dx\right\|_1\right]+$$

$$\left[\left\|\int_{x_0}^{b} A_1(x)\left(\int_{g(x_0)}^{g(x)} (g(x)-z)^{(n+1)\alpha-1}\left(\left(D_{x_0+;g}^{(n+1)\alpha} A_2\right)\circ g^{-1}\right)(z)\,dz\right)dx\right\|_1\right]\Bigg\} \le$$

$$\frac{1}{\Gamma((n+1)\alpha)}$$

$$\Bigg\{\left[\int_a^{x_0}\left\|A_2(x)\left(\int_{g(x)}^{g(x_0)} (z-g(x))^{(n+1)\alpha-1}\left(\left(D_{x_0-;g}^{(n+1)\alpha} A_1\right)\circ g^{-1}\right)(z)\,dz\right)\right\|_1 dx\right]+ \tag{8.78}$$

$$\left[\int_{x_0}^{b}\left\|A_2(x)\left(\int_{g(x_0)}^{g(x)} (g(x)-z)^{(n+1)\alpha-1}\left(\left(D_{x_0+;g}^{(n+1)\alpha} A_1\right)\circ g^{-1}\right)(z)\,dz\right)\right\|_1 dx\right]+$$

$$\left[\int_a^{x_0}\left\|A_1(x)\left(\int_{g(x)}^{g(x_0)} (z-g(x))^{(n+1)\alpha-1}\left(\left(D_{x_0-;g}^{(n+1)\alpha} A_2\right)\circ g^{-1}\right)(z)\,dz\right)\right\|_1 dx\right]+$$

$$\left[\int_{x_0}^{b}\left\|A_1(x)\left(\int_{g(x_0)}^{g(x)} (g(x)-z)^{(n+1)\alpha-1}\left(\left(D_{x_0+;g}^{(n+1)\alpha} A_2\right)\circ g^{-1}\right)(z)\,dz\right)\right\|_1 dx\right]\Bigg\} \le \tag{8.79}$$

(by using the p-Schatten norm and Hölder's type inequality (8.40) for $p, q > 1 : \frac{1}{p} + \frac{1}{q} = 1$)

$$\frac{1}{\Gamma((n+1)\alpha)}$$

$$\left\{\left[\int_a^{x_0} \|A_2(x)\|_q \left\| \left(\int_{g(x)}^{g(x_0)} (z-g(x))^{(n+1)\alpha-1} \left(\left(D_{x_0-;g}^{(n+1)\alpha} A_1 \right) \circ g^{-1} \right)(z)\, dz \right) \right\|_p dx \right] + \right.$$

$$\left[\int_{x_0}^{b} \|A_2(x)\|_q \left\| \left(\int_{g(x_0)}^{g(x)} (g(x)-z)^{(n+1)\alpha-1} \left(\left(D_{x_0+;g}^{(n+1)\alpha} A_1 \right) \circ g^{-1} \right)(z)\, dz \right) \right\|_p dx \right] +$$

$$\left[\int_a^{x_0} \|A_1(x)\|_p \left\| \left(\int_{g(x)}^{g(x_0)} (z-g(x))^{(n+1)\alpha-1} \left(\left(D_{x_0-;g}^{(n+1)\alpha} A_2 \right) \circ g^{-1} \right)(z)\, dz \right) \right\|_q dx \right] +$$

$$\left. \left[\int_{x_0}^{b} \|A_1(x)\|_p \left\| \left(\int_{g(x_0)}^{g(x)} (g(x)-z)^{(n+1)\alpha-1} \left(\left(D_{x_0+;g}^{(n+1)\alpha} A_2 \right) \circ g^{-1} \right)(z)\, dz \right) \right\|_q dx \right] \right\} \leq \tag{8.80}$$

$$\frac{1}{\Gamma((n+1)\alpha)}$$

$$\left\{\left[\int_a^{x_0} \|A_2(x)\|_q \left(\int_{g(x)}^{g(x_0)} (z-g(x))^{(n+1)\alpha-1} \left\| \left(\left(D_{x_0-;g}^{(n+1)\alpha} A_1 \right) \circ g^{-1} \right)(z) \right\|_p dz \right) dx \right] + \right.$$

$$\left[\int_{x_0}^{b} \|A_2(x)\|_q \left(\int_{g(x_0)}^{g(x)} (g(x)-z)^{(n+1)\alpha-1} \left(\left(D_{x_0+;g}^{(n+1)\alpha} A_1 \right) \circ g^{-1} \right)(z) \|_p dz \right) dx \right] +$$

$$\left[\int_a^{x_0} \|A_1(x)\|_p \left(\int_{g(x)}^{g(x_0)} (z-g(x))^{(n+1)\alpha-1} \left\| \left(\left(D_{x_0-;g}^{(n+1)\alpha} A_2 \right) \circ g^{-1} \right)(z) \right\|_q dz \right) dx \right] +$$

$$\left. \left[\int_{x_0}^{b} \|A_1(x)\|_p \left(\int_{g(x_0)}^{g(x)} (g(x)-z)^{(n+1)\alpha-1} \left\| \left(\left(D_{x_0+;g}^{(n+1)\alpha} A_2 \right) \circ g^{-1} \right)(z) \right\|_q dz \right) dx \right] \right\}. \tag{8.81}$$

We have proved, so far, that

$$\|\Omega(A_1, A_2)(x_0)\|_1 \leq \frac{1}{\Gamma((n+1)\alpha+1)}$$

$$\left\{\left[\int_a^{x_0} \|A_2(x)\|_q \left(\int_{g(x)}^{g(x_0)} (z-g(x))^{(n+1)\alpha-1} \left\| \left(\left(D_{x_0-;g}^{(n+1)\alpha} A_1 \right) \circ g^{-1} \right)(z) \right\|_p dz \right) dx \right] + \right.$$

$$\left[\int_{x_0}^{b} \|A_2(x)\|_q \left(\int_{g(x_0)}^{g(x)} (g(x)-z)^{(n+1)\alpha-1} \left\| \left(\left(D_{x_0+;g}^{(n+1)\alpha} A_1 \right) \circ g^{-1} \right)(z) \right\|_p dz \right) dx \right] +$$

$$\left[\int_a^{x_0}\|A_1(x)\|_p\left(\int_{g(x)}^{g(x_0)}(z-g(x))^{(n+1)\alpha-1}\left\|\left(\left(D_{x_0-;g}^{(n+1)\alpha}A_2\right)\circ g^{-1}\right)(z)\right\|_q dz\right)dx\right]+$$

$$\left[\int_{x_0}^{b}\|A_1(x)\|_p\left(\int_{g(x_0)}^{g(x)}(g(x)-z)^{(n+1)\alpha-1}\left\|\left(\left(D_{x_0+;g}^{(n+1)\alpha}A_2\right)\circ g^{-1}\right)(z)\right\|_q dz\right)dx\right]\Bigg\} =: (\lambda). \tag{8.82}$$

Let now $\gamma, \delta > 1$ such that $\frac{1}{\gamma}+\frac{1}{\delta}=1$, and we apply the usual Hölder's inequality in (8.82); $\frac{1}{\delta(n+1)} < \alpha < 1$. Then we have that

$$\|\Omega(A_1,A_2)(x_0)\|_1 \le (\lambda) \le \frac{1}{\Gamma((n+1)\alpha)(\gamma((n+1)\alpha-1)+1)^{\frac{1}{\gamma}}}$$

$$\Bigg\{\left[\int_a^{x_0}\|A_2(x)\|_q(g(x_0)-g(x))^{\frac{\gamma((n+1)\alpha-1)+1}{\gamma}}\left(\int_{g(x)}^{g(x_0)}\left\|\left(\left(D_{x_0-;g}^{(n+1)\alpha}A_1\right)\circ g^{-1}\right)(z)\right\|_p^{\delta}dz\right)^{\frac{1}{\delta}}dx\right]+$$

$$\left[\int_{x_0}^{b}\|A_2(x)\|_q(g(x)-g(x_0))^{\frac{\gamma((n+1)\alpha-1)+1}{\gamma}}\left(\int_{g(x_0)}^{g(x)}\left\|\left(\left(D_{x_0+;g}^{(n+1)\alpha}A_1\right)\circ g^{-1}\right)(z)\right\|_p^{\delta}dz\right)^{\frac{1}{\delta}}dx\right]+$$

$$\left[\int_a^{x_0}\|A_1(x)\|_p(g(x_0)-g(x))^{\frac{\gamma((n+1)\alpha-1)+1}{\gamma}}\left(\int_{g(x)}^{g(x_0)}\left\|\left(\left(D_{x_0-;g}^{(n+1)\alpha}A_2\right)\circ g^{-1}\right)(z)\right\|_q^{\delta}dz\right)^{\frac{1}{\delta}}dx\right]+ \tag{8.83}$$

$$\left[\int_{x_0}^{b}\|A_1(x)\|_p(g(x)-g(x_0))^{\frac{\gamma((n+1)\alpha-1)+1}{\gamma}}\left(\int_{g(x_0)}^{g(x)}\left\|\left(\left(D_{x_0+;g}^{(n+1)\alpha}A_2\right)\circ g^{-1}\right)(z)\right\|_q^{\delta}dz\right)^{\frac{1}{\delta}}dx\right]\Bigg\}$$

$$\le \frac{1}{\Gamma((n+1)\alpha)(\gamma((n+1)\alpha-1)+1)^{\frac{1}{\gamma}}}$$

$$\Bigg\{\left[\left\|\left\|\left(D_{x_0-;g}^{(n+1)\alpha}A_1\right)\circ g^{-1}\right\|_p\right\|_{\delta,[g(a),g(x_0)]}\int_a^{x_0}\|A_2(x)\|_q(g(x_0)-g(x))^{(n+1)\alpha-\frac{1}{\delta}}dx\right]+$$

$$\left[\left\|\left\|\left(D_{x_0+;g}^{(n+1)\alpha}A_1\right)\circ g^{-1}\right\|_p\right\|_{\delta,[g(x_0),g(b)]}\int_{x_0}^{b}\|A_2(x)\|_q(g(x)-g(x_0))^{(n+1)\alpha-\frac{1}{\delta}}dx\right]+ \tag{8.84}$$

$$\left[\left\|\left\|\left(D_{x_0-;g}^{(n+1)\alpha}A_2\right)\circ g^{-1}\right\|_q\right\|_{\delta,[g(a),g(x_0)]}\int_a^{x_0}\|A_1(x)\|_p(g(x_0)-g(x))^{(n+1)\alpha-\frac{1}{\delta}}dx\right]+$$

$$\left[\left\|\left\|\left(D_{x_0+;g}^{(n+1)\alpha}A_2\right)\circ g^{-1}\right\|_q\right\|_{\delta,[g(x_0),g(b)]}\int_{x_0}^{b}\|A_1(x)\|_p(g(x)-g(x_0))^{(n+1)\alpha-\frac{1}{\delta}}dx\right]\Bigg\},$$

proving (8.74).

If $(n+1)\alpha \ge 1$, we obtain

$$\|\Omega(A_1, A_2)(x_0)\|_1 \leq (\lambda) \leq \frac{1}{\Gamma((n+1)\alpha)}$$

$$\left\{\left[\left\|\left\|\left(D_{x_0-;g}^{(n+1)\alpha} A_1\right) \circ g^{-1}\right\|_p\right\|_{L_1([g(a),g(x_0)])} \int_a^{x_0} \|A_2(x)\|_q (g(x_0) - g(x))^{(n+1)\alpha-1} dx\right] +\right.$$

$$\left[\left\|\left\|\left(D_{x_0+;g}^{(n+1)\alpha} A_1\right) \circ g^{-1}\right\|_p\right\|_{L_1([g(x_0),g(b)])} \int_{x_0}^{b} \|A_2(x)\|_q (g(x) - g(x_0))^{(n+1)\alpha-1} dx\right] +$$

$$\left[\left\|\left\|\left(D_{x_0-;g}^{(n+1)\alpha} A_2\right) \circ g^{-1}\right\|_q\right\|_{L_1([g(a),g(x_0)])} \int_a^{x_0} \|A_1(x)\|_p (g(x_0) - g(x))^{(n+1)\alpha-1} dx\right] +$$

$$\left.\left[\left\|\left\|\left(D_{x_0+;g}^{(n+1)\alpha} A_2\right) \circ g^{-1}\right\|_q\right\|_{L_1([g(x_0),g(b)])} \int_{x_0}^{b} \|A_1(x)\|_p (g(x) - g(x_0))^{(n+1)\alpha-1} dx\right]\right\}, \tag{8.85}$$

proving (8.75).

At last we derive

$$\|\Omega(A_1, A_2)(x_0)\|_1 \leq (\lambda) \leq \frac{1}{\Gamma((n+1)\alpha+1)}$$

$$\left\{\left[\left\|\left\|\left(D_{x_0-;g}^{(n+1)\alpha} A_1\right) \circ g^{-1}\right\|_p\right\|_{\infty,[g(a),g(x_0)]} \int_a^{x_0} \|A_2(x)\|_q (g(x_0) - g(x))^{(n+1)\alpha} dx\right] +\right.$$

$$\left[\left\|\left\|\left(D_{x_0+;g}^{(n+1)\alpha} A_1\right) \circ g^{-1}\right\|_p\right\|_{\infty,[g(x_0),g(b)]} \int_{x_0}^{b} \|A_2(x)\|_q (g(x) - g(x_0))^{(n+1)\alpha} dx\right] +$$

$$\left[\left\|\left\|\left(D_{x_0-;g}^{(n+1)\alpha} A_2\right) \circ g^{-1}\right\|_q\right\|_{\infty,[g(a),g(x_0)]} \int_a^{x_0} \|A_1(x)\|_p (g(x_0) - g(x))^{(n+1)\alpha} dx\right] +$$

$$\left.\left[\left\|\left\|\left(D_{x_0+;g}^{(n+1)\alpha} A_2\right) \circ g^{-1}\right\|_q\right\|_{\infty,[g(x_0),g(b)]} \int_{x_0}^{b} \|A_1(x)\|_p (g(x) - g(x_0))^{(n+1)\alpha} dx\right]\right\}, \tag{8.86}$$

proving (8.76).

The theorem is proved. ■

When $r = 2$ and $n = 1$ we obtain the following special operator related sequential fractional Ostrowski type inequalities.

Corollary 8.31 (to Theorem 8.30) *Let $p, q > 1 : \frac{1}{p} + \frac{1}{q} = 1$, and let the $*$-ideals $\mathcal{B}_p(H)$, $\mathcal{B}_q(H)$, for which $\left(\mathcal{B}_p(H), \|\cdot\|_p\right)$, $\left(\mathcal{B}_q(H), \|\cdot\|_q\right)$ are Banach algebras; $x_0 \in [a, b] \subset \mathbb{R}$, $0 < \alpha < 1$, $A_1 \in C^1\left([a, b], \mathcal{B}_p(H)\right)$, $A_2 \in C^1\left([a, b], \mathcal{B}_q(H)\right)$; $g \in C^1([a, b])$, strictly increasing, such that $g^{-1} \in C^1([g(a), g(b)])$. Assume that $F_{ki}^{x_0} := D_{x_0-;g}^{k\alpha} A_i$, for $k = 1, 2$, $i = 1, 2$, fulfill $F_{1,1}^{x_0} \in C^1\left([a, x_0], \mathcal{B}_p(H)\right)$, $F_{2,1}^{x_0} \in$*

$C\left([a,x_0],\mathcal{B}_p(H)\right)$; $F_{1,2}^{x_0}\in C^1\left([a,x_0],\mathcal{B}_q(H)\right)$, $F_{2,2}^{x_0}\in C\left([a,x_0],\mathcal{B}_q(H)\right)$. *Similarly, we assume that* $G_{ki}^{x_0}:=D_{x_0+;g}^{k\alpha}A_i$, $k=1,2$, $i=1,2$, *fulfill* $G_{1,1}^{x_0}\in C^1\left([x_0,b],\mathcal{B}_p(H)\right)$, $G_{2,1}^{x_0}\in C\left([x_0,b],\mathcal{B}_p(H)\right)$, $G_{1,2}^{x_0}\in C^1\left([x_0,b],\mathcal{B}_q(H)\right)$, $G_{2,2}^{x_0}\in C\left([x_0,b],\mathcal{B}_q(H)\right)$.

Then

(1) it holds

$$\Omega(A_1,A_2)(x_0):=\int_a^b A_2(x)A_1(x)\,dx+\int_a^b A_1(x)A_2(x)\,dx-$$
$$\left(\int_a^b A_2(x)\,dx\right)A_1(x_0)-\left(\int_a^b A_1(x)\,dx\right)A_2(x_0)=$$
$$\frac{1}{\Gamma(2\alpha)}\left\{\left[\int_a^{x_0}A_2(x)\left(\int_{g(x)}^{g(x_0)}(z-g(x))^{2\alpha-1}\left(\left(D_{x_0-;g}^{2\alpha}A_1\right)\circ g^{-1}\right)(z)\,dz\right)dx\right]+\right.$$
$$\left[\int_{x_0}^{b}A_2(x)\left(\int_{g(x_0)}^{g(x)}(g(x)-z)^{2\alpha-1}\left(\left(D_{x_0+;g}^{2\alpha}A_1\right)\circ g^{-1}\right)(z)\,dz\right)dx\right]+$$
$$\left[\int_a^{x_0}A_1(x)\left(\int_{g(x)}^{g(x_0)}(z-g(x))^{2\alpha-1}\left(\left(D_{x_0-;g}^{2\alpha}A_2\right)\circ g^{-1}\right)(z)\,dz\right)dx\right]+$$
$$\left.\left[\int_{x_0}^{b}A_1(x)\left(\int_{g(x_0)}^{g(x)}(g(x)-z)^{2\alpha-1}\left(\left(D_{x_0+;g}^{2\alpha}A_2\right)\circ g^{-1}\right)(z)\,dz\right)dx\right]\right\}, \tag{8.87}$$

(2) for $\gamma,\delta>1:\frac{1}{\gamma}+\frac{1}{\delta}=1$, *with* $\frac{1}{2\delta}<\alpha<1$, *we have that*

$$\|\Omega(A_1,A_2)(x_0)\|_1\le\frac{1}{\Gamma(2\alpha)(\gamma(2\alpha-1)+1)^{\frac{1}{\gamma}}}$$
$$\left\{\left[\left\|\left\|\left(D_{x_0-;g}^{2\alpha}A_1\right)\circ g^{-1}\right\|_p\right\|_{\delta,[g(a),g(x_0)]}\int_a^{x_0}\|A_2(x)\|_q(g(x_0)-g(x))^{2\alpha-\frac{1}{\delta}}dx\right]+\right.$$
$$\left[\left\|\left\|\left(D_{x_0+;g}^{2\alpha}A_1\right)\circ g^{-1}\right\|_p\right\|_{\delta,[g(x_0),g(b)]}\int_{x_0}^{b}\|A_2(x)\|_q(g(x)-g(x_0))^{2\alpha-\frac{1}{\delta}}dx\right]+ \tag{8.88}$$
$$\left[\left\|\left\|\left(D_{x_0-;g}^{2\alpha}A_2\right)\circ g^{-1}\right\|_q\right\|_{\delta,[g(a),g(x_0)]}\int_a^{x_0}\|A_1(x)\|_p(g(x_0)-g(x))^{2\alpha-\frac{1}{\delta}}dx\right]+$$
$$\left.\left[\left\|\left\|\left(D_{x_0+;g}^{2\alpha}A_2\right)\circ g^{-1}\right\|_q\right\|_{\delta,[g(x_0),g(b)]}\int_{x_0}^{b}\|A_1(x)\|_p(g(x)-g(x_0))^{2\alpha-\frac{1}{\delta}}dx\right]\right\}$$
$$=:\Lambda_7(x_0),$$

(3) if $\alpha \geq \frac{1}{2}$, we obtain

$$\|\Omega(A_1, A_2)(x_0)\|_1 \leq \frac{1}{\Gamma(2\alpha)}$$

$$\Big\{\Big[\Big\|\big\|\big(D^{2\alpha}_{x_0-;g}A_1\big)\circ g^{-1}\big\|_p\Big\|_{L_1([g(a),g(x_0)])}\int_a^{x_0}\|A_2(x)\|_q\,(g(x_0)-g(x))^{2\alpha-1}\,dx\Big]+$$

$$\Big[\Big\|\big\|\big(D^{2\alpha}_{x_0+;g}A_1\big)\circ g^{-1}\big\|_p\Big\|_{L_1([g(x_0),g(b)])}\int_{x_0}^{b}\|A_2(x)\|_q\,(g(x)-g(x_0))^{2\alpha-1}\,dx\Big]+$$

$$\Big[\Big\|\big\|\big(D^{2\alpha}_{x_0-;g}A_2\big)\circ g^{-1}\big\|_q\Big\|_{L_1([g(a),g(x_0)])}\int_a^{x_0}\|A_1(x)\|_p\,(g(x_0)-g(x))^{2\alpha-1}\,dx\Big]+ \tag{8.89}$$

$$\Big[\Big\|\big\|\big(D^{2\alpha}_{x_0+;g}A_2\big)\circ g^{-1}\big\|_q\Big\|_{L_1([g(x_0),g(b)])}\int_{x_0}^{b}\|A_1(x)\|_p\,(g(x)-g(x_0))^{2\alpha-1}\,dx\Big]\Big\}$$

$$=: \Lambda_8(x_0),$$

and

(4)

$$\|\Omega(A_1, A_2)(x_0)\|_1 \leq \frac{1}{\Gamma(2\alpha+1)}$$

$$\Big\{\Big[\Big\|\big\|\big(D^{2\alpha}_{x_0-;g}A_1\big)\circ g^{-1}\big\|_p\Big\|_{\infty,[g(a),g(x_0)]}\int_a^{x_0}\|A_2(x)\|_q\,(g(x_0)-g(x))^{2\alpha}\,dx\Big]+$$

$$\Big[\Big\|\big\|\big(D^{2\alpha}_{x_0+;g}A_1\big)\circ g^{-1}\big\|_p\Big\|_{\infty,[g(x_0),g(b)]}\int_{x_0}^{b}\|A_2(x)\|_q\,(g(x)-g(x_0))^{2\alpha}\,dx\Big]+$$

$$\Big[\Big\|\big\|\big(D^{2\alpha}_{x_0-;g}A_2\big)\circ g^{-1}\big\|_q\Big\|_{\infty,[g(a),g(x_0)]}\int_a^{x_0}\|A_1(x)\|_p\,(g(x_0)-g(x))^{2\alpha}\,dx\Big]+$$

$$\Big[\Big\|\big\|\big(D^{2\alpha}_{x_0+;g}A_2\big)\circ g^{-1}\big\|_q\Big\|_{\infty,[g(x_0),g(b)]}\int_{x_0}^{b}\|A_1(x)\|_p\,(g(x)-g(x_0))^{2\alpha}\,dx\Big]\Big\} \tag{8.90}$$

$$=: \Lambda_9(x_0).$$

Proof By Theorem 8.30. ∎

We make

Remark 8.32 Let $\Omega(A_1, \ldots, A_r)(x_0)$ as in (8.41). Denote by

$$\Delta\left(A_1,...,A_r\right):=\int_a^b \Omega\left(A_1,...,A_r\right)(x_0)\,dx_0=$$

$$\sum_{i=1}^{r}\left[(b-a)\int_a^b\left(\prod_{\substack{j=1\\ j\neq i}}^{r}A_j(x)\right)A_i(x)\,dx-\left(\int_a^b\left(\prod_{\substack{j=1\\ j\neq i}}^{r}A_j(x)\right)dx\right)\left(\int_a^b A_i(x)\,dx\right)\right],\quad (8.91)$$

$r\in\mathbb{N}-\{1\}$.

In particular, we have that

$$\Delta\left(A_1,A_2\right):=\int_a^b \Omega\left(A_1,A_2\right)(x_0)\,dx_0=$$

$$(b-a)\left(\int_a^b A_2(x)\,A_1(x)\,dx+\int_a^b A_1(x)\,A_2(x)\,dx\right)-\quad (8.92)$$

$$\left(\int_a^b A_2(x)\,dx\right)\left(\int_a^b A_1(x)\,dx\right)-\left(\int_a^b A_1(x)\,dx\right)\left(\int_a^b A_2(x)\,dx\right),$$

Clearly, it holds that

$$\left\|\Delta\left(A_1,...,A_r\right)\right\|_p\leq\int_a^b\left\|\Omega\left(A_1,...,A_r\right)(x_0)\right\|_p dx_0,\quad (8.93)$$

$\forall\, p\geq 1$.

We need

Remark 8.33 (i) Call and assume

$$W_1\left(A_1,...,A_r\right):=\max_{i=1,...,r}\left\{\sup_{x_0\in[a,b]}\left\|\left\|\left(D_{x_0-;g}^{2\alpha}A_i\right)\right\|_2\right\|_{\infty,[a,x_0]},\right.$$

$$\left.\sup_{x_0\in[a,b]}\left\|\left\|\left(D_{x_0+;g}^{2\alpha}A_i\right)\right\|_2\right\|_{\infty,[x_0,b]}\right\}<\infty.\quad (8.94)$$

Hence by (8.63) we obtain

$$\left\|\Omega\left(A_1,...,A_r\right)(x_0)\right\|_1\leq\Lambda_1(x_0)\leq\quad (8.95)$$

$$\frac{W_1\left(A_1,...,A_r\right)(g(b)-g(a))^{2\alpha}}{\Gamma(2\alpha+1)}\sum_{i=1}^{r}\left[\int_a^b\left(\prod_{\substack{j=1\\ j\neq i}}^{r}\left\|A_j(x)\right\|_2\right)dx\right],$$

and

$$\|\Delta(A_1, ..., A_r)\|_1 \le$$

$$\frac{W_1(A_1, ..., A_r)(g(b) - g(a))^{2\alpha}(b-a)}{\Gamma(2\alpha+1)} \sum_{i=1}^{r} \left[\int_a^b \left(\prod_{\substack{j=1 \\ j \neq i}}^{r} \|A_j(x)\|_2 \right) dx \right]. \tag{8.96}$$

(ii) Here $\frac{1}{2} \le \alpha < 1$. Call and assume

$$W_2(A_1, ..., A_r) := \max_{i=1,...,r} \left\{ \sup_{x_0 \in [a,b]} \left\| \left\| \left(D_{x_0-;g}^{2\alpha} A_i \right) \right\|_2 \right\|_{L_1([a,x_0])}, \right.$$

$$\left. \sup_{x_0 \in [a,b]} \left\| \left\| \left(D_{x_0+;g}^{2\alpha} A_i \right) \right\|_2 \right\|_{L_1([x_0,b])} \right\} < \infty. \tag{8.97}$$

Hence by (8.64) we obtain

$$\|\Omega(A_1, ..., A_r)(x_0)\|_1 \le \Lambda_2(x_0) \le$$

$$\frac{\|g'\|_{\infty,[a,b]}}{\Gamma(2\alpha)} W_2(A_1, ..., A_r)(g(b) - g(a))^{2\alpha-1} \sum_{i=1}^{r} \left[\int_a^b \left(\prod_{\substack{j=1 \\ j \neq i}}^{r} \|A_j(x)\|_2 \right) dx \right], \tag{8.98}$$

and

$$\|\Delta(A_1, ..., A_r)\|_1 \le \frac{\|g'\|_{\infty,[a,b]}}{\Gamma(2\alpha)}$$

$$W_2(A_1, ..., A_r)(g(b) - g(a))^{2\alpha-1}(b-a) \sum_{i=1}^{r} \left[\int_a^b \left(\prod_{\substack{j=1 \\ j \neq i}}^{r} \|A_j(x)\|_2 \right) dx \right]. \tag{8.99}$$

(iii) Here $\gamma, \delta > 1 : \frac{1}{\gamma} + \frac{1}{\delta} = 1$, and $\frac{1}{2\delta} < \alpha < 1$. Call and assume

$$W_3(A_1, ..., A_r) := \max_{i=1,...,r} \left\{ \sup_{x_0 \in [a,b]} \left\| \left\| \left(D_{x_0-;g}^{2\alpha} A_i \right) \circ g^{-1} \right\|_2 \right\|_{\delta,[g(a),g(x_0)]}, \right.$$

$$\left. \sup_{x_0 \in [a,b]} \left\| \left\| \left(D_{x_0+;g}^{2\alpha} A_i \right) \circ g^{-1} \right\|_2 \right\|_{\delta,[g(x_0),g(b)]} \right\} < \infty. \tag{8.100}$$

Therefore by (8.65) we get

$$\|\Omega(A_1, ..., A_r)(x_0)\|_1 \leq \Lambda_3(x_0) \leq$$

$$\frac{W_3(A_1, ..., A_r)(g(b) - g(a))^{2\alpha - \frac{1}{\delta}}}{\Gamma(2\alpha)(\gamma(2\alpha - 1) + 1)^{\frac{1}{\gamma}}} \sum_{i=1}^{r} \left[\int_a^b \left(\prod_{\substack{j=1 \\ j \neq i}}^{r} \|A_j(x)\|_2 \right) dx \right], \quad (8.101)$$

and

$$\|\Delta(A_1, ..., A_r)\|_1 \leq$$

$$\frac{W_3(A_1, ..., A_r)(g(b) - g(a))^{2\alpha - \frac{1}{\delta}}(b - a)}{\Gamma(2\alpha)(\gamma(2\alpha - 1) + 1)^{\frac{1}{\gamma}}} \sum_{i=1}^{r} \left[\int_a^b \left(\prod_{\substack{j=1 \\ j \neq i}}^{r} \|A_j(x)\|_2 \right) dx \right]. \quad (8.102)$$

We need

Remark 8.34 (i) Call and assume

$$W_4(A_1, ..., A_r) := \max_{i=1,...,r} \left\{ \sup_{x_0 \in [a,b]} \left\| \left\| \left(D_{x_0-;g}^{2\alpha} A_i \right) \right\|_p \right\|_{\infty,[a,x_0]}, \right.$$

$$\left. \sup_{x_0 \in [a,b]} \left\| \left\| \left(D_{x_0+;g}^{2\alpha} A_i \right) \right\|_p \right\|_{\infty,[x_0,b]} \right\} < \infty. \quad (8.103)$$

Hence by (8.70) we obtain

$$\|\Omega(A_1, ..., A_r)(x_0)\|_p \leq \Lambda_4(x_0) \leq \quad (8.104)$$

$$\frac{W_4(A_1, ..., A_r)(g(b) - g(a))^{2\alpha}}{\Gamma(2\alpha + 1)} \sum_{i=1}^{r} \left[\int_a^b \left(\prod_{\substack{j=1 \\ j \neq i}}^{r} \|A_j(x)\|_p \right) dx \right],$$

and

$$\|\Delta(A_1, ..., A_r)\|_p \leq$$

$$\frac{W_4(A_1, ..., A_r)(g(b) - g(a))^{2\alpha}(b - a)}{\Gamma(2\alpha + 1)} \sum_{i=1}^{r} \left[\int_a^b \left(\prod_{\substack{j=1 \\ j \neq i}}^{r} \|A_j(x)\|_p \right) dx \right], \quad (8.105)$$

(ii) Here $\frac{1}{2} \leq \alpha < 1$. Call and assume

$$W_5(A_1, ..., A_r) := \max_{i=1,...,r} \left\{ \sup_{x_0 \in [a,b]} \left\| \left\| \left(D^{2\alpha}_{x_0-;g} A_i \right) \right\|_p \right\|_{L_1([a,x_0])}, \right.$$

$$\left. \sup_{x_0 \in [a,b]} \left\| \left\| \left(D^{2\alpha}_{x_0+;g} A_i \right) \right\|_p \right\|_{L_1([x_0,b])} \right\} < \infty. \tag{8.106}$$

Hence by (8.71) we obtain

$$\|\Omega(A_1, ..., A_r)(x_0)\|_p \leq \Lambda_5(x_0) \leq$$

$$\frac{\|g'\|_{\infty,[a,b]}}{\Gamma(2\alpha)} W_5(A_1, ..., A_r)(g(b) - g(a))^{2\alpha-1} \sum_{i=1}^{r} \left[\int_a^b \left(\prod_{\substack{j=1 \\ j \neq i}}^{r} \|A_j(x)\|_p \right) dx \right], \tag{8.107}$$

and

$$\|\Delta(A_1, ..., A_r)\|_p \leq \frac{\|g'\|_{\infty,[a,b]}}{\Gamma(2\alpha)}$$

$$W_5(A_1, ..., A_r)(g(b) - g(a))^{2\alpha-1}(b-a) \sum_{i=1}^{r} \left[\int_a^b \left(\prod_{\substack{j=1 \\ j \neq i}}^{r} \|A_j(x)\|_p \right) dx \right]. \tag{8.108}$$

(iii) Here $\gamma, \delta > 1 : \frac{1}{\gamma} + \frac{1}{\delta} = 1$, and $\frac{1}{2\delta} < \alpha < 1$. Call and assume

$$W_6(A_1, ..., A_r) := \max_{i=1,...,r} \left\{ \sup_{x_0 \in [a,b]} \left\| \left\| \left(D^{2\alpha}_{x_0-;g} A_i \right) \circ g^{-1} \right\|_p \right\|_{\delta,[g(a),g(x_0)]}, \right.$$

$$\left. \sup_{x_0 \in [a,b]} \left\| \left\| \left(D^{2\alpha}_{x_0+;g} A_i \right) \circ g^{-1} \right\|_p \right\|_{\delta,[g(x_0),g(b)]} \right\} < \infty. \tag{8.109}$$

Therefore by (8.72) we get

$$\|\Omega(A_1, ..., A_r)(x_0)\|_p \leq \Lambda_6(x_0) \leq$$

$$\frac{W_6(A_1, ..., A_r)(g(b) - g(a))^{2\alpha - \frac{1}{\delta}}}{\Gamma(2\alpha)(\gamma(2\alpha-1)+1)^{\frac{1}{\gamma}}} \sum_{i=1}^{r} \left[\int_a^b \left(\prod_{\substack{j=1 \\ j \neq i}}^{r} \|A_j(x)\|_p \right) dx \right], \tag{8.110}$$

and

$$\|\Delta(A_1, ..., A_r)\|_p \leq$$

$$\frac{W_6(A_1, ..., A_r)(g(b) - g(a))^{2\alpha - \frac{1}{\delta}}(b-a)}{\Gamma(2\alpha)(\gamma(2\alpha - 1) + 1)^{\frac{1}{\gamma}}} \sum_{i=1}^{r} \left[\int_a^b \left(\prod_{\substack{j=1 \\ j \neq i}}^{r} \|A_j(x)\|_p\right) dx\right]. \tag{8.111}$$

We also need

Remark 8.35 (i) Call and assume

$$W_7(A_1, A_2) :=$$

$$\max \left\{ \sup_{x_0 \in [a,b]} \left\| \left\| \left(D^{2\alpha}_{x_0-;g} A_1 \right) \circ g^{-1} \right\|_p \right\|_{\delta, [g(a), g(x_0)]}, \right.$$

$$\sup_{x_0 \in [a,b]} \left\| \left\| \left(D^{2\alpha}_{x_0+;g} A_1 \right) \circ g^{-1} \right\|_p \right\|_{\delta, [g(x_0), g(b)]},$$

$$\sup_{x_0 \in [a,b]} \left\| \left\| \left(D^{2\alpha}_{x_0-;g} A_2 \right) \circ g^{-1} \right\|_q \right\|_{\delta, [g(a), g(x_0)]},$$

$$\left. \sup_{x_0 \in [a,b]} \left\| \left\| \left(D^{2\alpha}_{x_0+;g} A_2 \right) \circ g^{-1} \right\|_q \right\|_{\delta, [g(x_0), g(b)]} \right\} < \infty. \tag{8.112}$$

Let $\gamma, \delta > 1 : \frac{1}{\gamma} + \frac{1}{\delta} = 1$, with $\frac{1}{2\delta} < \alpha < 1$.
Then, by (8.88), we get

$$\|\Omega(A_1, A_2)(x_0)\|_1 \leq \Lambda_7(x_0) \leq$$

$$\frac{W_7(A_1, A_2)(g(b) - g(a))^{2\alpha - \frac{1}{\delta}}}{\Gamma(2\alpha)(\gamma(2\alpha - 1) + 1)^{\frac{1}{\gamma}}} \left[\int_a^b \left(\|A_1(x)\|_p + \|A_2(x)\|_q\right) dx\right], \tag{8.113}$$

and

$$\|\Delta(A_1, A_2)\|_1 \leq$$

$$\frac{W_7(A_1, A_2)(g(b) - g(a))^{2\alpha - \frac{1}{\delta}}(b-a)}{\Gamma(2\alpha)(\gamma(2\alpha - 1) + 1)^{\frac{1}{\gamma}}} \left[\int_a^b \left(\|A_1(x)\|_p + \|A_2(x)\|_q\right) dx\right]. \tag{8.114}$$

(ii) Call and assume

$$W_8(A_1, A_2) :=$$

$$\max\left\{ \sup_{x_0\in[a,b]} \left\| \left\| \left(D^{2\alpha}_{x_0-;g} A_1\right) \circ g^{-1} \right\|_p \right\|_{L_1([g(a), g(x_0)])}, \right.$$

$$\sup_{x_0\in[a,b]} \left\| \left\| \left(D^{2\alpha}_{x_0+;g} A_1\right) \circ g^{-1} \right\|_p \right\|_{L_1([g(x_0), g(b)])},$$

$$\sup_{x_0\in[a,b]} \left\| \left\| \left(D^{2\alpha}_{x_0-;g} A_2\right) \circ g^{-1} \right\|_q \right\|_{L_1([g(a), g(x_0)])},$$

$$\left. \sup_{x_0\in[a,b]} \left\| \left\| \left(D^{2\alpha}_{x_0+;g} A_2\right) \circ g^{-1} \right\|_q \right\|_{L_1([g(x_0), g(b)])} \right\} < \infty. \tag{8.115}$$

If $\alpha \geq \frac{1}{2}$, by (8.89), we get

$$\|\Omega(A_1, A_2)(x_0)\|_1 \leq \Lambda_8(x_0) \leq$$

$$\frac{W_8(A_1, A_2)(g(b) - g(a))^{2\alpha-1}}{\Gamma(2\alpha)} \left[\int_a^b \left(\|A_1(x)\|_p + \|A_2(x)\|_q \right) dx \right], \tag{8.116}$$

and

$$\|\Delta(A_1, A_2)\|_1 \leq$$

$$\frac{W_8(A_1, A_2)(g(b) - g(a))^{2\alpha-1}(b-a)}{\Gamma(2\alpha)} \left[\int_a^b \left(\|A_1(x)\|_p + \|A_2(x)\|_q \right) dx \right]. \tag{8.117}$$

(iii) Call and assume

$$W_9(A_1, A_2) :=$$

$$\max\left\{ \sup_{x_0\in[a,b]} \left\| \left\| \left(D^{2\alpha}_{x_0-;g} A_1\right) \circ g^{-1} \right\|_p \right\|_{\infty,[g(a), g(x_0)]}, \right.$$

$$\sup_{x_0\in[a,b]} \left\| \left\| \left(D^{2\alpha}_{x_0+;g} A_1\right) \circ g^{-1} \right\|_p \right\|_{\infty,[g(x_0), g(b)]},$$

$$\sup_{x_0\in[a,b]} \left\| \left\| \left(D^{2\alpha}_{x_0-;g} A_2\right) \circ g^{-1} \right\|_q \right\|_{\infty,[g(a), g(x_0)]},$$

$$\left. \sup_{x_0\in[a,b]} \left\| \left\| \left(D^{2\alpha}_{x_0+;g} A_2\right) \circ g^{-1} \right\|_q \right\|_{\infty,[g(x_0), g(b)]} \right\} < \infty. \tag{8.118}$$

Therefore, by (8.90), we get

$$\|\Omega(A_1, A_2)(x_0)\|_1 \leq \Lambda_9(x_0) \leq$$

$$\frac{W_9(A_1, A_2)(g(b) - g(a))^{2\alpha}}{\Gamma(2\alpha + 1)} \left[\int_a^b \left(\|A_1(x)\|_p + \|A_2(x)\|_q\right) dx\right], \tag{8.119}$$

and

$$\|\Delta(A_1, A_2)\|_1 \leq$$

$$\frac{W_9(A_1, A_2)(g(b) - g(a))^{2\alpha}(b - a)}{\Gamma(2\alpha + 1)} \left[\int_a^b \left(\|A_1(x)\|_p + \|A_2(x)\|_q\right) dx\right]. \tag{8.120}$$

Based on the Remarks 8.32–8.35 we formulate the following generalized sequential fractional Grüss type inequalities results.

First come 1-Schatten norm generalized sequential fractional Grüss type inequalities involving several functions taking values in the Banach algebra $\mathcal{B}_2(H) \subset \mathcal{B}(H)$:

Theorem 8.36 *Let the $*$-ideal $\mathcal{B}_2(H)$, $0 < \alpha < 1$; $A_i \in C^1([a, b], \mathcal{B}_2(H))$, $i = 1, ..., r \in \mathbb{N} - \{1\}$; $g \in C^1([a, b])$, strictly increasing such that $g^{-1} \in C^1([g(a), g(b)])$. Assume $\forall\ x_0 \in [a, b]$ that $D^{\alpha}_{x_0-;g} A_i \in C^1([a, x_0], \mathcal{B}_2(H))$ and $D^{2\alpha}_{x_0-;g} A_i \in C([a, x_0], \mathcal{B}_2(H))$; $i = 1, ..., r$. Similarly, assume $\forall\ x_0 \in [a, b]$ that $D^{\alpha}_{x_0+;g} A_i \in C^1([x_0, b], \mathcal{B}_2(H))$ and $D^{2\alpha}_{x_0+;g} A_i \in C([x_0, b], \mathcal{B}_2(H))$; $i = 1, ..., r$. Here $\Delta(A_1, ..., A_r)$ is as in (8.91), $W_1(A_1, ..., A_r)$ as in (8.94), $W_2(A_1, ..., A_r)$ is as in (8.97), and $W_3(A_1, ..., A_r)$ is as in (8.100).*

Then

(i)

$$\|\Delta(A_1, ..., A_r)\|_1 \leq \frac{W_1(A_1, ..., A_r)(g(b) - g(a))^{2\alpha}(b - a)}{\Gamma(2\alpha + 1)}$$

$$\sum_{i=1}^{r} \left[\int_a^b \left(\prod_{\substack{j=1 \\ j \neq i}}^{r} \|A_j(x)\|_2\right) dx\right], \tag{8.121}$$

(ii) if $\frac{1}{2} \leq \alpha < 1$, we have

$$\|\Delta(A_1, ..., A_r)\|_1 \leq \frac{\|g'\|_{\infty,[a,b]}}{\Gamma(2\alpha)}$$

$$W_2(A_1, ..., A_r)(g(b) - g(a))^{2\alpha - 1}(b - a) \sum_{i=1}^{r} \left[\int_a^b \left(\prod_{\substack{j=1 \\ j \neq i}}^{r} \|A_j(x)\|_2\right) dx\right], \tag{8.122}$$

(iii) if $\gamma, \delta > 1 : \frac{1}{\gamma} + \frac{1}{\delta} = 1$; $\frac{1}{2\delta} < \alpha < 1$, *we have*

$$\|\Delta(A_1, ..., A_r)\|_1 \leq$$

$$\frac{W_3(A_1, ..., A_r)(g(b) - g(a))^{2\alpha - \frac{1}{\delta}}(b-a)}{\Gamma(2\alpha)(\gamma(2\alpha - 1) + 1)^{\frac{1}{\gamma}}} \sum_{i=1}^{r} \left[\int_a^b \left(\prod_{\substack{j=1 \\ j \neq i}}^{r} \|A_j(x)\|_2 \right) dx \right]. \tag{8.123}$$

We continue with p-Schatten norm generalized sequential fractional Grüss type inequalities involving several functions taking values in the Banach algebra $\mathcal{B}_p(H) \subset \mathcal{B}(H)$, $p > 1$.

Theorem 8.37 *Let* $p > 1$, *the* $*$-*ideal* $\mathcal{B}_p(H)$, $0 < \alpha < 1$; $A_i \in C^1([a, b], \mathcal{B}_p(H))$, $i = 1, ..., r \in \mathbb{N} - \{1\}$; $g \in C^1([a, b])$, *strictly increasing such that* $g^{-1} \in C^1([g(a), g(b)])$. *Assume* $\forall\ x_0 \in [a, b]$ *that* $D^{\alpha}_{x_0-;g} A_i \in C^1([a, x_0], \mathcal{B}_p(H))$ *and* $D^{2\alpha}_{x_0-;g} A_i \in C([a, x_0], \mathcal{B}_p(H))$; $i = 1, ..., r$. *Similarly, assume* $\forall\ x_0 \in [a, b]$ *that* $D^{\alpha}_{x_0+;g} A_i \in C^1([x_0, b], \mathcal{B}_p(H))$ *and* $D^{2\alpha}_{x_0+;g} A_i \in C([x_0, b], \mathcal{B}_p(H))$; $i = 1, ..., r$. *Here* $\Delta(A_1, ..., A_r)$ *is as in (8.91),* $W_4(A_1, ..., A_r)$ *as in (8.103),* $W_5(A_1, ..., A_r)$ *is as in (8.106), and* $W_6(A_1, ..., A_r)$ *is as in (8.109).*

Then

(i)

$$\|\Delta(A_1, ..., A_r)\|_p \leq \frac{W_4(A_1, ..., A_r)(g(b) - g(a))^{2\alpha}(b-a)}{\Gamma(2\alpha + 1)}$$

$$\sum_{i=1}^{r} \left[\int_a^b \left(\prod_{\substack{j=1 \\ j \neq i}}^{r} \|A_j(x)\|_p \right) dx \right], \tag{8.124}$$

(ii) if $\frac{1}{2} \leq \alpha < 1$, *we have*

$$\|\Delta(A_1, ..., A_r)\|_p \leq \frac{\|g'\|_{\infty,[a,b]}}{\Gamma(2\alpha)}$$

$$W_5(A_1, ..., A_r)(g(b) - g(a))^{2\alpha - 1}(b-a) \sum_{i=1}^{r} \left[\int_a^b \left(\prod_{\substack{j=1 \\ j \neq i}}^{r} \|A_j(x)\|_p \right) dx \right], \tag{8.125}$$

(iii) if $\gamma, \delta > 1 : \frac{1}{\gamma} + \frac{1}{\delta} = 1$; $\frac{1}{2\delta} < \alpha < 1$, *we have*

$$\|\Delta (A_1, ..., A_r)\|_p \leq$$

$$\frac{W_6 (A_1, ..., A_r) (g (b) - g (a))^{2\alpha - \frac{1}{\delta}} (b - a)}{\Gamma (2\alpha) (\gamma (2\alpha - 1) + 1)^{\frac{1}{\gamma}}} \sum_{i=1}^{r} \left[\int_a^b \left(\prod_{\substack{j=1 \\ j \neq i}}^{r} \|A_j (x)\|_p \right) dx \right]. \tag{8.126}$$

When $r = 2$ we obtain the following operator related Grüss type sequential fractional inequalities.

Theorem 8.38 *Let* $p, q > 1 : \frac{1}{p} + \frac{1}{q} = 1$, *and let the* $*$*-ideals* $\mathcal{B}_p (H)$, $\mathcal{B}_q (H)$, *for which* $\left(\mathcal{B}_p (H), \|\cdot\|_p\right)$, $\left(\mathcal{B}_q (H), \|\cdot\|_q\right)$ *are Banach algebras;* $0 < \alpha < 1$, $A_1 \in C^1 \left([a, b], \mathcal{B}_p (H)\right)$, $A_2 \in C^1 \left([a, b], \mathcal{B}_q (H)\right)$; $g \in C^1 ([a, b])$, *strictly increasing, such that* $g^{-1} \in C^1 ([g (a), g (b)])$ *. Assume* $\forall\, x_0 \in [a, b]$ *that* $F_{ki}^{x_0} := D_{x_0-;g}^{k\alpha} A_i$, *for* $k = 1, 2$, $i = 1, 2$, *fulfill* $F_{1,1}^{x_0} \in C^1 \left([a, x_0], \mathcal{B}_p (H)\right)$, $F_{2,1}^{x_0} \in C \left([a, x_0], \mathcal{B}_p (H)\right)$; $F_{1,2}^{x_0} \in C^1 \left([a, x_0], \mathcal{B}_q (H)\right)$, $F_{2,2}^{x_0} \in C \left([a, x_0], \mathcal{B}_q (H)\right)$. *Similarly, we assume* $\forall\, x_0 \in [a, b]$ *that* $G_{ki}^{x_0} := D_{x_0+;g}^{k\alpha} A_i$, $k = 1, 2$, $i = 1, 2$, *fulfill* $G_{1,1}^{x_0} \in C^1 \left([x_0, b], \mathcal{B}_p (H)\right)$, $G_{2,1}^{x_0} \in C \left([x_0, b], \mathcal{B}_p (H)\right)$, $G_{1,2}^{x_0} \in C^1 \left([x_0, b], \mathcal{B}_q (H)\right)$, $G_{2,2}^{x_0} \in C \left([x_0, b], \mathcal{B}_q (H)\right)$. *Here* $\Delta (A_1, A_2)$ *is as in (8.92),* $W_7 (A_1, A_2)$ *is as in (8.112),* $W_8 (A_1, A_2)$ *is as in (8.115), and* $W_9 (A_1, A_2)$ *as in (8.118).*

Then

(i) if $\gamma, \delta > 1 : \frac{1}{\gamma} + \frac{1}{\delta} = 1$; $\frac{1}{2\delta} < \alpha < 1$, *we have*

$$\|\Delta (A_1, A_2)\|_1 \leq$$

$$\frac{W_7 (A_1, A_2) (g (b) - g (a))^{2\alpha - \frac{1}{\delta}} (b - a)}{\Gamma (2\alpha) (\gamma (2\alpha - 1) + 1)^{\frac{1}{\gamma}}} \left[\int_a^b \left(\|A_1 (x)\|_p + \|A_2 (x)\|_q\right) dx \right], \tag{8.127}$$

(ii) if $\frac{1}{2} \leq \alpha < 1$, *we have*

$$\|\Delta (A_1, A_2)\|_1 \leq$$

$$\frac{W_8 (A_1, A_2) (g (b) - g (a))^{2\alpha - 1} (b - a)}{\Gamma (2\alpha)} \left[\int_a^b \left(\|A_1 (x)\|_p + \|A_2 (x)\|_q\right) dx \right], \tag{8.128}$$

and

(iii)

$$\|\Delta (A_1, A_2)\|_1 \leq$$

$$\frac{W_9 (A_1, A_2) (g (b) - g (a))^{2\alpha} (b - a)}{\Gamma (2\alpha + 1)} \left[\int_a^b \left(\|A_1 (x)\|_p + \|A_2 (x)\|_q\right) dx \right]. \tag{8.129}$$

8.6 Applications

We give the following sequential fractional Ostrowski inequalities:

Corollary 8.39 (to Corollary 8.31) *All as in Corollary 8.31, with* $g(t) = e^t$. *Then*
(1) for $\gamma, \delta > 1 : \frac{1}{\gamma} + \frac{1}{\delta} = 1$; $\frac{1}{2\delta} < \alpha < 1$, *we have*

$$\|\Omega(A_1, A_2)(x_0)\|_1 \leq \frac{1}{\Gamma(2\alpha)(\gamma(2\alpha - 1) + 1)^{\frac{1}{\gamma}}}$$

$$\left\{\left[\left\|\left\|\left(D^{2\alpha}_{x_0-;e^t} A_1\right) \circ \log\right\|_p\right\|_{\delta,[e^a,e^{x_0}]} \int_a^{x_0} \|A_2(x)\|_q \left(e^{x_0} - e^x\right)^{2\alpha - \frac{1}{\delta}} dx\right] + \right. \tag{8.130}$$

$$\left[\left\|\left\|\left(D^{2\alpha}_{x_0+;e^t} A_1\right) \circ \log\right\|_p\right\|_{\delta,[e^{x_0},e^b]} \int_{x_0}^{b} \|A_2(x)\|_q \left(e^x - e^{x_0}\right)^{2\alpha - \frac{1}{\delta}} dx\right] +$$

$$\left[\left\|\left\|\left(D^{2\alpha}_{x_0-;e^t} A_2\right) \circ \log\right\|_q\right\|_{\delta,[e^a,e^{x_0}]} \int_a^{x_0} \|A_1(x)\|_p \left(e^{x_0} - e^x\right)^{2\alpha - \frac{1}{\delta}} dx\right] +$$

$$\left.\left[\left\|\left\|\left(D^{2\alpha}_{x_0+;e^t} A_2\right) \circ \log\right\|_q\right\|_{\delta,[e^{x_0},e^b]} \int_{x_0}^{b} \|A_1(x)\|_p \left(e^x - e^{x_0}\right)^{2\alpha - \frac{1}{\delta}} dx\right]\right\},$$

(2) if $\frac{1}{2} \leq \alpha < 1$, *we have*

$$\|\Omega(A_1, A_2)(x_0)\|_1 \leq \frac{1}{\Gamma(2\alpha)}$$

$$\left\{\left[\left\|\left\|\left(D^{2\alpha}_{x_0-;e^t} A_1\right) \circ \log\right\|_p\right\|_{L_1([e^a,e^{x_0}])} \int_a^{x_0} \|A_2(x)\|_q \left(e^{x_0} - e^x\right)^{2\alpha - 1} dx\right] + \right. \tag{8.131}$$

$$\left[\left\|\left\|\left(D^{2\alpha}_{x_0+;e^t} A_1\right) \circ \log\right\|_p\right\|_{L_1([e^{x_0},e^b])} \int_{x_0}^{b} \|A_2(x)\|_q \left(e^x - e^{x_0}\right)^{2\alpha - 1} dx\right] +$$

$$\left[\left\|\left\|\left(D^{2\alpha}_{x_0-;e^t} A_2\right) \circ \log\right\|_q\right\|_{L_1([e^a,e^{x_0}])} \int_a^{x_0} \|A_1(x)\|_p \left(e^{x_0} - e^x\right)^{2\alpha - 1} dx\right] +$$

$$\left.\left[\left\|\left\|\left(D^{2\alpha}_{x_0+;e^t} A_2\right) \circ \log\right\|_q\right\|_{L_1([e^{x_0},e^b])} \int_{x_0}^{b} \|A_1(x)\|_p \left(e^x - e^{x_0}\right)^{2\alpha - 1} dx\right]\right\},$$

and

(3)

$$\|\Omega(A_1, A_2)(x_0)\|_1 \leq \frac{1}{\Gamma(2\alpha+1)}$$

$$\left\{\left[\left\|\left\|\left(D^{2\alpha}_{x_0-;e^t} A_1\right)\circ \log\right\|_p\right\|_{\infty,[e^a,e^{x_0}]} \int_a^{x_0} \|A_2(x)\|_q \left(e^{x_0}-e^x\right)^{2\alpha} dx\right]+\right. \tag{8.132}$$

$$\left[\left\|\left\|\left(D^{2\alpha}_{x_0+;e^t} A_1\right)\circ \log\right\|_p\right\|_{\infty,[e^{x_0},e^b]} \int_{x_0}^{b} \|A_2(x)\|_q \left(e^{x}-e^{x_0}\right)^{2\alpha} dx\right]+$$

$$\left[\left\|\left\|\left(D^{2\alpha}_{x_0-;e^t} A_2\right)\circ \log\right\|_q\right\|_{\infty,[e^a,e^{x_0}]} \int_a^{x_0} \|A_1(x)\|_p \left(e^{x_0}-e^x\right)^{2\alpha} dx\right]+$$

$$\left.\left[\left\|\left\|\left(D^{2\alpha}_{x_0+;e^t} A_2\right)\circ \log\right\|_q\right\|_{\infty,[e^{x_0},e^b]} \int_{x_0}^{b} \|A_1(x)\|_p \left(e^{x}-e^{x_0}\right)^{2\alpha} dx\right]\right\}.$$

We continue with

Corollary 8.40 (to Corollary 8.31) *All as in Corollary 8.31, with $g(t) = t$. Then (1) for $\gamma, \delta > 1 : \frac{1}{\gamma} + \frac{1}{\delta} = 1$; $\frac{1}{2\delta} < \alpha < 1$, we have*

$$\|\Omega(A_1, A_2)(x_0)\|_1 \leq \frac{1}{\Gamma(2\alpha)\left(\gamma(2\alpha-1)+1\right)^{\frac{1}{\gamma}}}$$

$$\left\{\left[\left\|\left\|D^{2\alpha}_{x_0-} A_1\right\|_p\right\|_{\delta,[a,x_0]} \int_a^{x_0} \|A_2(x)\|_q (x_0-x)^{2\alpha-\frac{1}{\delta}} dx\right]+\right. \tag{8.133}$$

$$\left[\left\|\left\|D^{2\alpha}_{*x_0} A_1\right\|_p\right\|_{\delta,[x_0,b]} \int_{x_0}^{b} \|A_2(x)\|_q (x-x_0)^{2\alpha-\frac{1}{\delta}} dx\right]+$$

$$\left[\left\|\left\|D^{2\alpha}_{x_0-} A_2\right\|_q\right\|_{\delta,[a,x_0]} \int_a^{x_0} \|A_1(x)\|_p (x_0-x)^{2\alpha-\frac{1}{\delta}} dx\right]+$$

$$\left.\left[\left\|\left\|D^{2\alpha}_{*x_0} A_2\right\|_q\right\|_{\delta,[x_0,b]} \int_{x_0}^{b} \|A_1(x)\|_p (x-x_0)^{2\alpha-\frac{1}{\delta}} dx\right]\right\},$$

(2) if $\frac{1}{2} \leq \alpha < 1$, we have

$$\|\Omega(A_1, A_2)(x_0)\|_1 \leq \frac{1}{\Gamma(2\alpha)}$$

$$\left\{\left[\left\|\left\|D^{2\alpha}_{x_0-} A_1\right\|_p\right\|_{L_1([a,x_0])} \int_a^{x_0} \|A_2(x)\|_q (x_0-x)^{2\alpha-1} dx\right]+\right. \tag{8.134}$$

$$\left[\left\| \left\| D_{*x_0}^{2\alpha} A_1 \right\|_p \right\|_{L_1([x_0,b])} \int_{x_0}^{b} \|A_2(x)\|_q (x - x_0)^{2\alpha-1} dx\right] +$$

$$\left[\left\| \left\| D_{x_0-}^{2\alpha} A_2 \right\|_q \right\|_{L_1([a,x_0])} \int_{a}^{x_0} \|A_1(x)\|_p (x_0 - x)^{2\alpha-1} dx\right] +$$

$$\left[\left\| \left\| D_{*x_0}^{2\alpha} A_2 \right\|_q \right\|_{L_1([x_0,b])} \int_{x_0}^{b} \|A_1(x)\|_p (x - x_0)^{2\alpha-1} dx\right]\Bigg\},$$

and

(3)

$$\|\Omega(A_1, A_2)(x_0)\|_1 \leq \frac{1}{\Gamma(2\alpha+1)}$$

$$\Bigg\{\left[\left\| \left\| D_{x_0-}^{2\alpha} A_1 \right\|_p \right\|_{\infty,[a,x_0]} \int_{a}^{x_0} \|A_2(x)\|_q (x_0 - x)^{2\alpha} dx\right] + \tag{8.135}$$

$$\left[\left\| \left\| D_{*x_0}^{2\alpha} A_1 \right\|_p \right\|_{\infty,[x_0,b]} \int_{x_0}^{b} \|A_2(x)\|_q (x - x_0)^{2\alpha} dx\right] +$$

$$\left[\left\| \left\| D_{x_0-}^{2\alpha} A_2 \right\|_q \right\|_{\infty,[a,x_0]} \int_{a}^{x_0} \|A_1(x)\|_p (x_0 - x)^{2\alpha} dx\right] +$$

$$\left[\left\| \left\| D_{*x_0}^{2\alpha} A_2 \right\|_q \right\|_{\infty,[x_0,b]} \int_{x_0}^{b} \|A_1(x)\|_p (x - x_0)^{2\alpha} dx\right]\Bigg\}.$$

Next come sequential fractional Grüss inequalities:

Corollary 8.41 (to Theorem 8.38) *All as in Theorem 8.38, with $g(t) = e^t$. Then*

(1) if $\gamma, \delta > 1 : \frac{1}{\gamma} + \frac{1}{\delta} = 1$, $\frac{1}{2\delta} < \alpha < 1$, we have

$$\|\Delta(A_1, A_2)\|_1 \leq \frac{W_7(A_1, A_2)\left(e^b - e^a\right)^{2\alpha - \frac{1}{\delta}}(b-a)}{\Gamma(2\alpha)(\gamma(2\alpha-1)+1)^{\frac{1}{\gamma}}}$$

$$\left[\int_a^b \left(\|A_1(x)\|_p + \|A_2(x)\|_q\right) dx\right], \tag{8.136}$$

(2) if $\frac{1}{2} \leq \alpha < 1$, we have

$$\|\Delta(A_1, A_2)\|_1 \leq \frac{W_8(A_1, A_2)\left(e^b - e^a\right)^{2\alpha-1}(b-a)}{\Gamma(2\alpha)}$$

$$\left[\int_a^b \left(\|A_1(x)\|_p + \|A_2(x)\|_q\right) dx\right], \tag{8.137}$$

and

(3)

$$\|\Delta(A_1, A_2)\|_1 \leq \frac{W_9(A_1, A_2)\left(e^b - e^a\right)^{2\alpha}(b-a)}{\Gamma(2\alpha+1)}$$

$$\left[\int_a^b \left(\|A_1(x)\|_p + \|A_2(x)\|_q\right) dx\right], \tag{8.138}$$

We finish with

Corollary 8.42 (to Theorem 8.38) *All as in Theorem 8.38, with $g(t) = t$. Then*

(1) if $\gamma, \delta > 1 : \frac{1}{\gamma} + \frac{1}{\delta} = 1$, $\frac{1}{2\delta} < \alpha < 1$, we have

$$\|\Delta(A_1, A_2)\|_1 \leq \frac{W_7(A_1, A_2)(b-a)^{2\alpha+\frac{1}{\gamma}}}{\Gamma(2\alpha)(\gamma(2\alpha-1)+1)^{\frac{1}{\gamma}}}$$

$$\left[\int_a^b \left(\|A_1(x)\|_p + \|A_2(x)\|_q\right) dx\right], \tag{8.139}$$

(2) if $\frac{1}{2} \leq \alpha < 1$, we have

$$\|\Delta(A_1, A_2)\|_1 \leq \frac{W_8(A_1, A_2)(b-a)^{2\alpha}}{\Gamma(2\alpha)}$$

$$\left[\int_a^b \left(\|A_1(x)\|_p + \|A_2(x)\|_q\right) dx\right], \tag{8.140}$$

and

(3)

$$\|\Delta(A_1, A_2)\|_1 \leq \frac{W_9(A_1, A_2)(b-a)^{2\alpha+1}}{\Gamma(2\alpha+1)}$$

$$\left[\int_a^b \left(\|A_1(x)\|_p + \|A_2(x)\|_q\right) dx\right]. \tag{8.141}$$

References

1. Aliprantis, C., Border, K.: Infinite Dimensional Analysis, 3rd edn. Springer, New York (2006)
2. Anastassiou, G.A.: Fractional Differentiation Inequalities. Research Monograph. Springer, New York (2009)
3. Anastassiou, G.A.: Advances on Fractional Inequalities. Research Monograph. Springer, New York (2011)

4. Anastassiou, G.A.: Intelligent Comparisons: Analytic Inequalities. Springer, Heidelberg (2016)
5. Anastassiou, G.A.: Intelligent Computations: Abstract Fractional Calculus, Inequalities, Approximations. Springer, Heidelberg (2018)
6. Anastassiou, G.A.: p-Schatten norm sequential generalized fractional Ostrowski and Grüss type inequalities for several functions. Submitted (2021)
7. Bellman, R.: Some inequalities for positive definite matrices. In: Beckenbach, E.F. (ed.), General Inequalities 2, Proceedings of the 2nd International Conference on General Inequalities, pp. 89–90. Birkhauser, Basel (1980)
8. Čebyšev: Sur les expressions approximatives des intégrales définies par les aures prises entre les mêmes limites. Proc. Math. Soc. Charkov **2**), 93–98 (1882
9. Chang, D.: A matrix trace inequality for products of Hermitian matrices. J. Math. Anal. Appl. **237**, 721–725 (1999)
10. Coop, I.D.: On matrix trace inequalities and related topics for products of Hermitian matrix. J. Math. Anal. Appl. **188**, 999–1001 (1994)
11. Dragomir, S.S.: p-Schatten norm inequalities of Ostrowski's type. RGMIA Res. Rep. Coll. **24**, Art. 108, 19 pp (2021)
12. Dragomir, S.S.: p-Schatten norm inequalities of Grüss type. RGMIA Res. Rep. Coll. **24**, Art. 115, 16 pp (2021)
13. Mikusinski, J.: The Bochner Integral. Academic, New York (1978)
14. Neudecker, H.: A matrix trace inequality. J. Math. Anal. Appl. **166**, 302–303 (1992)
15. Ostrowski, A.: Über die Absolutabweichung einer differentiabaren Funcktion von ihrem Integralmittelwert. Comment. Math. Helv. **10**, 226–227 (1938)
16. Rudin, W.: Functional Analysis, 2nd edn. McGraw-Hill Inc, New York (1991)
17. Simon, B.: Trace Ideals and Their Applications. Cambridge University Press, Cambridge (1979)
18. Zagrebvov, V.A.: Gibbs Semigroups. Operator Theory: Advances and Applications, vol. 273. Birkhauser (2019)

Chapter 9
p-Schatten Norm Generalized Canavati Fractional Ostrowski, Opial and Grüss Inequalities for Multiple Functions

Using generalized Canavati fractional left and right vectorial Taylor formulae we establish generalized fractional Ostrowski, Opial and Grüss inequalities for several functions that take values in the von Neumann-Schatten class $\mathcal{B}_p(H)$, $1 \leq p < \infty$. The estimates are with respect to all p-Schatten norms, $1 \leq p < \infty$. We finish with applications. It follows [7].

9.1 Introduction

The following results inspire this chapter.

Theorem 9.1 (1938, Ostrowski [17]) *Let $f : [a, b] \to \mathbb{R}$ be continuous on $[a, b]$ and differentiable on (a, b) whose derivative $f' : (a, b) \to \mathbb{R}$ is bounded on (a, b), i.e., $\|f'\|_{\infty}^{\sup} := \sup_{t \in (a,b)} |f'(t)| < +\infty$. Then*

$$\left| \frac{1}{b-a} \int_a^b f(t)\, dt - f(x) \right| \leq \left[\frac{1}{4} + \frac{\left(x - \frac{a+b}{2}\right)^2}{(b-a)^2} \right] (b-a) \|f'\|_{\infty}^{\sup}, \tag{9.1}$$

for any $x \in [a, b]$. The constant $\frac{1}{4}$ is the best possible.

Ostrowski type inequalities have great applications to integral approximations in Numerical Analysis.

We mention

Theorem 9.2 (1882, Čebyšev [9]) *Let $f, g : [a, b] \to \mathbb{R}$ be absolutely continuous functions with $f', g' \in L_{\infty}([a, b])$. Then*

G. A. Anastassiou, *Fractional Inequalities In Banach Algebras*,
Studies in Systems, Decision and Control 441,
https://doi.org/10.1007/978-3-031-05148-7_9

$$\left|\frac{1}{b-a}\int_a^b f(x)\,g(x)\,dx-\left(\frac{1}{b-a}\int_a^b f(x)\,dx\right)\left(\frac{1}{b-a}\int_a^b g(x)\,dx\right)\right|$$
$$\le\frac{1}{12}(b-a)^2\left\|f'\right\|_\infty\left\|g'\right\|_\infty. \tag{9.2}$$

The above integrals are assumed to exist.

The related Grüss type inequalities have many applications to Probability Theory. We presented also ([3], Chaps. 8 and 9) mixed fractional Ostrowski and Grüss-Cebysev type inequalities for several functions, acting to all possible directions. The estimates involve the left and right Caputo fractional derivatives. See also the monographs written by the author [1], Chaps. 24–26 and [2], Chaps. 2–6.

We are motivated also by Dragomir [12] recent work:

An operator $A\in\mathcal{B}(H)$ is said to belong to the von Neumann-Schatten class $\mathcal{B}_p(H)$, $1\le p<\infty$ if the p-Schatten norm is finite

$$\|A\|_p:=\left[tr\left(|A|^p\right)\right]^{\frac{1}{p}}<\infty.$$

Assume that $A:[a,b]\to\mathcal{B}_p(H)$, $B:[a,b]\to\mathcal{B}_q(H)$, $p,q>1$ with $\frac{1}{p}+\frac{1}{q}=1$, are continuous and B is strongly differentiable on (a,b), then

$$\left\|\int_a^b A(t)\,B(t)\,dt-\left(\int_a^b A(s)\,ds\right)B(u)\right\|_1\le$$

$$\sup_{t\in[a,b]}\left\|B'(t)\right\|_q\times\begin{cases}\left[\frac{1}{2}(b-a)+\left|u-\frac{a+b}{2}\right|\right]\int_a^b\|A(t)\|_p\,dt,\\[2ex]\left[\frac{(u-a)^{\beta+1}+(b-u)^{\beta+1}}{\beta+1}\right]^{\frac{1}{\beta}}\left(\int_a^b\|A(t)\|_p^\alpha\right)^{\frac{1}{\alpha}},\\ \text{for }\alpha,\beta>1\text{ with }\frac{1}{\alpha}+\frac{1}{\beta}=1\\[2ex]\left[\frac{1}{4}(b-a)^2+\left(u-\frac{a+b}{2}\right)^2\right]\sup_{t\in[a,b]}\|A(t)\|_p,\end{cases} \tag{9.3}$$

for all $u\in[a,b]$, an Ostrowski type inequality.

Further inspiration comes from Dragomir [13] recent work on Grüss inequalities:

For two continuous functions $A,B:[a,b]\to\mathcal{B}(H)$ we define the noncommutative Cebysev fractional

$$D(A,B):=(b-a)\int_a^b A(t)\,B(t)\,dt-\int_a^b A(t)\,dt\int_a^b B(t)\,dt.$$

If $p,q>1$ with $\frac{1}{p}+\frac{1}{q}=1$, let $A:[a,b]\to\mathcal{B}_p(H)$, $B:[a,b]\to\mathcal{B}_q(H)$ be strongly differentiable functions on the interval (a,b), then

$$\|D(A,B)\|_1 \le D\left(\int_a^b \|A'(u)\|_p du, \int_a^b \|B'(u)\|_q du\right) \le \tag{9.4}$$

$$\frac{1}{4}(b-a)^2 \int_a^b \|A'(u)\|_p du \int_a^b \|B'(u)\|_q du.$$

We are also inspired by Opial [16], 1960, famous inequality.

Theorem 9.3 *Let $x(t) \in C^1([0,h])$ be such that $x(0) = x(h) = 0$, and $x(t) > 0$ in $(0,h)$. Then*

$$\int_0^h |x(t)x'(t)|\,dt \le \frac{h}{4}\int_0^h (x'(t))^2\,dt. \tag{9.5}$$

In (9.5), the constant $\frac{h}{4}$ is the best possible. Inequality (9.5) holds as equality for the optimal function

$$x(t) = \begin{cases} ct, & 0 \le t \le \frac{h}{2}, \\ c(h-t), & \frac{h}{2} \le t \le h, \end{cases} \tag{9.6}$$

where $c > 0$ is an arbitrary constant.

Opial-type inequalities are used a lot in proving uniqueness of solutions to differential equations and also to give upper bounds to their solutions.

For an extensive study about fractional Opial inequalities see the author's monograph [1].

In this chapter we generalize [3], Ch. 8,9 for several Banach algebra $\mathcal{B}_p(H)$ valued functions, in the sense of developing fractional Ostrowski, Opial and Grüss type inequalities. Now our left and right generalized Canavati fractional derivatives are for Banach space valued functions and our integrals are of Bochner type [14]. Applications finish the chapter.

9.2 Background on Vectorial Generalized Canavati Fractional Calculus

All in this section come from [5], pp. 109–115 and [4].

Let $g : [a,b] \to \mathbb{R}$ be a strictly increasing function. such that $g \in C^1([a,b])$, and $g^{-1} \in C^n([g(a), g(b)])$, $n \in \mathbb{N}$, $(X, \|\cdot\|)$ is a Banach space. Let $f \in C^n([a,b], X)$, and call $l := f \circ g^{-1} : [g(a), g(b)] \to X$. It is clear that $l, l', \ldots, l^{(n)}$ are continuous functions from $[g(a), g(b)]$ into $f([a,b]) \subseteq X$.

Let $\nu \ge 1$ such that $[\nu] = n$, $n \in \mathbb{N}$ as above, where $[\cdot]$ is the integral part of the number.

Clearly when $0 < \nu < 1$, $[\nu] = 0$.

(I) Let $h \in C([g(a), g(b)], X)$, we define the left Riemann-Liouville Bochner fractional integral as

$$\left(J_{\nu}^{z_0}h\right)(z) := \frac{1}{\Gamma(\nu)}\int_{z_0}^{z}(z-t)^{\nu-1}h(t)\,dt, \tag{9.7}$$

for $g(a) \leq z_0 \leq z \leq g(b)$, where Γ is the gamma function; $\Gamma(\nu) = \int_0^{\infty} e^{-t}t^{\nu-1}dt$. We set $J_0^{z_0}h = h$.

Let $\alpha := \nu - [\nu]$ $(0 < \alpha < 1)$. We define the subspace $C_{g(x_0)}^{\nu}([g(a), g(b)], X)$ of $C^{[\nu]}([g(a), g(b)], X)$, where $x_0 \in [a, b]$ as:

$$C_{g(x_0)}^{\nu}([g(a), g(b)], X) =$$

$$\left\{h \in C^{[\nu]}([g(a), g(b)], X) : J_{1-\alpha}^{g(x_0)}h^{([\nu])} \in C^1([g(x_0), g(b)], X)\right\}. \tag{9.8}$$

So let $h \in C_{g(x_0)}^{\nu}([g(a), g(b)], X)$, we define the left g-generalized X-valued fractional derivative of h of order ν, of Canavati type, over $[g(x_0), g(b)]$ as

$$D_{g(x_0)}^{\nu}h := \left(J_{1-\alpha}^{g(x_0)}h^{([\nu])}\right)'. \tag{9.9}$$

Clearly, for $h \in C_{g(x_0)}^{\nu}([g(a), g(b)], X)$, there exists

$$\left(D_{g(x_0)}^{\nu}h\right)(z) = \frac{1}{\Gamma(1-\alpha)}\frac{d}{dz}\int_{g(x_0)}^{z}(z-t)^{-\alpha}h^{([\nu])}(t)\,dt, \tag{9.10}$$

for all $g(x_0) \leq z \leq g(b)$.

In particular, when $f \circ g^{-1} \in C_{g(x_0)}^{\nu}([g(a), g(b)], X)$, we have that

$$\left(D_{g(x_0)}^{\nu}\left(f \circ g^{-1}\right)\right)(z) = \frac{1}{\Gamma(1-\alpha)}\frac{d}{dz}\int_{g(x_0)}^{z}(z-t)^{-\alpha}\left(f \circ g^{-1}\right)^{([\nu])}(t)\,dt, \tag{9.11}$$

for all $g(x_0) \leq z \leq g(b)$. We have that $D_{g(x_0)}^{n}\left(f \circ g^{-1}\right) = \left(f \circ g^{-1}\right)^{(n)}$ and $D_{g(x_0)}^{0}\left(f \circ g^{-1}\right) = f \circ g^{-1}$, see [4].

By [4], we have for $f \circ g^{-1} \in C_{g(x_0)}^{\nu}([g(a), g(b)], X)$, where $x_0 \in [a, b]$ the following left generalized g-fractional, of Canavati type, X-valued Taylor's formula:

Theorem 9.4 *Let $f \circ g^{-1} \in C_{g(x_0)}^{\nu}([g(a), g(b)], X)$, where $x_0 \in [a, b]$ is fixed.*

(i) If $\nu \geq 1$, then

$$f(x) - f(x_0) = \sum_{k=1}^{[\nu]-1}\frac{\left(f \circ g^{-1}\right)^{(k)}(g(x_0))}{k!}(g(x) - g(x_0))^k +$$

$$\frac{1}{\Gamma(\nu)}\int_{g(x_0)}^{g(x)}(g(x)-t)^{\nu-1}\left(D_{g(x_0)}^{\nu}\left(f \circ g^{-1}\right)\right)(t)\,dt, \tag{9.12}$$

for all $x_0 \leq x \leq b$.

(ii) If $0 < \nu < 1$, *we get*

$$f(x) = \frac{1}{\Gamma(\nu)} \int_{g(x_0)}^{g(x)} (g(x) - t)^{\nu-1} \left(D_{g(x_0)}^{\nu} \left(f \circ g^{-1}\right)\right)(t)\, dt, \tag{9.13}$$

for all $x_0 \leq x \leq b$.

(II) Let $h \in C([g(a), g(b)], X)$, we define the right Riemann-Liouville Bochner fractional integral as

$$\left(J_{z_0-}^{\nu} h\right)(z) := \frac{1}{\Gamma(\nu)} \int_{z}^{z_0} (t - z)^{\nu-1} h(t)\, dt, \tag{9.14}$$

for $g(a) \leq z \leq z_0 \leq g(b)$. We set $J_{z_0-}^{0} h = h$.

Let $\alpha := \nu - [\nu]$ $(0 < \alpha < 1)$. We define the subspace $C_{g(x_0)-}^{\nu}([g(a)[, g(b)], X)$ of $C^{[\nu]}([g(a), g(b)], X)$, where $x_0 \in [a, b]$ as:

$$C_{g(x_0)-}^{\nu}([g(a), g(b)], X) :=$$

$$\left\{h \in C^{[\nu]}([g(a), g(b)], X) : J_{g(x_0)-}^{1-\alpha} h^{([\nu])} \in C^1([g(a), g(x_0)], X)\right\}. \tag{9.15}$$

So let $h \in C_{g(x_0)-}^{\nu}([g(a), g(b)], X)$, we define the right g-generalized X-valued fractional derivative of h of order ν, of Canavati type, over $[g(a), g(x_0)]$ as

$$D_{g(x_0)-}^{\nu} h := (-1)^{n-1} \left(J_{g(x_0)-}^{1-\alpha} h^{([\nu])}\right)'. \tag{9.16}$$

Clearly, for $h \in C_{g(x_0)-}^{\nu}([g(a), g(b)], X)$, there exists

$$\left(D_{g(x_0)-}^{\nu} h\right)(z) = \frac{(-1)^{n-1}}{\Gamma(1-\alpha)} \frac{d}{dz} \int_{z}^{g(x_0)} (t - z)^{-\alpha} h^{([\nu])}(t)\, dt, \tag{9.17}$$

for all $g(a) \leq z \leq g(x_0) \leq g(b)$.

In particular, when $f \circ g^{-1} \in C_{g(x_0)-}^{\nu}([g(a), g(b)], X)$, we have that

$$\left(D_{g(x_0)-}^{\nu} \left(f \circ g^{-1}\right)\right)(z) = \frac{(-1)^{n-1}}{\Gamma(1-\alpha)} \frac{d}{dz} \int_{z}^{g(x_0)} (t - z)^{-\alpha} \left(f \circ g^{-1}\right)^{([\nu])}(t)\, dt, \tag{9.18}$$

for all $g(a) \leq z \leq g(x_0) \leq g(b)$.

We get that

$$\left(D_{g(x_0)-}^{n} \left(f \circ g^{-1}\right)\right)(z) = (-1)^n \left(f \circ g^{-1}\right)^{(n)}(z) \tag{9.19}$$

and $\left(D_{g(x_0)-}^{0} \left(f \circ g^{-1}\right)\right)(z) = \left(f \circ g^{-1}\right)(z)$, all $z \in [g(a), g(b)]$, see [4].

By [4], we have for $f \circ g^{-1} \in C^{\nu}_{g(x_0)-}([g(a), g(b)], X)$, where $x_0 \in [a, b]$ is fixed, the following right generalized g-fractional, of Canavati type, X-valued Taylor's formula:

Theorem 9.5 *Let $f \circ g^{-1} \in C^{\nu}_{g(x_0)-}([g(a), g(b)], X)$, where $x_0 \in [a, b]$ is fixed.*
(i) If $\nu \geq 1$, then

$$f(x) - f(x_0) = \sum_{k=1}^{[\nu]-1} \frac{\left(f \circ g^{-1}\right)^{(k)}(g(x_0))}{k!} (g(x) - g(x_0))^k +$$

$$\frac{1}{\Gamma(\nu)} \int_{g(x)}^{g(x_0)} (t - g(x))^{\nu-1} \left(D^{\nu}_{g(x_0)-}\left(f \circ g^{-1}\right)\right)(t)\, dt, \tag{9.20}$$

for all $a \leq x \leq x_0$,
(ii) If $0 < \nu < 1$, we get

$$f(x) = \frac{1}{\Gamma(\nu)} \int_{g(x)}^{g(x_0)} (t - g(x))^{\nu-1} \left(D^{\nu}_{g(x_0)-}\left(f \circ g^{-1}\right)\right)(t)\, dt, \tag{9.21}$$

all $a \leq x \leq x_0$.

(III) Denote by

$$D^{m\nu}_{g(x_0)} = D^{\nu}_{g(x_0)} D^{\nu}_{g(x_0)} ... D^{\nu}_{g(x_0)} \text{ (m-times)}, m \in \mathbb{N}. \tag{9.22}$$

We mention the following modified and generalized left X-valued fractional Taylor's formula of Canavati type:

Theorem 9.6 *Let $f \in C^1([a, b], X)$, $g \in C^1([a, b])$, strictly increasing: $g^{-1} \in C^1([g(a), g(b)])$. Assume that $\left(D^{i\nu}_{g(x_0)}\left(f \circ g^{-1}\right)\right) \in C^{\nu}_{g(x_0)}([g(a), g(b)], X)$, $0 < \nu < 1$, $x_0 \in [a, b]$, for $i = 0, 1, ..., m$. Then*

$$f(x) = \frac{1}{\Gamma((m+1)\nu)} \int_{g(x_0)}^{g(x)} (g(x) - z)^{(m+1)\nu-1} \left(D^{(m+1)\nu}_{g(x_0)}\left(f \circ g^{-1}\right)\right)(z)\, dz, \tag{9.23}$$

all $x_0 \leq x \leq b$.

(IV) Denote by

$$D^{m\nu}_{g(x_0)-} = D^{\nu}_{g(x_0)-} D^{\nu}_{g(x_0)-} ... D^{\nu}_{g(x_0)-} \text{ (m times)}, m \in \mathbb{N}. \tag{9.24}$$

We mention the following modified and generalized right X-valued fractional Taylor's formula of Canavati type:

Theorem 9.7 *Let* $f \in C^1([a,b], X)$, $g \in C^1([a,b])$, *strictly increasing:* $g^{-1} \in C^1([g(a), g(b)])$. *Assume that* $\left(D^{i\nu}_{g(x_0)-}\left(f \circ g^{-1}\right)\right) \in C^{\nu}_{g(x_0)-}([g(a), g(b)], X)$, $0 < \nu < 1$, $x_0 \in [a,b]$, *for all* $i = 0, 1, ..., m$. *Then*

$$f(x) = \frac{1}{\Gamma((m+1)\nu)} \int_{g(x)}^{g(x_0)} (z - g(x))^{(m+1)\nu - 1} \left(D^{(m+1)\nu}_{g(x_0)-}\left(f \circ g^{-1}\right)\right)(z)\, dz, \tag{9.25}$$

all $a \leq x \leq x_0 \leq b$.

9.3 Basic Banach Algebras Background

All here come from [18].

We need

Definition 9.8 ([18], p. 245) A complex algebra is a vector space A over the complex filed $\mathbb{C}$ in which a multiplication is defined that satisfies

$$x(yz) = (xy)z, \tag{9.26}$$

$$(x+y)z = xz + yz, \quad x(y+z) = xy + xz, \tag{9.27}$$

and

$$\alpha(xy) = (\alpha x)y = x(\alpha y), \tag{9.28}$$

for all x, y and z in A and for all scalars α.

Additionally if A is a Banach space with respect to a norm that satisfies the multiplicative inequality

$$\|xy\| \leq \|x\| \|y\| \quad (x \in A, y \in A) \tag{9.29}$$

and if A contains a unit element e such that

$$xe = ex = x \quad (x \in A) \tag{9.30}$$

and

$$\|e\| = 1, \tag{9.31}$$

then A is called a Banach algebra.

A is commutative iff $xy = yx$ for all $x, y \in A$.

We make

Remark 9.9 Commutativity of A is explicited stated when needed.

There exists at most one $e \in A$ that satisfies (9.30).

Inequality (9.29) makes multiplication to be continuous, more precisely left and right continuous, see [18], p. 246.

Multiplication in A is not necessarily the numerical multiplication, it is something more general and it is defined abstractly, that is for $x, y \in A$ we have $xy \in A$, e.g. composition or convolution, etc.

For nice examples about Banach algebras see [18], pp. 247–248, Sect. 10.3.

We also make

Remark 9.10 Next we mention about integration of A-valued functions, see [18], p. 259, Sect. 10.22:

If A is a Banach algebra and f is a continuous A-valued function on some compact Hausdorff space Q on which a complex Borel measure μ is defined, then $\int f d\mu$ exists and has all the properties that were discussed in Chap. 3 of [18], simply because A is a Banach space. However, an additional property can be added to these, namely: If $x \in A$, then

$$x \int_Q f \, d\mu = \int_Q xf(p) \, d\mu(p) \tag{9.32}$$

and

$$\left(\int_Q f \, d\mu \right) x = \int_Q f(p) \, x \, d\mu(p). \tag{9.33}$$

The Bochner integrals we will involve in our chapter follow (9.32) and (9.33). Also, let $f \in C([a, b], X)$, where $[a, b] \subset \mathbb{R}$, $(X, \|\cdot\|)$ is a Banach space. By [5], p. 3, f is Bochner integrable.

9.4 p-Schatten Norms Background

In this advanced section all come from [12].

Let $(H, \langle \cdot, \cdot \rangle)$ be a complex Hilbert space and $\mathcal{B}(H)$ the Banach algebra of all bounded linear operators on H. If $\{e_i\}_{i \in I}$ an orthonormal basis of H, we say that $A \in \mathcal{B}(H)$ is of trace class if

$$\|A\|_1 := \sum_{i \in I} \langle |A| e_i, e_i \rangle < \infty. \tag{9.34}$$

The definition of $\|A\|_1$ does not depend on the choice of the orthornormal basis $\{e_i\}_{i \in I}$. We denote by $\mathcal{B}_1(H)$ the set of trace class operators in $\mathcal{B}(H)$.

We define the trace of a trace class operator $A \in \mathcal{B}_1(H)$ to be

$$tr(A) := \sum_{i \in I} \langle Ae_i, e_i \rangle, \tag{9.35}$$

where $\{e_i\}_{i\in I}$ an orthonormal basis of H. Note that this coincides with the usual definition of the trace if H is finite-dimensional. We observe that the series (9.35) converges absolutely and it is independent from the choice of basis.

The following result collects some properties of the trace:

Theorem 9.11 *We have:*

(i) If $A \in \mathcal{B}_1(H)$ *then* $A^* \in \mathcal{B}_1(H)$ *and*

$$tr\left(A^*\right) = \overline{tr\left(A\right)}; \tag{9.36}$$

(ii) If $A \in \mathcal{B}_1(H)$ *and* $T \in \mathcal{B}(H)$, *then* $AT, TA \in \mathcal{B}_1(H)$ *and*

$$tr\left(AT\right) = tr\left(TA\right) \text{ and } \left|tr\left(AT\right)\right| \leq \|A\|_1 \|T\|; \tag{9.37}$$

(iii) $tr(\cdot)$ *is a bounded linear functional on* $\mathcal{B}_1(H)$ *with* $\|tr\| = 1$;
(iv) If $A, B \in \mathcal{B}_2(H)$ *then* $AB, BA \in \mathcal{B}_1(H)$ *and* $tr(AB) = tr(BA)$;
(v) $\mathcal{B}_{fin}(H)$ *(finite rank operators) is a dense subspace of* $\mathcal{B}_1(H)$.

An operator $A \in \mathcal{B}(H)$ is said to belong to the von Neumann-Schatten class $\mathcal{B}_p(H)$, $1 \leq p < \infty$ if the p-Schatten norm is finite [20, pp. 60–64]

$$\|A\|_p := \left[tr\left(|A|^p\right)\right]^{\frac{1}{p}} < \infty,$$

$|A|^p$ is an operator notation and not a power.

For $1 < p < q < \infty$ we have that

$$\mathcal{B}_1(H) \subset \mathcal{B}_p(H) \subset \mathcal{B}_q(H) \subset \mathcal{B}(H) \tag{9.38}$$

and

$$\|A\|_1 \geq \|A\|_p \geq \|A\|_q \geq \|A\|. \tag{9.39}$$

For $p \geq 1$ the functional $\|\cdot\|_p$ is a norm on the $*$-ideal $\mathcal{B}_p(H)$, which is a Banach algebra, and $\left(\mathcal{B}_p(H), \|\cdot\|_p\right)$ is a Banach space.

Also, see for instance [20, p. 60–64], for $p \geq 1$,

$$\|A\|_p = \left\|A^*\right\|_p, \; A \in \mathcal{B}_p(H) \tag{9.40}$$

$$\|AB\|_p \leq \|A\|_p \|B\|_p, \; A, B \in \mathcal{B}_p(H) \tag{9.41}$$

and

$$\|AB\|_p \leq \|A\|_p \|B\|, \; \|BA\|_p \leq \|B\| \|A\|_p, \; A \in \mathcal{B}_p(H), B \in \mathcal{B}(H). \tag{9.42}$$

This implies that

$$\|CAB\|_p \leq \|C\| \, \|A\|_p \, \|B\|, \ A \in \mathcal{B}_p(H), B, C \in \mathcal{B}(H). \tag{9.43}$$

In terms of p-Schatten norm we have the Hölder inequality for $p, q > 1$ with $\frac{1}{p} + \frac{1}{q} = 1$:

$$(|tr(AB)| \leq) \, \|AB\|_1 \leq \|A\|_p \, \|B\|_q, \quad A \in \mathcal{B}_p(H), B \in \mathcal{B}_q(H). \tag{9.44}$$

For the theory of trace functionals and their applications the interested reader is referred to [19, 20].

For some classical trace inequalities see [10, 11, 15], which are continuations of the work of Bellman [8].

9.5 Main Results

We start with 1-Schatten norm weighted mixed generalized Canavati fractional Ostrowski type inequalities involving several functions taking values in the Banach algebra $\mathcal{B}_2(H) \subset \mathcal{B}(H)$:

Theorem 9.12 *Let the $*$-ideal $\mathcal{B}_2(H)$, which $(\mathcal{B}_2(H), \|\cdot\|_2)$ is a Banach algebra; $x_0 \in [a,b] \subset \mathbb{R}$, $\nu \geq 1$, $n = [\nu]$; $f_i \in C^n([a,b], \mathcal{B}_2(H))$, $i = 1, ..., r \in \mathbb{N} - \{1\}$; $g \in C^1([a,b])$, strictly increasing such that $g^{-1} \in C^n([g(a), g(b)])$, with $(f_i \circ g^{-1})^{(k)}(g(x_0)) = 0$, $k = 1, ..., n-1$; $i = 1, ..., r$. Assume further that $f_i \circ g^{-1} \in C^{\nu}_{g(x_0)-}([g(a), g(b)], \mathcal{B}_2(H)) \cap C^{\nu}_{g(x_0)}([g(a), g(b)], \mathcal{B}_2(H))$, $i = 1, ..., r$.*

Denote by

$$K(f_1, ..., f_r)(x_0) :=$$

$$\sum_{i=1}^{r} \left[\int_a^b \left(\prod_{\substack{j=1 \\ j \neq i}}^{r} f_j(x) \right) f_i(x) \, dx - \left(\int_a^b \left(\prod_{\substack{j=1 \\ j \neq i}}^{r} f_j(x) \right) dx \right) f_i(x_0) \right]. \tag{9.45}$$

Then

$$\|K(f_1, ..., f_r)(x_0)\|_1 \leq \frac{1}{\Gamma(\nu+1)} \sum_{i=1}^{r} \left[\left[\left\| \left\| D^{\nu}_{g(x_0)-} \left(f_i \circ g^{-1} \right) \right\|_2 \right\|_{\infty, [g(a), g(x_0)]} \right. \right.$$

$$\left. (g(x_0) - g(a))^{\nu} \left(\int_a^{x_0} \left(\prod_{\substack{j=1 \\ j \neq i}}^{r} \|f_j(x)\|_2 \right) dx \right) \right] + \tag{9.46}$$

$$\left[\left\| \left\| D^{\nu}_{g(x_0)} \left(f_i \circ g^{-1} \right) \right\|_2 \right\|_{\infty,[g(x_0),g(b)]} (g(b) - g(x_0))^{\nu} \left(\int_{x_0}^{b} \left(\prod_{\substack{j=1 \\ j \neq i}}^{r} \| f_j(x) \|_2 \right) dx \right) \right] \right] .$$

Proof Since $\left(f_i \circ g^{-1} \right)^{(k)} (g(x_0)) = 0$, $k = 1, ..., [\nu] - 1$; $i = 1, ..., r$; we have by Theorem 9.4 that

$$f_i(x) - f_i(x_0) = \frac{1}{\Gamma(\nu)} \int_{g(x_0)}^{g(x)} (g(x) - t)^{\nu-1} \left(D^{\nu}_{g(x_0)} \left(f_i \circ g^{-1} \right) \right)(t)\, dt, \quad (9.47)$$

$\forall\, x \in [x_0, b]$,
and by Theorem 9.5 that

$$f_i(x) - f_i(x_0) = \frac{1}{\Gamma(\nu)} \int_{g(x)}^{g(x_0)} (t - g(x))^{\nu-1} \left(D^{\nu}_{g(x_0)-} \left(f_i \circ g^{-1} \right) \right)(t)\, dt, \quad (9.48)$$

$\forall\, x \in [a, x_0]$, for all $i = 1, ..., r$.

Left multiplying (9.47) and (9.48) with $\left(\prod_{\substack{j=1 \\ j \neq i}}^{r} f_j(x) \right)$ we get, respectively,

$$\left(\prod_{\substack{j=1 \\ j \neq i}}^{r} f_j(x) \right) f_i(x) - \left(\prod_{\substack{j=1 \\ j \neq i}}^{r} f_j(x) \right) f_i(x_0) =$$

$$\frac{\left(\prod_{\substack{j=1 \\ j \neq i}}^{r} f_j(x) \right)}{\Gamma(\nu)} \int_{g(x_0)}^{g(x)} (g(x) - t)^{\nu-1} \left(D^{\nu}_{g(x_0)} \left(f_i \circ g^{-1} \right) \right)(t)\, dt, \quad (9.49)$$

$\forall\, x \in [x_0, b]$,
and

$$\left(\prod_{\substack{j=1 \\ j \neq i}}^{r} f_j(x) \right) f_i(x) - \left(\prod_{\substack{j=1 \\ j \neq i}}^{r} f_j(x) \right) f_i(x_0) = \quad (9.50)$$

$$\frac{\left(\prod_{\substack{j=1 \\ j \neq i}}^{r} f_j(x) \right)}{\Gamma(\nu)} \int_{g(x)}^{g(x_0)} (t - g(x))^{\nu-1} \left(D^{\nu}_{g(x_0)-} \left(f_i \circ g^{-1} \right) \right)(t)\, dt,$$

$\forall\, x \in [a, x_0]$, for all $i = 1, ..., r$.

Adding (9.49) and (9.50) as separate groups, we obtain

$$\sum_{i=1}^{r}\left(\prod_{\substack{j=1\\j\neq i}}^{r}f_{j}(x)\right)f_{i}(x)-\sum_{i=1}^{r}\left(\prod_{\substack{j=1\\j\neq i}}^{r}f_{j}(x)\right)f_{i}(x_{0})=$$

$$\frac{1}{\Gamma(\nu)}\sum_{i=1}^{r}\left(\prod_{\substack{j=1\\j\neq i}}^{r}f_{j}(x)\right)\int_{g(x_{0})}^{g(x)}(g(x)-t)^{\nu-1}\left(D_{g(x_{0})}^{\nu}\left(f_{i}\circ g^{-1}\right)\right)(t)\,dt, \quad (9.51)$$

$\forall\, x\in[x_{0},b]$,

and

$$\sum_{i=1}^{r}\left(\prod_{\substack{j=1\\j\neq i}}^{r}f_{j}(x)\right)f_{i}(x)-\sum_{i=1}^{r}\left(\prod_{\substack{j=1\\j\neq i}}^{r}f_{j}(x)\right)f_{i}(x_{0})=$$

$$\frac{1}{\Gamma(\nu)}\sum_{i=1}^{r}\left(\prod_{\substack{j=1\\j\neq i}}^{r}f_{j}(x)\right)\int_{g(x)}^{g(x_{0})}(t-g(x))^{\nu-1}\left(D_{g(x_{0})-}^{\nu}\left(f_{i}\circ g^{-1}\right)\right)(t)\,dt, \quad (9.52)$$

$\forall\, x\in[a,x_{0}]$.

Next, we integrate (9.51) and (9.52) with respect to $x\in[a,b]$. We have

$$\sum_{i=1}^{r}\int_{x_{0}}^{b}\left(\prod_{\substack{j=1\\j\neq i}}^{r}f_{j}(x)\right)f_{i}(x)\,dx-\sum_{i=1}^{r}\left(\int_{x_{0}}^{b}\left(\prod_{\substack{j=1\\j\neq i}}^{r}f_{j}(x)\right)dx\right)f_{i}(x_{0})=$$

$$\frac{1}{\Gamma(\nu)}\sum_{i=1}^{r}\left[\int_{x_{0}}^{b}\left(\prod_{\substack{j=1\\j\neq i}}^{r}f_{j}(x)\right)\left(\int_{g(x_{0})}^{g(x)}(g(x)-t)^{\nu-1}\left(D_{g(x_{0})}^{\nu}\left(f_{i}\circ g^{-1}\right)\right)(t)\,dt\right)dx\right], \quad (9.53)$$

and

$$\sum_{i=1}^{r}\int_{a}^{x_{0}}\left(\prod_{\substack{j=1\\j\neq i}}^{r}f_{j}(x)\right)f_{i}(x)\,dx-\sum_{i=1}^{r}\left(\int_{a}^{x_{0}}\left(\prod_{\substack{j=1\\j\neq i}}^{r}f_{j}(x)\right)dx\right)f_{i}(x_{0})=$$

$$\frac{1}{\Gamma(\nu)}\sum_{i=1}^{r}\left[\int_a^{x_0}\left(\prod_{\substack{j=1\\ j\neq i}}^{r} f_j(x)\right)\left(\int_{g(x)}^{g(x_0)}(t-g(x))^{\nu-1}\left(D_{g(x_0)-}^{\nu}\left(f_i\circ g^{-1}\right)\right)(t)\,dt\right)dx\right], \tag{9.54}$$

Finally, adding (9.53) and (9.54) we obtain the useful identity

$$K(f_1,\ldots,f_r)(x_0) :=$$

$$\sum_{i=1}^{r}\left[\int_a^b\left(\prod_{\substack{j=1\\ j\neq i}}^{r} f_j(x)\right)f_i(x)\,dx-\left(\int_a^b\left(\prod_{\substack{j=1\\ j\neq i}}^{r} f_j(x)\right)dx\right)f_i(x_0)\right]=$$

$$\frac{1}{\Gamma(\nu)}\sum_{i=1}^{r}\left[\left[\int_a^{x_0}\left(\prod_{\substack{j=1\\ j\neq i}}^{r} f_j(x)\right)\left(\int_{g(x)}^{g(x_0)}(t-g(x))^{\nu-1}\left(D_{g(x_0)-}^{\nu}\left(f_i\circ g^{-1}\right)\right)(t)\,dt\right)dx\right]\right.$$

$$\left.+\left[\int_{x_0}^b\left(\prod_{\substack{j=1\\ j\neq i}}^{r} f_j(x)\right)\left(\int_{g(x_0)}^{g(x)}(g(x)-t)^{\nu-1}\left(D_{g(x_0)}^{\nu}\left(f_i\circ g^{-1}\right)\right)(t)\,dt\right)dx\right]\right]. \tag{9.55}$$

Therefore, we get that

$$\|K(f_1,\ldots,f_r)(x_0)\|_1 =$$

$$\left\|\sum_{i=1}^{r}\left[\int_a^b\left(\prod_{\substack{j=1\\ j\neq i}}^{r} f_j(x)\right)f_i(x)\,dx-\left(\int_a^b\left(\prod_{\substack{j=1\\ j\neq i}}^{r} f_j(x)\right)dx\right)f_i(x_0)\right]\right\|_1\leq\frac{1}{\Gamma(\nu)}$$

$$\sum_{i=1}^{r}\left[\left\|\left[\int_a^{x_0}\left(\prod_{\substack{j=1\\ j\neq i}}^{r} f_j(x)\right)\left(\int_{g(x)}^{g(x_0)}(t-g(x))^{\nu-1}\left(D_{g(x_0)-}^{\nu}\left(f_i\circ g^{-1}\right)\right)(t)\,dt\right)dx\right]\right\|_1\right. \tag{9.56}$$

$$\left.+\left\|\left[\int_{x_0}^b\left(\prod_{\substack{j=1\\ j\neq i}}^{r} f_j(x)\right)\left(\int_{g(x_0)}^{g(x)}(g(x)-t)^{\nu-1}\left(D_{g(x_0)}^{\nu}\left(f_i\circ g^{-1}\right)\right)(t)\,dt\right)dx\right]\right\|_1\right]\leq$$

$$\frac{1}{\Gamma(\nu)}\sum_{i=1}^{r}\left[\left[\int_{a}^{x_0}\left\|\left(\prod_{\substack{j=1\\j\neq i}}^{r} f_j(x)\right)\left(\int_{g(x)}^{g(x_0)}(t-g(x))^{\nu-1}\left(D_{g(x_0)-}^{\nu}\left(f_i\circ g^{-1}\right)\right)(t)\,dt\right)\right\|_1 dx\right]\right. \tag{9.57}$$

$$\left.+\left[\int_{x_0}^{b}\left\|\left(\prod_{\substack{j=1\\j\neq i}}^{r} f_j(x)\right)\left(\int_{g(x_0)}^{g(x)}(g(x)-t)^{\nu-1}\left(D_{g(x_0)}^{\nu}\left(f_i\circ g^{-1}\right)\right)(t)\,dt\right)\right\|_1 dx\right]\right]\leq$$

$$\frac{1}{\Gamma(\nu)}\sum_{i=1}^{r}\left[\left[\int_{a}^{x_0}\left(\prod_{\substack{j=1\\j\neq i}}^{r}\|f_j(x)\|_2\right)\left(\int_{g(x)}^{g(x_0)}(t-g(x))^{\nu-1}\left\|\left(D_{g(x_0)-}^{\nu}\left(f_i\circ g^{-1}\right)\right)(t)\right\|_2 dt\right)dx\right]\right. \tag{9.58}$$

$$\left.+\left[\int_{x_0}^{b}\left(\prod_{\substack{j=1\\j\neq i}}^{r}\|f_j(x)\|_2\right)\left(\int_{g(x_0)}^{g(x)}(g(x)-t)^{\nu-1}\left\|\left(D_{g(x_0)}^{\nu}\left(f_i\circ g^{-1}\right)\right)(t)\right\|_2 dt\right)dx\right]\right]=:(*).$$

Hence it holds

$$\|K(f_1,\ldots,f_r)(x_0)\|_1\leq(*). \tag{9.59}$$

We have that

$$(*)\leq\frac{1}{\Gamma(\nu+1)}$$

$$\sum_{i=1}^{r}\left[\left[\left\|\left\|\left(D_{g(x_0)-}^{\nu}\left(f_i\circ g^{-1}\right)\right)\right\|_2\right\|_{\infty,[g(a),g(x_0)]}\int_{a}^{x_0}\left(\prod_{\substack{j=1\\j\neq i}}^{r}\|f_j(x)\|_2\right)(g(x_0)-g(x))^{\nu}dx\right]\right. \tag{9.60}$$

$$\left.+\left[\left\|\left\|\left(D_{g(x_0)}^{\nu}\left(f_i\circ g^{-1}\right)\right)\right\|_2\right\|_{\infty,[g(x_0),g(b)]}\int_{x_0}^{b}\left(\prod_{\substack{j=1\\j\neq i}}^{r}\|f_j(x)\|_2\right)(g(x)-g(x_0))^{\nu}dx\right]\right]\leq$$

$$\frac{1}{\Gamma(\nu+1)}\sum_{i=1}^{r}\left[\left[\left\|\left\|\left(D_{g(x_0)-}^{\nu}\left(f_i\circ g^{-1}\right)\right)\right\|_2\right\|_{\infty,[g(a),g(x_0)]}\right.\right.$$

$$\left.(g(x_0)-g(a))^{\nu}\left(\int_{a}^{x_0}\left(\prod_{\substack{j=1\\j\neq i}}^{r}\|f_j(x)\|_2\right)dx\right)\right]+ \tag{9.61}$$

$$\left[\left\|\left\|\left(D^{\nu}_{g(x_0)}\left(f_i \circ g^{-1}\right)\right)\right\|_2\right\|_{\infty,[g(x_0),g(b)]} (g(b) - g(x_0))^{\nu} \left(\int_{x_0}^{b} \left(\prod_{\substack{j=1\\j\neq i}}^{r} \|f_j(x)\|_2\right) dx\right)\right]\right],$$

proving (9.46). ■

Next comes an L_1 estimate.

Theorem 9.13 *All as in Theorem 9.12. Then*

$$\left\|\left|K(f_1, ..., f_r)(x_0)\right\|_1 \le \frac{1}{\Gamma(\nu)}\right.$$

$$\sum_{i=1}^{r}\left[\left[\left\|\left\|\left(D^{\nu}_{g(x_0)-}\left(f_i \circ g^{-1}\right)\right)\right\|_2\right\|_{L_1([g(a),g(x_0)])} \int_a^{x_0} \left(\prod_{\substack{j=1\\j\neq i}}^{r} \|f_j(x)\|_2\right) (g(x_0) - g(x))^{\nu-1} dx\right]\right.$$

$$+\left[\left\|\left\|\left(D^{\nu}_{g(x_0)}\left(f_i \circ g^{-1}\right)\right)\right\|_2\right\|_{L_1([g(x_0),g(b)])} \int_{x_0}^{b} \left(\prod_{\substack{j=1\\j\neq i}}^{r} \|f_j(x)\|_2\right) (g(x) - g(x_0))^{\nu-1} dx\right]\right]. \tag{9.62}$$

Proof We observe that (by (9.58), (9.59))

$$(*) \le \frac{1}{\Gamma(\nu)}$$

$$\sum_{i=1}^{r}\left[\left[\left\|\left\|\left(D^{\nu}_{g(x_0)-}\left(f_i \circ g^{-1}\right)\right)\right\|_2\right\|_{L_1([g(a),g(x_0)])} \int_a^{x_0} \left(\prod_{\substack{j=1\\j\neq i}}^{r} \|f_j(x)\|_2\right) (g(x_0) - g(x))^{\nu-1} dx\right]\right.$$

$$+\left[\left\|\left\|\left(D^{\nu}_{g(x_0)}\left(f_i \circ g^{-1}\right)\right)\right\|_2\right\|_{L_1([g(x_0),g(b)])} \int_{x_0}^{b} \left(\prod_{\substack{j=1\\j\neq i}}^{r} \|f_j(x)\|_2\right) (g(x) - g(x_0))^{\nu-1} dx\right]\right], \tag{9.63}$$

proving (9.62). ■

An L_p estimate follows.

Theorem 9.14 *All as in Theorem 9.12. Let now* $p, q > 1 : \frac{1}{p} + \frac{1}{q} = 1$. *Then*

$$\|K(f_1, ..., f_r)(x_0)\|_1 \le \frac{1}{(p(\nu-1)+1)^{\frac{1}{p}}\,\Gamma(\nu)}$$

$$\sum_{i=1}^{r}\left[\left[\left\|\left\|\left(D^{\nu}_{g(x_0)-}\left(f_i\circ g^{-1}\right)\right)\right\|_2\right\|_{q,[g(a),g(x_0)]}\left(\int_a^{x_0}\left(g\left(x_0\right)-g\left(x\right)\right)^{\nu-\frac{1}{q}}\left(\prod_{\substack{j=1\\j\neq i}}^{r}\left\|f_j\left(x\right)\right\|_2\right)dx\right)\right]\right.$$

$$\left.+\left[\left\|\left\|\left(D^{\nu}_{g(x_0)}\left(f_i\circ g^{-1}\right)\right)\right\|_2\right\|_{q,[g(x_0),g(b)]}\left(\int_{x_0}^{b}\left(g\left(x\right)-g\left(x_0\right)\right)^{\nu-\frac{1}{q}}\left(\prod_{\substack{j=1\\j\neq i}}^{r}\left\|f_j\left(x\right)\right\|_2\right)dx\right)\right]\right]. \tag{9.64}$$

Proof We have that (by (9.58), (9.59))

$$(*)\leq\frac{1}{\Gamma(\nu)}\sum_{i=1}^{r}\left[\left[\int_a^{x_0}\left(\prod_{\substack{j=1\\j\neq i}}^{r}\left\|f_j\left(x\right)\right\|_2\right)\left(\int_{g(x)}^{g(x_0)}\left(t-g\left(x\right)\right)^{p(\nu-1)}dt\right)^{\frac{1}{p}}\right.\right.$$

$$\left.\left(\int_{g(x)}^{g(x_0)}\left\|\left(D^{\nu}_{g(x_0)-}\left(f_i\circ g^{-1}\right)\right)(t)\right\|_2^q dt\right)^{\frac{1}{q}}dx\right]+$$

$$\left[\int_{x_0}^{b}\left(\prod_{\substack{j=1\\j\neq i}}^{r}\left\|f_j\left(x\right)\right\|_2\right)\left(\int_{g(x_0)}^{g(x)}\left(g\left(x\right)-t\right)^{p(\nu-1)}dt\right)^{\frac{1}{p}}\right.$$

$$\left.\left.\left(\int_{g(x)}^{g(x_0)}\left\|\left(D^{\nu}_{g(x_0)}\left(f_i\circ g^{-1}\right)\right)(t)\right\|_2^q dt\right)^{\frac{1}{q}}dx\right]\right]\leq \tag{9.65}$$

$$\frac{1}{\Gamma(\nu)}\sum_{i=1}^{r}\left[\left[\int_a^{x_0}\left(\prod_{\substack{j=1\\j\neq i}}^{r}\left\|f_j\left(x\right)\right\|_2\right)\frac{\left(g\left(x_0\right)-g\left(x\right)\right)^{\nu-1+\frac{1}{p}}}{\left(p\left(\nu-1\right)+1\right)^{\frac{1}{p}}}\left\|\left\|\left(D^{\nu}_{g(x_0)-}\left(f_i\circ g^{-1}\right)\right)\right\|_2\right\|_{q,[g(a),g(x_0)]}dx\right]\right.$$

$$\left.+\left[\int_{x_0}^{b}\left(\prod_{\substack{j=1\\j\neq i}}^{r}\left\|f_j\left(x\right)\right\|_2\right)\frac{\left(g\left(x\right)-g\left(x_0\right)\right)^{\nu-1+\frac{1}{p}}}{\left(p\left(\nu-1\right)+1\right)^{\frac{1}{p}}}\left\|\left\|\left(D^{\nu}_{g(x_0)}\left(f_i\circ g^{-1}\right)\right)(z)\right\|_2\right\|_{q,[g(x_0),g(b)]}dx\right]\right]$$

$$=\frac{1}{\left(p\left(\nu-1\right)+1\right)^{\frac{1}{p}}\Gamma(\nu)}$$

$$\sum_{i=1}^{r}\left[\left\|\left\|\left(D^{\nu}_{g(x_0)-}\left(f_i\circ g^{-1}\right)\right)\right\|_2\right\|_{q,[g(a),g(x_0)]}\left(\int_a^{x_0}\left(g\left(x_0\right)-g\left(x\right)\right)^{\nu-\frac{1}{q}}\left(\prod_{\substack{j=1\\j\neq i}}^{r}\left\|f_j\left(x\right)\right\|_2\right)dx\right)\right. \tag{9.66}$$

$$+\left\| \left\| \left(D^{\nu}_{g(x_0)} \left(f_i \circ g^{-1} \right) \right) \right\|_2 \right\|_{q,[g(x_0),g(b)]} \left(\int_{x_0}^{b} (g(x) - g(x_0))^{\nu - \frac{1}{q}} \left(\prod_{\substack{j=1 \\ j \neq i}}^{r} \| f_j(x) \|_2 \right) dx \right) \Bigg],$$

proving (9.64). ■

We continue with γ-Schatten norm related Ostrowski fractional inequalities:

Theorem 9.15 *Let $\gamma \geq 1$, the $*$-ideal $\mathcal{B}_\gamma(H)$, which $\left(\mathcal{B}_\gamma(H), \|\cdot\|_\gamma\right)$ is a Banach algebra; $x_0 \in [a,b] \subset \mathbb{R}$, $\nu \geq 1$, $n = [\nu]$; $f_i \in C^n\left([a,b], \mathcal{B}_\gamma(H)\right)$, $i = 1, ..., r \in \mathbb{N} - \{1\}$; $g \in C^1([a,b])$, strictly increasing such that $g^{-1} \in C^n([g(a), g(b)])$, with $\left(f_i \circ g^{-1}\right)^{(k)}(g(x_0)) = 0$, $k = 1, ..., n-1$; $i = 1, ..., r$. Assume further that $f_i \circ g^{-1} \in C^{\nu}_{g(x_0)-}\left([g(a), g(b)], \mathcal{B}_\gamma(H)\right) \cap C^{\nu}_{g(x_0)}\left([g(a), g(b)], \mathcal{B}_\gamma(H)\right)$, $i = 1, ..., r$.*

Here $K(f_1, ..., f_r)(x_0)$ is as in (9.45). Then

$$\|K(f_1, ..., f_r)(x_0)\|_\gamma \leq \frac{1}{\Gamma(\nu+1)} \sum_{i=1}^{r} \Bigg[\Bigg[\left\| \left\| D^{\nu}_{g(x_0)-} \left(f_i \circ g^{-1} \right) \right\|_\gamma \right\|_{\infty,[g(a),g(x_0)]}$$

$$(g(x_0) - g(a))^{\nu} \left(\int_a^{x_0} \left(\prod_{\substack{j=1 \\ j \neq i}}^{r} \| f_j(x) \|_\gamma \right) dx \right) \Bigg] + \quad (9.67)$$

$$\left[\left\| \left\| D^{\nu}_{g(x_0)} \left(f_i \circ g^{-1} \right) \right\|_\gamma \right\|_{\infty,[g(x_0),g(b)]} (g(b) - g(x_0))^{\nu} \left(\int_{x_0}^{b} \left(\prod_{\substack{j=1 \\ j \neq i}}^{r} \| f_j(x) \|_\gamma \right) dx \right) \right] \Bigg].$$

Proof As similar to Theorem 9.12 is omitted. Use of (9.41). ■

An L_1 estimate follows:

Theorem 9.16 *All as in Theorem 9.15. Then*

$$\||K(f_1, ..., f_r)(x_0)\|_\gamma \leq \frac{1}{\Gamma(\nu)}$$

$$\sum_{i=1}^{r} \Bigg[\Bigg[\left\| \left\| \left(D^{\nu}_{g(x_0)-} \left(f_i \circ g^{-1} \right) \right) \right\|_\gamma \right\|_{L_1([g(a),g(x_0)])} \int_a^{x_0} \left(\prod_{\substack{j=1 \\ j \neq i}}^{r} \| f_j(x) \|_\gamma \right) (g(x_0) - g(x))^{\nu-1} dx \Bigg] \quad (9.68)$$

$$+ \left[\left\| \left\| \left(D^{\nu}_{g(x_0)} \left(f_i \circ g^{-1} \right) \right) \right\|_\gamma \right\|_{L_1([g(x_0),g(b)])} \int_{x_0}^{b} \left(\prod_{\substack{j=1 \\ j \neq i}}^{r} \| f_j(x) \|_\gamma \right) (g(x) - g(x_0))^{\nu-1} dx \right] \Bigg].$$

Proof As similar to Theorem 9.13 is omitted. ■

An L_p estimate follows.

Theorem 9.17 *All as in Theorem 9.15. Let now* $p, q > 1 : \frac{1}{p} + \frac{1}{q} = 1$. *Then*

$$\|K(f_1, ..., f_r)(x_0)\|_\gamma \leq \frac{1}{(p(\nu-1)+1)^{\frac{1}{p}} \Gamma(\nu)}$$

$$\sum_{i=1}^{r} \left[\left[\left\|\left\|\left(D^{\nu}_{g(x_0)-}\left(f_i \circ g^{-1}\right)\right)\right\|_\gamma\right\|_{q,[g(a),g(x_0)]} \left(\int_a^{x_0} (g(x_0) - g(x))^{\nu - \frac{1}{q}} \left(\prod_{\substack{j=1 \\ j \neq i}}^{r} \|f_j(x)\|_\gamma\right) dx\right)\right]\right.$$

$$\left. + \left[\left\|\left\|\left(D^{\nu}_{g(x_0)}\left(f_i \circ g^{-1}\right)\right)\right\|_\gamma\right\|_{q,[g(x_0),g(b)]} \left(\int_{x_0}^{b} (g(x) - g(x_0))^{\nu - \frac{1}{q}} \left(\prod_{\substack{j=1 \\ j \neq i}}^{r} \|f_j(x)\|_\gamma\right) dx\right)\right]\right]. \quad (9.69)$$

Proof As similar to Theorem 9.14 is omitted. ■

When $r = 2$ we derive the following p-Schatten norm operator related Ostrowski type Canavati fractional inequalities.

Theorem 9.18 *Let* $p, q > 1 : \frac{1}{p} + \frac{1}{q} = 1$, *and let the* $*$*-ideals* $\mathcal{B}_p(H)$, $\mathcal{B}_q(H)$, *for which* $\left(\mathcal{B}_p(H), \|\cdot\|_p\right)$, $\left(\mathcal{B}_q(H), \|\cdot\|_q\right)$ *are Banach algebras;* $x_0 \in [a, b] \subset \mathbb{R}$, $\alpha \geq 1$, $n = [\alpha]$; $A_1 \in C^n\left([a, b], \mathcal{B}_p(H)\right)$, $A_2 \in C^n\left([a, b], \mathcal{B}_q(H)\right)$; $g \in C^1([a, b])$, *strictly increasing, such that* $g^{-1} \in C^n([g(a), g(b)])$, *with* $\left(A_i \circ g^{-1}\right)^{(k)}(g(x_0)) = 0$, $k = 1, ..., n-1$; $i = 1, 2$. *Assume further that* $A_1 \circ g^{-1} \in C^{\alpha}_{g(x_0)-}\left([g(a), g(b)], \mathcal{B}_p(H)\right) \cap C^{\alpha}_{g(x_0)}\left([g(a), g(b)], \mathcal{B}_p(H)\right)$, *and* $A_2 \circ g^{-1} \in C^{\alpha}_{g(x_0)-}\left([g(a), g(b)], \mathcal{B}_q(H)\right) \cap C^{\alpha}_{g(x_0)}\left([g(a), g(b)], \mathcal{B}_q(H)\right)$. *Then*

(1) it holds

$$\Phi(A_1, A_2)(x_0) := \int_a^b A_2(x) A_1(x)\, dx + \int_a^b A_1(x) A_2(x)\, dx -$$

$$\left(\int_a^b A_2(x)\, dx\right) A_1(x_0) - \left(\int_a^b A_1(x)\, dx\right) A_2(x_0) =$$

$$\frac{1}{\Gamma(\alpha)} \left\{\left[\int_a^{x_0} A_2(x) \left(\int_{g(x)}^{g(x_0)} (z - g(x))^{\alpha-1} \left(D^{\alpha}_{g(x_0)-}\left(A_1 \circ g^{-1}\right)\right)(z)\, dz\right) dx\right] + \right.$$

$$\left[\int_{x_0}^{b} A_2(x) \left(\int_{g(x_0)}^{g(x)} (g(x) - z)^{\alpha-1} \left(D^{\alpha}_{g(x_0)}\left(A_1 \circ g^{-1}\right)\right)(z)\, dz\right) dx\right] + \quad (9.70)$$

$$\left[\int_a^{x_0} A_1(x)\left(\int_{g(x)}^{g(x_0)} (z-g(x))^{\alpha-1}\left(D_{g(x_0)-}^{\alpha}\left(A_2\circ g^{-1}\right)\right)(z)\,dz\right)dx\right]+$$

$$\left[\int_{x_0}^{b} A_1(x)\left(\int_{g(x_0)}^{g(x)} (g(x)-z)^{\alpha-1}\left(D_{g(x_0)}^{\alpha}\left(A_2\circ g^{-1}\right)\right)(z)\,dz\right)dx\right]\Bigg\},$$

(2) for $\gamma,\delta>1:\frac{1}{\gamma}+\frac{1}{\delta}=1$, we have that

$$\|\Phi(A_1,A_2)(x_0)\|_1\le\frac{1}{\Gamma(\alpha)\left(\gamma(\alpha-1)+1\right)^{\frac{1}{\gamma}}}$$

$$\Bigg\{\left[\left\|\left\|D_{g(x_0)-}^{\alpha}\left(A_1\circ g^{-1}\right)\right\|_p\right\|_{\delta,[g(a),g(x_0)]}\int_a^{x_0}\|A_2(x)\|_q\,(g(x_0)-g(x))^{\alpha-\frac{1}{\delta}}\,dx\right]+$$

$$\left[\left\|\left\|D_{g(x_0)}^{\alpha}\left(A_1\circ g^{-1}\right)\right\|_p\right\|_{\delta,[g(x_0),g(b)]}\int_{x_0}^{b}\|A_2(x)\|_q\,(g(x)-g(x_0))^{\alpha-\frac{1}{\delta}}\,dx\right]+ \tag{9.71}$$

$$\left[\left\|\left\|D_{g(x_0)-}^{\alpha}\left(A_2\circ g^{-1}\right)\right\|_q\right\|_{\delta,[g(a),g(x_0)]}\int_a^{x_0}\|A_1(x)\|_p\,(g(x_0)-g(x))^{\alpha-\frac{1}{\delta}}\,dx\right]+$$

$$\left[\left\|\left\|D_{g(x_0)}^{\alpha}\left(A_2\circ g^{-1}\right)\right\|_q\right\|_{\delta,[g(x_0),g(b)]}\int_{x_0}^{b}\|A_1(x)\|_p\,(g(x)-g(x_0))^{\alpha-\frac{1}{\delta}}\,dx\right]\Bigg\},$$

(3) we also obtain

$$\|\Phi(A_1,A_2)(x_0)\|_1\le\frac{1}{\Gamma(\alpha)}$$

$$\Bigg\{\left[\left\|\left\|D_{g(x_0)-}^{\alpha}\left(A_1\circ g^{-1}\right)\right\|_p\right\|_{L_1([g(a),g(x_0)])}\int_a^{x_0}\|A_2(x)\|_q\,(g(x_0)-g(x))^{\alpha-1}\,dx\right]+$$

$$\left[\left\|\left\|D_{g(x_0)}^{\alpha}\left(A_1\circ g^{-1}\right)\right\|_p\right\|_{L_1([g(x_0),g(b)])}\int_{x_0}^{b}\|A_2(x)\|_q\,(g(x)-g(x_0))^{\alpha-1}\,dx\right]+$$

$$\left[\left\|\left\|D_{g(x_0)-}^{\alpha}\left(A_2\circ g^{-1}\right)\right\|_q\right\|_{L_1([g(a),g(x_0)])}\int_a^{x_0}\|A_1(x)\|_p\,(g(x_0)-g(x))^{\alpha-1}\,dx\right]+ \tag{9.72}$$

$$\left[\left\|\left\|D_{g(x_0)}^{\alpha}\left(A_2\circ g^{-1}\right)\right\|_q\right\|_{L_1([g(x_0),g(b)])}\int_{x_0}^{b}\|A_1(x)\|_p\,(g(x)-g(x_0))^{\alpha-1}\,dx\right]\Bigg\},$$

and

(4)

$$\|\Phi(A_1,A_2)(x_0)\|_1\le\frac{1}{\Gamma(\alpha+1)}$$

$$\Big\{\Big[\Big\|\big\|D^{\alpha}_{g(x_0)-}\left(A_1\circ g^{-1}\right)\big\|_p\Big\|_{\infty,[g(a),g(x_0)]}\int_a^{x_0}\|A_2(x)\|_q\,(g(x_0)-g(x))^{\alpha}\,dx\Big]+$$

$$\Big[\Big\|\big\|D^{\alpha}_{g(x_0)}\left(A_1\circ g^{-1}\right)\big\|_p\Big\|_{\infty,[g(x_0),g(b)]}\int_{x_0}^{b}\|A_2(x)\|_q\,(g(x)-g(x_0))^{\alpha}\,dx\Big]+$$

$$\Big[\Big\|\big\|D^{\alpha}_{g(x_0)-}\left(A_2\circ g^{-1}\right)\big\|_q\Big\|_{\infty,[g(a),g(x_0)]}\int_a^{x_0}\|A_1(x)\|_p\,(g(x_0)-g(x))^{\alpha}\,dx\Big]+ \tag{9.73}$$

$$\Big[\Big\|\big\|D^{\alpha}_{g(x_0)}\left(A_2\circ g^{-1}\right)\big\|_q\Big\|_{\infty,[g(x_0),g(b)]}\int_{x_0}^{b}\|A_1(x)\|_p\,(g(x)-g(x_0))^{\alpha}\,dx\Big]\Big\}.$$

Proof Here we have that (acting as in the proof of Theorem 9.12 for $r=2$)

$$\Phi(A_1,A_2)(x_0):=\int_a^b A_2(x)\,A_1(x)\,dx+\int_a^b A_1(x)\,A_2(x)\,dx-$$

$$\left(\int_a^b A_2(x)\,dx\right)A_1(x_0)-\left(\int_a^b A_1(x)\,dx\right)A_2(x_0)\overset{(9.55)}{=}$$

$$\frac{1}{\Gamma(\alpha)}\Big\{\Big[\int_a^{x_0}A_2(x)\left(\int_{g(x)}^{g(x_0)}(z-g(x))^{\alpha-1}\left(D^{\alpha}_{g(x_0)-}\left(A_1\circ g^{-1}\right)\right)(z)\,dz\right)dx\Big]+$$

$$\Big[\int_{x_0}^{b}A_2(x)\left(\int_{g(x_0)}^{g(x)}(g(x)-z)^{\alpha-1}\left(D^{\alpha}_{g(x_0)}\left(A_1\circ g^{-1}\right)\right)(z)\,dz\right)dx\Big]+$$

$$\Big[\int_a^{x_0}A_1(x)\left(\int_{g(x)}^{g(x_0)}(z-g(x))^{\alpha-1}\left(D^{\alpha}_{g(x_0)-}\left(A_2\circ g^{-1}\right)\right)(z)\,dz\right)dx\Big]+$$

$$\Big[\int_{x_0}^{b}A_1(x)\left(\int_{g(x_0)}^{g(x)}(g(x)-z)^{\alpha-1}\left(D^{\alpha}_{g(x_0)}\left(A_2\circ g^{-1}\right)\right)(z)\,dz\right)dx\Big]\Big\}. \tag{9.74}$$

Therefore it holds by taking the 1-Schatten norm that

$$\|\Phi(A_1,A_2)(x_0)\|_1=\Big\|\int_a^b A_2(x)\,A_1(x)\,dx+\int_a^b A_1(x)\,A_2(x)\,dx-$$

$$\left(\int_a^b A_2(x)\,dx\right)A_1(x_0)-\left(\int_a^b A_1(x)\,dx\right)A_2(x_0)\Big\|_1\leq$$

$$\frac{1}{\Gamma(\alpha)}\Big\{\Big[\Big\|\int_a^{x_0}A_2(x)\left(\int_{g(x)}^{g(x_0)}(z-g(x))^{\alpha-1}\left(D^{\alpha}_{g(x_0)-}\left(A_1\circ g^{-1}\right)\right)(z)\,dz\right)dx\Big\|_1\Big]+$$

$$\left[\left\| \int_{x_0}^{b} A_2(x) \left(\int_{g(x_0)}^{g(x)} (g(x) - z)^{\alpha-1} \left(D_{g(x_0)}^{\alpha} \left(A_1 \circ g^{-1} \right) \right)(z)\, dz \right) dx \right\|_1 \right] +$$

$$\left[\left\| \int_{a}^{x_0} A_1(x) \left(\int_{g(x)}^{g(x_0)} (z - g(x))^{\alpha-1} \left(D_{g(x_0)-}^{\alpha} \left(A_2 \circ g^{-1} \right) \right)(z)\, dz \right) dx \right\|_1 \right] +$$

$$\left[\left\| \int_{x_0}^{b} A_1(x) \left(\int_{g(x_0)}^{g(x)} (g(x) - z)^{\alpha-1} \left(D_{g(x_0)}^{\alpha} \left(A_2 \circ g^{-1} \right) \right)(z)\, dz \right) dx \right\|_1 \right] \Bigg\} \leq \tag{9.75}$$

$$\frac{1}{\Gamma(\alpha)} \Bigg\{ \left[\int_{a}^{x_0} \left\| A_2(x) \left(\int_{g(x)}^{g(x_0)} (z - g(x))^{\alpha-1} \left(D_{g(x_0)-}^{\alpha} \left(A_1 \circ g^{-1} \right) \right)(z)\, dz \right) \right\|_1 dx \right] +$$

$$\left[\int_{x_0}^{b} \left\| A_2(x) \left(\int_{g(x_0)}^{g(x)} (g(x) - z)^{\alpha-1} \left(D_{g(x_0)}^{\alpha} \left(A_1 \circ g^{-1} \right) \right)(z)\, dz \right) \right\|_1 dx \right] +$$

$$\left[\int_{a}^{x_0} \left\| A_1(x) \left(\int_{g(x)}^{g(x_0)} (z - g(x))^{\alpha-1} \left(D_{g(x_0)-}^{\alpha} \left(A_2 \circ g^{-1} \right) \right)(z)\, dz \right) \right\|_1 dx \right] +$$

$$\left[\int_{x_0}^{b} \left\| A_1(x) \left(\int_{g(x_0)}^{g(x)} (g(x) - z)^{\alpha-1} \left(D_{g(x_0)}^{\alpha} \left(A_2 \circ g^{-1} \right) \right)(z)\, dz \right) \right\|_1 dx \right] \Bigg\} \leq \tag{9.76}$$

(by using the p-Schatten norm and Hölder's type inequality (9.44) for $p, q > 1 : \frac{1}{p} + \frac{1}{q} = 1$)

$$\frac{1}{\Gamma(\alpha)} \Bigg\{ \left[\int_{a}^{x_0} \|A_2(x)\|_q \left\| \left(\int_{g(x)}^{g(x_0)} (z - g(x))^{\alpha-1} \left(D_{g(x_0)-}^{\alpha} \left(A_1 \circ g^{-1} \right) \right)(z)\, dz \right) \right\|_p dx \right] +$$

$$\left[\int_{x_0}^{b} \|A_2(x)\|_q \left\| \left(\int_{g(x_0)}^{g(x)} (g(x) - z)^{\alpha-1} \left(D_{g(x_0)}^{\alpha} \left(A_1 \circ g^{-1} \right) \right)(z)\, dz \right) \right\|_p dx \right] +$$

$$\left[\int_{a}^{x_0} \|A_1(x)\|_p \left\| \left(\int_{g(x)}^{g(x_0)} (z - g(x))^{\alpha-1} \left(D_{g(x_0)-}^{\alpha} \left(A_2 \circ g^{-1} \right) \right)(z)\, dz \right) \right\|_q dx \right] +$$

$$\left[\int_{x_0}^{b} \|A_1(x)\|_p \left\| \left(\int_{g(x_0)}^{g(x)} (g(x) - z)^{\alpha-1} \left(D_{g(x_0)}^{\alpha} \left(A_2 \circ g^{-1} \right) \right)(z)\, dz \right) \right\|_q dx \right] \Bigg\} \leq \tag{9.77}$$

$$\frac{1}{\Gamma(\alpha)} \Bigg\{ \left[\int_{a}^{x_0} \|A_2(x)\|_q \left(\int_{g(x)}^{g(x_0)} (z - g(x))^{\alpha-1} \left\| \left(D_{g(x_0)-}^{\alpha} \left(A_1 \circ g^{-1} \right) \right)(z) \right\|_p dz \right) dx \right] +$$

$$\left[\int_{x_0}^{b} \|A_2(x)\|_q \left(\int_{g(x_0)}^{g(x)} (g(x)-z)^{\alpha-1} \left\|\left(D_{g(x_0)}^{\alpha}\left(A_1 \circ g^{-1}\right)\right)(z)\right\|_p dz\right) dx\right]+$$

$$\left[\int_{a}^{x_0} \|A_1(x)\|_p \left(\int_{g(x)}^{g(x_0)} (z-g(x))^{\alpha-1} \left\|\left(D_{g(x_0)-}^{\alpha}\left(A_2 \circ g^{-1}\right)\right)(z)\right\|_q dz\right) dx\right]+$$

$$\left[\int_{x_0}^{b} \|A_1(x)\|_p \left(\int_{g(x_0)}^{g(x)} (g(x)-z)^{\alpha-1} \left\|\left(D_{g(x_0)}^{\alpha}\left(A_2 \circ g^{-1}\right)\right)(z)\right\|_q dz\right) dx\right]\Bigg\}. \tag{9.78}$$

We have proved, so far, that

$$\|\Phi(A_1, A_2)(x_0)\|_1 \leq$$

$$\frac{1}{\Gamma(\alpha)}\Bigg\{\left[\int_{a}^{x_0} \|A_2(x)\|_q \left(\int_{g(x)}^{g(x_0)} (z-g(x))^{\alpha-1} \left\|\left(D_{g(x_0)-}^{\alpha}\left(A_1 \circ g^{-1}\right)\right)(z)\right\|_p dz\right) dx\right]+$$

$$\left[\int_{x_0}^{b} \|A_2(x)\|_q \left(\int_{g(x_0)}^{g(x)} (g(x)-z)^{\alpha-1} \left\|\left(D_{g(x_0)}^{\alpha}\left(A_1 \circ g^{-1}\right)\right)(z)\right\|_p dz\right) dx\right]+$$

$$\left[\int_{a}^{x_0} \|A_1(x)\|_p \left(\int_{g(x)}^{g(x_0)} (z-g(x))^{\alpha-1} \left\|\left(D_{g(x_0)-}^{\alpha}\left(A_2 \circ g^{-1}\right)\right)(z)\right\|_q dz\right) dx\right]+$$

$$\left[\int_{x_0}^{b} \|A_1(x)\|_p \left(\int_{g(x_0)}^{g(x)} (g(x)-z)^{\alpha-1} \left\|\left(D_{g(x_0)}^{\alpha}\left(A_2 \circ g^{-1}\right)\right)(z)\right\|_q dz\right) dx\right]\Bigg\} =: (\lambda). \tag{9.79}$$

Let now $\gamma, \delta > 1$ such that $\frac{1}{\gamma}+\frac{1}{\delta}=1$, and we apply the usual Hölder's inequality in (9.79). Then we have that

$$\|\Phi(A_1, A_2)(x_0)\|_1 \leq (\lambda) \leq \frac{1}{\Gamma(\alpha)(\gamma(\alpha-1)+1)^{\frac{1}{\gamma}}}$$

$$\Bigg\{\left[\int_{a}^{x_0} \|A_2(x)\|_q (g(x_0)-g(x))^{\frac{\gamma(\alpha-1)+1}{\gamma}} \left(\int_{g(x)}^{g(x_0)} \left\|\left(D_{g(x_0)-}^{\alpha}\left(A_1 \circ g^{-1}\right)\right)(z)\right\|_p^{\delta} dz\right)^{\frac{1}{\delta}} dx\right]+$$

$$\left[\int_{x_0}^{b} \|A_2(x)\|_q (g(x)-g(x_0))^{\frac{\gamma(\alpha-1)+1}{\gamma}} \left(\int_{g(x_0)}^{g(x)} \left\|\left(D_{g(x_0)}^{\alpha}\left(A_1 \circ g^{-1}\right)\right)(z)\right\|_p^{\delta} dz\right)^{\frac{1}{\delta}} dx\right]+$$

$$\left[\int_{a}^{x_0} \|A_1(x)\|_p (g(x_0)-g(x))^{\frac{\gamma(\alpha-1)+1}{\gamma}} \left(\int_{g(x)}^{g(x_0)} \left\|\left(D_{g(x_0)-}^{\alpha}\left(A_2 \circ g^{-1}\right)\right)(z)\right\|_q^{\delta} dz\right)^{\frac{1}{\delta}} dx\right]+ \tag{9.80}$$

$$\left[\int_{x_0}^{b}\|A_1(x)\|_p\,(g(x)-g(x_0))^{\frac{\gamma(\alpha-1)+1}{\gamma}}\left(\int_{g(x_0)}^{g(x)}\left\|\left(D_{g(x_0)}^{\alpha}\left(A_2\circ g^{-1}\right)\right)(z)\right\|_q^{\delta}dz\right)^{\frac{1}{\delta}}dx\right]\Bigg\}$$

$$\leq\frac{1}{\Gamma(\alpha)\,(\gamma(\alpha-1)+1)^{\frac{1}{\gamma}}}$$

$$\Bigg\{\left[\left\|\left\|D_{g(x_0)-}^{\alpha}\left(A_1\circ g^{-1}\right)\right\|_p\right\|_{\delta,[g(a),g(x_0)]}\int_{a}^{x_0}\|A_2(x)\|_q\,(g(x_0)-g(x))^{\alpha-\frac{1}{\delta}}dx\right]+$$

$$\left[\left\|\left\|D_{g(x_0)}^{\alpha}\left(A_1\circ g^{-1}\right)\right\|_p\right\|_{\delta,[g(x_0),g(b)]}\int_{x_0}^{b}\|A_2(x)\|_q\,(g(x)-g(x_0))^{\alpha-\frac{1}{\delta}}dx\right]+$$

$$\left[\left\|\left\|D_{g(x_0)-}^{\alpha}\left(A_2\circ g^{-1}\right)\right\|_q\right\|_{\delta,[g(a),g(x_0)]}\int_{a}^{x_0}\|A_1(x)\|_p\,(g(x_0)-g(x))^{\alpha-\frac{1}{\delta}}dx\right]+ \tag{9.81}$$

$$\left[\left\|\left\|D_{g(x_0)}^{\alpha}\left(A_2\circ g^{-1}\right)\right\|_q\right\|_{\delta,[g(x_0),g(b)]}\int_{x_0}^{b}\|A_1(x)\|_p\,(g(x)-g(x_0))^{\alpha-\frac{1}{\delta}}dx\right]\Bigg\},$$

proving (9.71).

We also obtain

$$\|\Phi(A_1,A_2)(x_0)\|_1\leq(\lambda)\leq\frac{1}{\Gamma(\alpha)}$$

$$\Bigg\{\left[\left\|\left\|D_{g(x_0)-}^{\alpha}\left(A_1\circ g^{-1}\right)\right\|_p\right\|_{L_1([g(a),g(x_0)])}\int_{a}^{x_0}\|A_2(x)\|_q\,(g(x_0)-g(x))^{\alpha-1}dx\right]+$$

$$\left[\left\|\left\|D_{g(x_0)}^{\alpha}\left(A_1\circ g^{-1}\right)\right\|_p\right\|_{L_1([g(x_0),g(b)])}\int_{x_0}^{b}\|A_2(x)\|_q\,(g(x)-g(x_0))^{\alpha-1}dx\right]+$$

$$\left[\left\|\left\|D_{g(x_0)-}^{\alpha}\left(A_2\circ g^{-1}\right)\right\|_q\right\|_{L_1([g(a),g(x_0)])}\int_{a}^{x_0}\|A_1(x)\|_p\,(g(x_0)-g(x))^{\alpha-1}dx\right]+ \tag{9.82}$$

$$\left[\left\|\left\|D_{g(x_0)}^{\alpha}\left(A_2\circ g^{-1}\right)\right\|_q\right\|_{L_1([g(x_0),g(b)])}\int_{x_0}^{b}\|A_1(x)\|_p\,(g(x)-g(x_0))^{\alpha-1}dx\right]\Bigg\},$$

proving (9.72).

At last we derive

$$\|\Phi(A_1,A_2)(x_0)\|_1\leq(\lambda)\leq\frac{1}{\Gamma(\alpha+1)}$$

$$\Bigg\{\left[\left\|\left\|D_{g(x_0)-}^{\alpha}\left(A_1\circ g^{-1}\right)\right\|_p\right\|_{\infty,[g(a),g(x_0)]}\int_{a}^{x_0}\|A_2(x)\|_q\,(g(x_0)-g(x))^{\alpha}dx\right]+$$

$$\left[\left\|\left\|D_{g(x_0)}^{\alpha}\left(A_1 \circ g^{-1}\right)\right\|_p\right\|_{\infty,[g(x_0),g(b)]} \int_{x_0}^{b} \|A_2(x)\|_q \left(g(x) - g(x_0)\right)^{\alpha} dx\right] +$$

$$\left[\left\|\left\|D_{g(x_0)-}^{\alpha}\left(A_2 \circ g^{-1}\right)\right\|_q\right\|_{\infty,[g(a),g(x_0)]} \int_{a}^{x_0} \|A_1(x)\|_p \left(g(x_0) - g(x)\right)^{\alpha} dx\right] +$$

$$\left[\left\|\left\|D_{g(x_0)}^{\alpha}\left(A_2 \circ g^{-1}\right)\right\|_q\right\|_{\infty,[g(x_0),g(b)]} \int_{x_0}^{b} \|A_1(x)\|_p \left(g(x) - g(x_0)\right)^{\alpha} dx\right]\Bigg\}, \tag{9.83}$$

proving (9.73).

The theorem is proved. ■

Next we present p-Schatten left and right generalized Canavati fractional Opial type inequalities:

Theorem 9.19 *Let the $*$-ideal $\mathcal{B}_2(H)$, which $(\mathcal{B}_2(H), \|\cdot\|_2)$ is a Banach algebra; $x_0 \in [a,b] \subset \mathbb{R}$, $\nu \geq 1$, $n = [\nu]$; $f \in C^n([a,b], \mathcal{B}_2(H))$, $g \in C^1([a,b])$, strictly increasing such that $g^{-1} \in C^n([g(a), g(b)])$, with $\left(f \circ g^{-1}\right)^{(k)}(g(x_0)) = 0$, $k = 0, 1, ..., n-1$. Assume further that $f \circ g^{-1} \in C_{g(x_0)}^{\nu}([g(a), g(b)], \mathcal{B}_2(H))$.*

Let also $p, q > 1 : \frac{1}{p} + \frac{1}{q} = 1$. Then

$$\int_{g(x_0)}^{z} \left\|\left(\left(f \circ g^{-1}\right)(w)\right)\left(\left(D_{g(x_0)}^{\nu}\left(f \circ g^{-1}\right)\right)(w)\right)\right\|_1 dw \leq \tag{9.84}$$

$$\frac{2^{-\frac{1}{q}}\left(z - g(x_0)\right)^{\nu + \frac{1}{p} - \frac{1}{q}}}{\Gamma(\nu)\left[\left(p(\nu-1)+1\right)\left(p(\nu-1)+2\right)\right]^{\frac{1}{p}}} \left(\int_{g(x_0)}^{z} \left\|\left(D_{g(x_0)}^{\nu}\left(f \circ g^{-1}\right)\right)(w)\right\|_2^q dw\right)^{\frac{2}{q}},$$

for all $g(x_0) \leq z \leq g(b)$.

Proof Very similar to the proof of Theorem 13 of [6]. Use of (9.44) for $p = q = 2$. ■

A similar result comex next:

Theorem 9.20 *Let $\gamma \geq 1$, the $*$-ideal $\mathcal{B}_{\gamma}(H)$, which $\left(\mathcal{B}_{\gamma}(H), \|\cdot\|_{\gamma}\right)$ is a Banach algebra; $x_0 \in [a,b] \subset \mathbb{R}$, $\nu \geq 1$, $n = [\nu]$; $f \in C^n\left([a,b], \mathcal{B}_{\gamma}(H)\right)$, $g \in C^1([a,b])$, strictly increasing such that $g^{-1} \in C^n([g(a), g(b)])$, with $\left(f \circ g^{-1}\right)^{(k)}(g(x_0)) = 0$, $k = 0, 1, ..., n-1$. Assume further that $f \circ g^{-1} \in C_{g(x_0)}^{\nu}\left([g(a), g(b)], \mathcal{B}_{\gamma}(H)\right)$.*

Let also $p, q > 1 : \frac{1}{p} + \frac{1}{q} = 1$. Then

$$\int_{g(x_0)}^{z} \left\|\left(\left(f \circ g^{-1}\right)(w)\right)\left(\left(D_{g(x_0)}^{\nu}\left(f \circ g^{-1}\right)\right)(w)\right)\right\|_{\gamma} dw \leq \tag{9.85}$$

$$\frac{2^{-\frac{1}{q}}\left(z-g\left(x_{0}\right)\right)^{\nu+\frac{1}{p}-\frac{1}{q}}}{\Gamma\left(\nu\right)\left[\left(p\left(\nu-1\right)+1\right)\left(p\left(\nu-1\right)+2\right)\right]^{\frac{1}{p}}}\left(\int_{g(x_{0})}^{z}\left\|\left(D_{g(x_{0})}^{\nu}\left(f\circ g^{-1}\right)\right)(w)\right\|_{\gamma}^{q}dw\right)^{\frac{2}{q}},$$

for all $g\left(x_{0}\right)\leq z\leq g\left(b\right)$.

Proof Very similar to the proof of Theorem 13 of [6]. Use of (9.41) for $p=\gamma$. ■

It follows the corresponding right side fractional Opial type inequalities:

Theorem 9.21 *All as in Theorem 9.19, however now it is* $f\circ g^{-1}\in C_{g(x_{0})-}^{\nu}\left(\left[g\left(a\right),g\left(b\right)\right],\mathcal{B}_{2}\left(H\right)\right)$. *Then*

$$\int_{z}^{g(x_{0})}\left\|\left(\left(f\circ g^{-1}\right)(w)\right)\left(\left(D_{g(x_{0})-}^{\nu}\left(f\circ g^{-1}\right)\right)(w)\right)\right\|_{1}dw\leq$$

$$\frac{2^{-\frac{1}{q}}\left(g\left(x_{0}\right)-z\right)^{\nu+\frac{1}{p}-\frac{1}{q}}}{\Gamma\left(\nu\right)\left[\left(p\left(\nu-1\right)+1\right)\left(p\left(\nu-1\right)+2\right)\right]^{\frac{1}{p}}}\left(\int_{z}^{g(x_{0})}\left\|\left(D_{g(x_{0})-}^{\nu}\left(f\circ g^{-1}\right)\right)(t)\right\|_{2}^{q}dt\right)^{\frac{2}{q}},\tag{9.86}$$

for all $g\left(a\right)\leq z\leq g\left(x_{0}\right)$.

Proof Based on (9.20), and as similar to the proof of Theorem 9.19 is omitted. ■

Next comes another right ride fractional Opial type inequality:

Theorem 9.22 *All as in Theorem 9.20, however now it is* $f\circ g^{-1}\in C_{g(x_{0})-}^{\nu}\left(\left[g\left(a\right),g\left(b\right)\right],\mathcal{B}_{\gamma}\left(H\right)\right)$. *Then*

$$\int_{z}^{g(x_{0})}\left\|\left(\left(f\circ g^{-1}\right)(w)\right)\left(\left(D_{g(x_{0})-}^{\nu}\left(f\circ g^{-1}\right)\right)(w)\right)\right\|_{\gamma}dw\leq$$

$$\frac{2^{-\frac{1}{q}}\left(g\left(x_{0}\right)-z\right)^{\nu+\frac{1}{p}-\frac{1}{q}}}{\Gamma\left(\nu\right)\left[\left(p\left(\nu-1\right)+1\right)\left(p\left(\nu-1\right)+2\right)\right]^{\frac{1}{p}}}\left(\int_{z}^{g(x_{0})}\left\|\left(D_{g(x_{0})-}^{\nu}\left(f\circ g^{-1}\right)\right)(t)\right\|_{\gamma}^{q}dt\right)^{\frac{2}{q}},\tag{9.87}$$

for all $g\left(a\right)\leq z\leq g\left(x_{0}\right)$.

Proof Based on (9.20), and as similar to the proof of Theorem 9.19 is omitted. ■

It follows the modified generalized left $\mathcal{B}_{2}\left(H\right)$-valued fractional Opial inequality:

Theorem 9.23 *All as in Theorem 9.6, where* $X=\mathcal{B}_{2}\left(H\right)$ *and let* $p,q>1:\frac{1}{p}+\frac{1}{q}=1$. *Here we assume that* $\frac{1}{(m+1)q}<\nu<1$. *Then*

$$\int_{g(x_{0})}^{z}\left\|\left(\left(f\circ g^{-1}\right)(w)\right)\left(\left(D_{g(x_{0})}^{(m+1)\nu}\left(f\circ g^{-1}\right)\right)(w)\right)\right\|_{1}dw\leq\tag{9.88}$$

$$\frac{2^{-\frac{1}{q}}(z-g(x_0))^{(m+1)\nu+\frac{1}{p}-\frac{1}{q}}}{\Gamma((m+1)\nu)[(p((m+1)\nu-1)+1)(p((m+1)\nu-1)+2)]^{\frac{1}{p}}}$$

$$\left(\int_{g(x_0)}^{z}\left\|\left(D_{g(x_0)}^{(m+1)\nu}\left(f\circ g^{-1}\right)\right)(t)\right\|_2^q dt\right)^{\frac{2}{q}},$$

for all $g(x_0)\leq z\leq g(b)$.

Proof As in Theorem 9.19. ■

Next comes another modified generalized left $\mathcal{B}_\gamma(H)$-valued fractional Opial inequality:

Theorem 9.24 *All as in Theorem 9.6, where* $X=\mathcal{B}_\gamma(H)$ *and let* $p,q>1:\frac{1}{p}+\frac{1}{q}=1$. *Here we assume that* $\frac{1}{(m+1)q}<\nu<1$. *Then*

$$\int_{g(x_0)}^{z}\left\|\left(\left(f\circ g^{-1}\right)(w)\right)\left(\left(D_{g(x_0)}^{(m+1)\nu}\left(f\circ g^{-1}\right)\right)(w)\right)\right\|_\gamma dw\leq \tag{9.89}$$

$$\frac{2^{-\frac{1}{q}}(z-g(x_0))^{(m+1)\nu+\frac{1}{p}-\frac{1}{q}}}{\Gamma((m+1)\nu)[(p((m+1)\nu-1)+1)(p((m+1)\nu-1)+2)]^{\frac{1}{p}}}$$

$$\left(\int_{g(x_0)}^{z}\left\|\left(D_{g(x_0)}^{(m+1)\nu}\left(f\circ g^{-1}\right)\right)(t)\right\|_\gamma^q dt\right)^{\frac{2}{q}},$$

for all $g(x_0)\leq z\leq g(b)$.

Proof As in Theorem 9.19. ■

The corresponding modified generalized right $\mathcal{B}_2(H)$-valued fractional Opial inequality comes next:

Theorem 9.25 *All as in Theorem 9.7, where* $X=\mathcal{B}_2(H)$ *and let* $p,q>1:\frac{1}{p}+\frac{1}{q}=1$. *Here we assume that* $\frac{1}{(m+1)q}<\nu<1$. *Then*

$$\int_{z}^{g(x_0)}\left\|\left(\left(f\circ g^{-1}\right)(w)\right)\left(\left(D_{g(x_0)-}^{(m+1)\nu}\left(f\circ g^{-1}\right)\right)(w)\right)\right\|_1 dw\leq \tag{9.90}$$

$$\frac{2^{-\frac{1}{q}}(g(x_0)-z)^{(m+1)\nu+\frac{1}{p}-\frac{1}{q}}}{\Gamma((m+1)\nu)[(p((m+1)\nu-1)+1)(p((m+1)\nu-1)+2)]^{\frac{1}{p}}}$$

$$\left(\int_{z}^{g(x_0)}\left\|\left(D_{g(x_0)-}^{(m+1)\nu}\left(f\circ g^{-1}\right)\right)(t)\right\|_2^q dt\right)^{\frac{2}{q}},$$

for all $g(a) \leq z \leq g(x_0)$.

Proof As in Theorem 9.19. ■

The corresponding modified generalized right $\mathcal{B}_\gamma(H)$-valued fractional Opial inequality comes next:

Theorem 9.26 *All as in Theorem 9.7, where* $X = \mathcal{B}_\gamma(H)$ *and let* $p, q > 1 : \frac{1}{p} + \frac{1}{q} = 1$. *Here we assume that* $\frac{1}{(m+1)q} < \nu < 1$. *Then*

$$\int_z^{g(x_0)} \left\| \left(\left(f \circ g^{-1} \right)(w) \right) \left(\left(D_{g(x_0)-}^{(m+1)\nu} \left(f \circ g^{-1} \right) \right)(w) \right) \right\|_\gamma dw \leq \tag{9.91}$$

$$\frac{2^{-\frac{1}{q}} \left(g(x_0) - z \right)^{(m+1)\nu + \frac{1}{p} - \frac{1}{q}}}{\Gamma\left((m+1)\nu\right) \left[\left(p\left((m+1)\nu - 1\right) + 1 \right) \left(p\left((m+1)\nu - 1\right) + 2 \right) \right]^{\frac{1}{p}}}$$

$$\left(\int_z^{g(x_0)} \left\| \left(D_{g(x_0)-}^{(m+1)\nu} \left(f \circ g^{-1} \right) \right)(t) \right\|_\gamma^q dt \right)^{\frac{2}{q}},$$

for all $g(a) \leq z \leq g(x_0)$.

Proof As in Theorem 9.19. ■

We make

Remark 9.27 (to Theorem 9.12)
Case of inequality (9.46):
Call and assume

$$M_1(f_1, ..., f_r) := \tag{9.92}$$

$$\max_{i=1,...,r} \left\{ \sup_{x_0 \in [a,b]} \left\| \left\| D_{g(x_0)-}^{\nu} \left(f_i \circ g^{-1} \right) \right\|_2 \right\|_{\infty, [g(a), g(x_0)]}, \right.$$

$$\left. \sup_{x_0 \in [a,b]} \left\| \left\| D_{g(x_0)}^{\gamma} \left(f_i \circ g^{-1} \right) \right\|_2 \right\|_{\infty, [g(x_0), g(b)]} \right\} < +\infty.$$

Then

$$\left\| K(f_1, ..., f_r)(x_0) \right\|_1 \leq \text{Right hand side } (9.46) \leq$$

$$\frac{M_1(f_1, ..., f_r) \left(g(b) - g(a) \right)^\nu}{\Gamma(\nu + 1)} \sum_{i=1}^r \left(\int_a^b \left(\prod_{\substack{j=1 \\ j \neq i}}^r \left\| f_j(x) \right\|_2 \right) dx \right). \tag{9.93}$$

We make

Remark 9.28 (to Theorem 9.13)
Case of inequality (9.62):
Call and assume

$$M_2 (f_1, ..., f_r) := \tag{9.94}$$

$$\max_{i=1,...,r} \left\{ \sup_{x_0 \in [a,b]} \left\| \left\| D^{\nu}_{g(x_0)-} \left(f_i \circ g^{-1} \right) \right\|_2 \right\|_{L_1([g(a), g(x_0)])}, \right.$$

$$\left. \sup_{x_0 \in [a,b]} \left\| \left\| D^{\nu}_{g(x_0)} \left(f_i \circ g^{-1} \right) \right\|_2 \right\|_{L_1([g(x_0), g(b)])} \right\} < +\infty.$$

Then

$$\| K (f_1, ..., f_r) (x_0) \|_1 \leq \text{Right hand side (9.62)} \leq$$

$$\frac{M_2 (f_1, ..., f_r) (g (b) - g (a))^{\nu-1}}{\Gamma (\nu)} \sum_{i=1}^{r} \left(\int_a^b \left(\prod_{\substack{j=1 \\ j \neq i}}^{r} \| f_j (x) \|_2 \right) dx \right). \tag{9.95}$$

We make

Remark 9.29 (to Theorem 9.14)
Case of inequality (9.64):
Call and assume ($p, q > 1 : \frac{1}{p} + \frac{1}{q} = 1$):

$$M_3 (f_1, ..., f_r) := \tag{9.96}$$

$$\max_{i=1,...,r} \left\{ \sup_{x_0 \in [a,b]} \left\| \left\| D^{\nu}_{g(x_0)-} \left(f_i \circ g^{-1} \right) \right\|_2 \right\|_{q,([g(a), g(x_0)])}, \right.$$

$$\left. \sup_{x_0 \in [a,b]} \left\| \left\| D^{\nu}_{g(x_0)} \left(f_i \circ g^{-1} \right) \right\|_2 \right\|_{q,([g(x_0), g(b)])} \right\} < +\infty.$$

Then

$$\| K (f_1, ..., f_r) (x_0) \|_1 \leq \text{Right hand side (9.64)} \leq$$

$$\frac{M_3 (f_1, ..., f_r) (g (b) - g (a))^{\nu-\frac{1}{q}}}{(p (\nu - 1) + 1)^{\frac{1}{p}} \Gamma (\nu)} \sum_{i=1}^{r} \left(\int_a^b \left(\prod_{\substack{j=1 \\ j \neq i}}^{r} \| f_j (x) \|_2 \right) dx \right). \tag{9.97}$$

We make

Remark 9.30 (to Theorem 9.15) ($\gamma \geq 1$)
Case of inequality (9.67):
Call and assume

$$M_1^{\gamma}\left(f_1, ..., f_r\right) := \tag{9.98}$$

$$\max_{i=1,...,r}\left\{\sup_{x_0\in[a,b]}\left\|\left\|D_{g(x_0)-}^{\nu}\left(f_i\circ g^{-1}\right)\right\|_{\gamma}\right\|_{\infty,[g(a),g(x_0)]},\right.$$

$$\left.\sup_{x_0\in[a,b]}\left\|\left\|D_{g(x_0)}^{\nu}\left(f_i\circ g^{-1}\right)\right\|_{\gamma}\right\|_{\infty,[g(x_0),g(b)]}\right\}<+\infty.$$

Then

$$\left\|K\left(f_1, ..., f_r\right)(x_0)\right\|_{\gamma} \leq \text{Right hand side (9.67)} \leq$$

$$\frac{M_1^{\gamma}\left(f_1, ..., f_r\right)\left(g\left(b\right)-g\left(a\right)\right)^{\nu}}{\Gamma\left(\nu+1\right)}\sum_{i=1}^{r}\left(\int_a^b\left(\prod_{\substack{j=1\\ j\neq i}}^{r}\left\|f_j\left(x\right)\right\|_{\gamma}\right)dx\right). \tag{9.99}$$

We make

Remark 9.31 (to Theorem 9.16) ($\gamma \geq 1$)
Case of inequality (9.68):
Call and assume:

$$M_2^{\gamma}\left(f_1, ..., f_r\right) := \tag{9.100}$$

$$\max_{i=1,...,r}\left\{\sup_{x_0\in[a,b]}\left\|\left\|D_{g(x_0)-}^{\nu}\left(f_i\circ g^{-1}\right)\right\|_{\gamma}\right\|_{L_1([g(a),g(x_0)])},\right.$$

$$\left.\sup_{x_0\in[a,b]}\left\|\left\|D_{g(x_0)}^{\nu}\left(f_i\circ g^{-1}\right)\right\|_{\gamma}\right\|_{L_1([g(x_0),g(b)])}\right\}<+\infty.$$

Then

$$\left\|K\left(f_1, ..., f_r\right)(x_0)\right\|_{\gamma} \leq \text{Right hand side (9.68)} \leq$$

$$\frac{M_2^{\gamma}\left(f_1, ..., f_r\right)\left(g\left(b\right)-g\left(a\right)\right)^{\nu-1}}{\Gamma\left(\nu\right)}\sum_{i=1}^{r}\left(\int_a^b\left(\prod_{\substack{j=1\\ j\neq i}}^{r}\left\|f_j\left(x\right)\right\|_{\gamma}\right)dx\right). \tag{9.101}$$

We make

Remark 9.32 (to Theorem 9.17) ($\gamma \geq 1$)

Case of inequality (9.69):
Call and assume ($p, q > 1 : \frac{1}{p} + \frac{1}{q} = 1$):

$$M_3^{\gamma}(f_1, ..., f_r) := \tag{9.102}$$

$$\max_{i=1,...,r} \left\{ \sup_{x_0 \in [a,b]} \left\| \left\| D^{\nu}_{g(x_0)-} \left(f_i \circ g^{-1} \right) \right\|_{\gamma} \right\|_{q,([g(a),g(x_0)])}, \right.$$

$$\left. \sup_{x_0 \in [a,b]} \left\| \left\| D^{\nu}_{g(x_0)} \left(f_i \circ g^{-1} \right) \right\|_{\gamma} \right\|_{q,([g(x_0),g(b)])} \right\} < +\infty.$$

Then

$$\| K(f_1, ..., f_r)(x_0) \|_{\gamma} \leq \text{Right hand side (9.69)} \leq$$

$$\frac{M_3^{\gamma}(f_1, ..., f_r)(g(b) - g(a))^{\nu - \frac{1}{q}}}{(p(\nu - 1) + 1)^{\frac{1}{p}} \Gamma(\nu)} \sum_{i=1}^{r} \left(\int_a^b \left(\prod_{\substack{j=1 \\ j \neq i}}^{r} \| f_j(x) \|_{\gamma} \right) dx \right). \tag{9.103}$$

Remark 9.33 (to Theorem 9.18)
(i) for $\gamma, \delta > 1 : \frac{1}{\gamma} + \frac{1}{\delta} = 1$, case of inequality (9.71):
Call and assume

$$N_1(A_1, A_2) :=$$

$$\max \left\{ \sup_{x_0 \in [a,b]} \left\| \left\| D^{\nu}_{g(x_0)-} \left(A_1 \circ g^{-1} \right) \right\|_p \right\|_{\delta,[g(a),g(x_0)]}, \right.$$

$$\sup_{x_0 \in [a,b]} \left\| \left\| D^{\nu}_{g(x_0)} \left(A_1 \circ g^{-1} \right) \right\|_p \right\|_{\delta,[g(x_0),g(b)]},$$

$$\sup_{x_0 \in [a,b]} \left\| \left\| D^{\nu}_{g(x_0)-} \left(A_2 \circ g^{-1} \right) \right\|_q \right\|_{\delta,[g(a),g(x_0)]},$$

$$\left. \sup_{x_0 \in [a,b]} \left\| \left\| D^{\nu}_{g(x_0)} \left(A_2 \circ g^{-1} \right) \right\|_q \right\|_{\delta,[g(x_0),g(b)]} \right\} < +\infty. \tag{9.104}$$

Then

$$\| \Phi(A_1, A_2)(x_0) \|_1 \leq \text{right hand side (9.71)} \leq$$

$$\frac{N_1(A_1, A_2)(g(b) - g(a))^{\alpha - \frac{1}{\delta}}}{\Gamma(\alpha)(\gamma(\alpha - 1) + 1)^{\frac{1}{\gamma}}} \left[\int_a^b \| A_1(x) \|_p \, dx + \int_a^b \| A_2(x) \|_q \, dx \right]. \tag{9.105}$$

(ii) case of inequality (9.72):
Call and assume

$$N_2(A_1, A_2) := \tag{9.106}$$

$$\max \left\{ \sup_{x_0 \in [a,b]} \left\| \left\| D^{\nu}_{g(x_0)-} \left(A_1 \circ g^{-1} \right) \right\|_p \right\|_{L_1([g(a), g(x_0)])}, \right.$$

$$\sup_{x_0 \in [a,b]} \left\| \left\| D^{\nu}_{g(x_0)} \left(A_1 \circ g^{-1} \right) \right\|_p \right\|_{L_1([g(x_0), g(b)])},$$

$$\sup_{x_0 \in [a,b]} \left\| \left\| D^{\nu}_{g(x_0)-} \left(A_2 \circ g^{-1} \right) \right\|_q \right\|_{L_1([g(a), g(x_0)])},$$

$$\left. \sup_{x_0 \in [a,b]} \left\| \left\| D^{\nu}_{g(x_0)} \left(A_2 \circ g^{-1} \right) \right\|_q \right\|_{L_1([g(x_0), g(b)])} \right\} < +\infty.$$

Then

$$\|\Phi(A_1, A_2)(x_0)\|_1 \leq \text{right hand side (9.72)} \leq$$

$$\frac{N_2(A_1, A_2)(g(b) - g(a))^{\alpha-1}}{\Gamma(\alpha)} \left[\int_a^b \|A_1(x)\|_p \, dx + \int_a^b \|A_2(x)\|_q \, dx \right]. \tag{9.107}$$

(iii) case of inequality (9.73):
Call and assume

$$N_3(A_1, A_2) := \tag{9.108}$$

$$\max \left\{ \sup_{x_0 \in [a,b]} \left\| \left\| D^{\nu}_{g(x_0)-} \left(A_1 \circ g^{-1} \right) \right\|_p \right\|_{\infty, [g(a), g(x_0)]}, \right.$$

$$\sup_{x_0 \in [a,b]} \left\| \left\| D^{\nu}_{g(x_0)} \left(A_1 \circ g^{-1} \right) \right\|_p \right\|_{\infty, [g(x_0), g(b)]},$$

$$\sup_{x_0 \in [a,b]} \left\| \left\| D^{\nu}_{g(x_0)-} \left(A_2 \circ g^{-1} \right) \right\|_q \right\|_{\infty, [g(a), g(x_0)]},$$

$$\left. \sup_{x_0 \in [a,b]} \left\| \left\| D^{\nu}_{g(x_0)} \left(A_2 \circ g^{-1} \right) \right\|_q \right\|_{\infty, [g(x_0), g(b)]} \right\} < +\infty.$$

Then

$$\|\Phi(A_1, A_2)(x_0)\|_1 \leq \text{right hand side (9.73)} \leq$$

$$\frac{N_3(A_1, A_2)(g(b) - g(a))^{\alpha}}{\Gamma(\alpha + 1)} \left[\int_a^b \|A_1(x)\|_p \, dx + \int_a^b \|A_2(x)\|_q \, dx \right]. \tag{9.109}$$

We need

Remark 9.34 (i) This is regarding Theorems 9.12-9.17. Here $K(f_1, ..., f_r)(x_0)$, $x_0 \in [a, b]$, is as in (9.45). Next we denote and have (case of $1 \leq \nu < 2$):

$$\Delta(f_1, ..., f_r) := \int_a^b K(f_1, ..., f_r)(x_0)\, dx_0 =$$

$$\sum_{i=1}^{r} \left[(b-a) \int_a^b \left(\prod_{\substack{j=1 \\ j \neq i}}^{r} f_j(x) \right) f_i(x)\, dx - \left(\int_a^b \left(\prod_{\substack{j=1 \\ j \neq i}}^{r} f_j(x) \right) dx \right) \left(\int_a^b f_i(x)\, dx \right) \right], \tag{9.110}$$

(ii) This is regarding Theorem 9.18. Here $\Phi(A_1, A_2)(x_0)$, $x_0 \in [a, b]$, is as in (9.70). Next we denote and have (case of $1 \leq \alpha < 2$):

$$\Delta(A_1, A_2) := \int_a^b \Phi(A_1, A_2)(x_0)\, dx_0 =$$

$$(b-a) \left(\int_a^b A_2(x) A_1(x)\, dx + \int_a^b A_1(x) A_2(x)\, dx \right) - \tag{9.111}$$

$$\left(\int_a^b A_2(x)\, dx \right) \left(\int_a^b A_1(x)\, dx \right) - \left(\int_a^b A_1(x)\, dx \right) \left(\int_a^b A_2(x)\, dx \right).$$

(iii) for $\gamma \geq 1$, it holds

$$\|\Delta(f_1, ..., f_r)\|_\gamma \leq \int_a^b \|K(f_1, ..., f_r)(x)\|_\gamma\, dx, \tag{9.112}$$

and

$$\|\Delta(A_1, A_2)\|_1 \leq \int_a^b \|\Phi(A_1, A_2)(x)\|_1\, dx. \tag{9.113}$$

We give the following set of γ-Schatten norm generalized Canavati type fractional Grüss type inequalities involving several functions over $\mathcal{B}_\gamma(H)$, $\gamma \geq 1$.

Theorem 9.35 *All as in Theorem 9.12, with $1 \leq \nu < 2$ (i.e. $n = 1$). Then*

(i)

$$\|\Delta(f_1, ..., f_r)\|_1 \leq \frac{M_1(f_1, ..., f_r)(g(b) - g(a))^\nu (b-a)^2}{\Gamma(\nu+1)}$$

$$\sum_{i=1}^{r}\left(\left\|\prod_{\substack{j=1\\j\neq i}}^{r}\|f_j(x)\|_2\right\|_{\infty,[a,b]}\right), \tag{9.114}$$

where $M_1(f_1, ..., f_r)$ *is as in (9.92),*

(ii)

$$\|\Delta(f_1, ..., f_r)\|_1 \leq \frac{M_2(f_1, ..., f_r)(g(b) - g(a))^{\nu-1}(b-a)^2}{\Gamma(\nu)}$$

$$\sum_{i=1}^{r}\left(\left\|\prod_{\substack{j=1\\j\neq i}}^{r}\|f_j(x)\|_2\right\|_{\infty,[a,b]}\right), \tag{9.115}$$

where $M_2(f_1, ..., f_r)$ *is as in (9.94),*

(iii) when $p, q > 1 : \frac{1}{p} + \frac{1}{q} = 1$, *we have*

$$\|\Delta(f_1, ..., f_r)\|_1 \leq \frac{M_3(f_1, ..., f_r)(g(b) - g(a))^{\nu-\frac{1}{q}}(b-a)^2}{(p(\nu-1)+1)^{\frac{1}{p}}\Gamma(\nu)}$$

$$\sum_{i=1}^{r}\left(\left\|\prod_{\substack{j=1\\j\neq i}}^{r}\|f_j(x)\|_2\right\|_{\infty,[a,b]}\right), \tag{9.116}$$

where $M_3(f_1, ..., f_r)$ *is as in (9.96).*

Proof By Remarks 9.34, 9.27-9.29 and that

$$\int_a^b\left(\prod_{\substack{j=1\\j\neq i}}^{r}\|f_j(x)\|_2\right)dx \leq (b-a)\left\|\prod_{\substack{j=1\\j\neq i}}^{r}\|f_j(x)\|_2\right\|_{\infty,[a,b]}.$$

■

We continue with

Theorem 9.36 *All as in Theorem 9.15, with* $1 \leq \nu < 2$ *(i.e.* $n = 1$*),* $\gamma \geq 1$*. Then*

(i)

$$\|\Delta(f_1, ..., f_r)\|_\gamma \leq \frac{M_1^\gamma(f_1, ..., f_r)(g(b) - g(a))^\nu(b-a)^2}{\Gamma(\nu+1)}$$

$$\sum_{i=1}^{r}\left(\left\|\prod_{\substack{j=1\\ j\neq i}}^{r}\left\|f_j(x)\right\|_\gamma\right\|_{\infty,[a,b]}\right), \tag{9.117}$$

where $M_1^\gamma(f_1, ..., f_r)$ *is as in (9.98),*

(ii)

$$\left\|\Delta(f_1, ..., f_r)\right\|_\gamma \leq \frac{M_2^\gamma(f_1, ..., f_r)(g(b)-g(a))^{\nu-1}(b-a)^2}{\Gamma(\nu)}$$

$$\sum_{i=1}^{r}\left(\left\|\prod_{\substack{j=1\\ j\neq i}}^{r}\left\|f_j(x)\right\|_\gamma\right\|_{\infty,[a,b]}\right), \tag{9.118}$$

where $M_2^\gamma(f_1, ..., f_r)$ *is as in (9.100),*

(iii) when $p, q > 1 : \frac{1}{p} + \frac{1}{q} = 1$, *we have*

$$\left\|\Delta(f_1, ..., f_r)\right\|_\gamma \leq \frac{M_3^\gamma(f_1, ..., f_r)(g(b)-g(a))^{\nu-\frac{1}{q}}(b-a)^2}{(p(\nu-1)+1)^{\frac{1}{p}}\Gamma(\nu)}$$

$$\sum_{i=1}^{r}\left(\left\|\prod_{\substack{j=1\\ j\neq i}}^{r}\left\|f_j(x)\right\|_\gamma\right\|_{\infty,[a,b]}\right), \tag{9.119}$$

where $M_3^\gamma(f_1, ..., f_r)$ *is as in (9.102).*

Proof By Remarks 9.34, 9.30-9.32 and that

$$\int_a^b\left(\prod_{\substack{j=1\\ j\neq i}}^{r}\left\|f_j(x)\right\|_\gamma\right)dx \leq (b-a)\left\|\prod_{\substack{j=1\\ j\neq i}}^{r}\left\|f_j(x)\right\|_\gamma\right\|_{\infty,[a,b]}.$$

■

Furthermore we have ($r = 2$ case of p-Schatten norm Grüss inequalities)

Theorem 9.37 *All as in Theorem 9.18, with* $1 \leq \alpha < 2$ *(i.e.* $[\alpha] = 1$*). Then*

(i) for $\gamma, \delta > 1 : \frac{1}{\gamma} + \frac{1}{\delta} = 1$, *we have*

$$\left\|\Delta(A_1, A_2)\right\|_1 \leq \frac{N_1(A_1, A_2)(g(b)-g(a))^{\alpha-\frac{1}{\delta}}(b-a)}{\Gamma(\alpha)(\gamma(\alpha-1)+1)^{\frac{1}{\gamma}}}$$

$$\left[\int_a^b \|A_1(x)\|_p\, dx + \int_a^b \|A_2(x)\|_q\, dx\right], \quad (9.120)$$

where $N_1(A_1, A_2)$ is as in (9.104),

(ii)

$$\|\Delta(A_1, A_2)\|_1 \leq \frac{N_2(A_1, A_2)(g(b) - g(a))^{\alpha-1}(b-a)}{\Gamma(\alpha)}$$

$$\left[\int_a^b \|A_1(x)\|_p\, dx + \int_a^b \|A_2(x)\|_q\, dx\right], \quad (9.121)$$

where $N_2(A_1, A_2)$ is as in (9.106),

and

(iii)

$$\|\Delta(A_1, A_2)\|_1 \leq \frac{N_3(A_1, A_2)(g(b) - g(a))^{\alpha}(b-a)}{\Gamma(\alpha+1)}$$

$$\left[\int_a^b \|A_1(x)\|_p\, dx + \int_a^b \|A_2(x)\|_q\, dx\right], \quad (9.122)$$

where $N_3(A_1, A_2)$ is as in (9.108).

Proof By Remarks 9.34 and 9.33. ■

9.6 Applications

We start with applications on Ostrowski type inequalities:

Corollary 9.38 (to Theorems 9.12–9.14) *All as in Theorem 9.12 for $g(t) = t$. Then*

(i)

$$\|K(f_1, ..., f_r)(x_0)\|_1 \leq \frac{1}{\Gamma(\nu+1)}$$

$$\sum_{i=1}^{r} \left[\left[\left\| \left\| \left(D_{x_0-}^{\nu} f_i\right) \right\|_2 \right\|_{\infty,[a,x_0]} (x_0 - a)^{\nu} \left(\int_a^{x_0} \left(\prod_{\substack{j=1 \\ j\neq i}}^{r} \|f_j(x)\|_2\right) dx\right)\right] + \right. \quad (9.123)$$

$$\left.\left[\left\| \left\| \left(D_{x_0}^{\nu} f_i\right) \right\|_2 \right\|_{\infty,[x_0,b]} (b - x_0)^{\nu} \left(\int_{x_0}^{b} \left(\prod_{\substack{j=1 \\ j\neq i}}^{r} \|f_j(x)\|_2\right) dx\right)\right]\right],$$

(ii)

$$\|K(f_1, ..., f_r)(x_0)\|_1 \leq \frac{1}{\Gamma(\nu)}$$

$$\sum_{i=1}^{r}\left[\left[\left\|\left\|\left(D_{x_0-}^{\nu} f_i\right)\right\|_2\right\|_{L_1([a,x_0])} \int_a^{x_0}\left(\prod_{\substack{j=1\\j\neq i}}^{r}\|f_j(x)\|_2\right)(x_0-x)^{\nu-1}dx\right]+\right.$$

$$\left.\left[\left\|\left\|\left(D_{x_0}^{\nu} f_i\right)\right\|_2\right\|_{L_1([x_0,b])} \int_{x_0}^{b}\left(\prod_{\substack{j=1\\j\neq i}}^{r}\|f_j(x)\|_2\right)(x-x_0)^{\nu-1}dx\right]\right], \tag{9.124}$$

(iii) when $p, q > 1 : \frac{1}{p} + \frac{1}{q} = 1$, we have

$$\|K(f_1, ..., f_r)(x_0)\|_1 \leq \frac{1}{(p(\nu-1)+1)^{\frac{1}{p}}\Gamma(\nu)}$$

$$\sum_{i=1}^{r}\left[\left[\left\|\left\|\left(D_{x_0-}^{\nu} f_i\right)\right\|_2\right\|_{q,[a,x_0]}\left(\int_a^{x_0}(x_0-x)^{\nu-\frac{1}{q}}\left(\prod_{\substack{j=1\\j\neq i}}^{r}\|f_j(x)\|_2\right)dx\right)\right]+\right.$$

$$\left.\left[\left\|\left\|\left(D_{x_0}^{\nu} f_i\right)\right\|_2\right\|_{q,[x_0,b]}\left(\int_{x_0}^{b}(x-x_0)^{\nu-\frac{1}{q}}\left(\prod_{\substack{j=1\\j\neq i}}^{r}\|f_j(x)\|_2\right)dx\right)\right]\right]. \tag{9.125}$$

It follows:

Corollary 9.39 (to Theorems 9.15–9.17) *All as in Theorem 9.15 for $g(t) = t$, $\gamma \geq 1$. Then*

(i)

$$\|K(f_1, ..., f_r)(x_0)\|_\gamma \leq \frac{1}{\Gamma(\nu+1)}$$

$$\sum_{i=1}^{r}\left[\left[\left\|\left\|\left(D_{x_0-}^{\nu} f_i\right)\right\|_\gamma\right\|_{\infty,[a,x_0]}(x_0-a)^{\nu}\left(\int_a^{x_0}\left(\prod_{\substack{j=1\\j\neq i}}^{r}\|f_j(x)\|_\gamma\right)dx\right)\right]+\right. \tag{9.126}$$

$$\left[\left\|\left\|\left(D_{x_0}^{\nu} f_i\right)\right\|_{\gamma}\right\|_{\infty,[x_0,b]} (b-x_0)^{\nu} \left(\int_{x_0}^{b} \left(\prod_{\substack{j=1\\ j\neq i}}^{r} \left\|f_j(x)\right\|_{\gamma}\right) dx\right)\right]\Bigg],$$

(ii)

$$\left\|K(f_1,...,f_r)(x_0)\right\|_{\gamma} \leq \frac{1}{\Gamma(\nu)}$$

$$\sum_{i=1}^{r}\left[\left[\left\|\left\|\left(D_{x_0-}^{\nu} f_i\right)\right\|_{\gamma}\right\|_{L_1([a,x_0])} \int_{a}^{x_0} \left(\prod_{\substack{j=1\\ j\neq i}}^{r} \left\|f_j(x)\right\|_{\gamma}\right) (x_0-x)^{\nu-1} dx\right]+\right.$$

$$\left.\left[\left\|\left\|\left(D_{x_0}^{\nu} f_i\right)\right\|_{\gamma}\right\|_{L_1([x_0,b])} \int_{x_0}^{b} \left(\prod_{\substack{j=1\\ j\neq i}}^{r} \left\|f_j(x)\right\|_{\gamma}\right) (x-x_0)^{\nu-1} dx\right]\right], \quad (9.127)$$

(iii) when $p,q>1: \frac{1}{p}+\frac{1}{q}=1$, we have

$$\left\|K(f_1,...,f_r)(x_0)\right\|_{\gamma} \leq \frac{1}{(p(\nu-1)+1)^{\frac{1}{p}}\,\Gamma(\nu)}$$

$$\sum_{i=1}^{r}\left[\left[\left\|\left\|\left(D_{x_0-}^{\nu} f_i\right)\right\|_{\gamma}\right\|_{q,[a,x_0]} \left(\int_{a}^{x_0} (x_0-x)^{\nu-\frac{1}{q}} \left(\prod_{\substack{j=1\\ j\neq i}}^{r} \left\|f_j(x)\right\|_{\gamma}\right) dx\right)\right]+\right.$$

$$\left.\left[\left\|\left\|\left(D_{x_0}^{\nu} f_i\right)\right\|_{\gamma}\right\|_{q,[x_0,b]} \left(\int_{x_0}^{b} (x-x_0)^{\nu-\frac{1}{q}} \left(\prod_{\substack{j=1\\ j\neq i}}^{r} \left\|f_j(x)\right\|_{\gamma}\right) dx\right)\right]\right]. \quad (9.128)$$

We continue with

Corollary 9.40 (to Theorem 9.18) *All as in Theorem 9.18, with $g(t)=e^t$. Then (i) for $\gamma,\delta>1: \frac{1}{\gamma}+\frac{1}{\delta}=1$, we have*

$$\left\|\Phi(A_1,A_2)(x_0)\right\|_1 \leq \frac{1}{\Gamma(\alpha)(\gamma(\alpha-1)+1)^{\frac{1}{\gamma}}}$$

$$\left\{\left[\left\|\left\|D_{e^{x_0}-}^{\alpha}(A_1\circ\log)\right\|_p\right\|_{\delta,[e^a,e^{x_0}]} \int_{a}^{x_0} \left\|A_2(x)\right\|_q \left(e^{x_0}-e^{x}\right)^{\alpha-\frac{1}{\delta}} dx\right]+\right.$$

$$\left[\left\|\left\|D^{\alpha}_{e^{x_0}}\left(A_1\circ\log\right)\right\|_p\right\|_{\delta,\left[e^{x_0},e^b\right]}\int_{x_0}^{b}\left\|A_2\left(x\right)\right\|_q\left(e^x-e^{x_0}\right)^{\alpha-\frac{1}{\delta}}dx\right]+$$

$$\left[\left\|\left\|D^{\alpha}_{e^{x_0}-}\left(A_2\circ\log\right)\right\|_q\right\|_{\delta,\left[e^{a},e^{x_0}\right]}\int_{a}^{x_0}\left\|A_1\left(x\right)\right\|_p\left(e^{x_0}-e^{x}\right)^{\alpha-\frac{1}{\delta}}dx\right]+$$

$$\left[\left\|\left\|D^{\alpha}_{e^{x_0}}\left(A_2\circ\log\right)\right\|_q\right\|_{\delta,\left[e^{x_0},e^b\right]}\int_{x_0}^{b}\left\|A_1\left(x\right)\right\|_p\left(e^x-e^{x_0}\right)^{\alpha-\frac{1}{\delta}}dx\right]\Bigg\}, \tag{9.129}$$

(ii) it holds

$$\left\|\Phi\left(A_1,A_2\right)\left(x_0\right)\right\|_1\leq\frac{1}{\Gamma\left(\alpha\right)}$$

$$\Bigg\{\left[\left\|\left\|D^{\alpha}_{e^{x_0}-}\left(A_1\circ\log\right)\right\|_p\right\|_{L_1\left(\left[e^{a},e^{x_0}\right]\right)}\int_{a}^{x_0}\left\|A_2\left(x\right)\right\|_q\left(e^{x_0}-e^{x}\right)^{\alpha-1}dx\right]+$$

$$\left[\left\|\left\|D^{\alpha}_{e^{x_0}}\left(A_1\circ\log\right)\right\|_p\right\|_{L_1\left(\left[e^{x_0},e^b\right]\right)}\int_{x_0}^{b}\left\|A_2\left(x\right)\right\|_q\left(e^x-e^{x_0}\right)^{\alpha-1}dx\right]+$$

$$\left[\left\|\left\|D^{\alpha}_{e^{x_0}-}\left(A_2\circ\log\right)\right\|_q\right\|_{L_1\left(\left[e^{a},e^{x_0}\right]\right)}\int_{a}^{x_0}\left\|A_1\left(x\right)\right\|_p\left(e^{x_0}-e^{x}\right)^{\alpha-1}dx\right]+ \tag{9.130}$$

$$\left[\left\|\left\|D^{\alpha}_{e^{x_0}}\left(A_2\circ\log\right)\right\|_q\right\|_{L_1\left(\left[e^{x_0},e^b\right]\right)}\int_{x_0}^{b}\left\|A_1\left(x\right)\right\|_p\left(e^x-e^{x_0}\right)^{\alpha-1}dx\right]\Bigg\},$$

and

(iii)

$$\left\|\Phi\left(A_1,A_2\right)\left(x_0\right)\right\|_1\leq\frac{1}{\Gamma\left(\alpha+1\right)}$$

$$\Bigg\{\left[\left\|\left\|D^{\alpha}_{e^{x_0}-}\left(A_1\circ\log\right)\right\|_p\right\|_{\infty,\left[e^{a},e^{x_0}\right]}\int_{a}^{x_0}\left\|A_2\left(x\right)\right\|_q\left(e^{x_0}-e^{x}\right)^{\alpha}dx\right]+$$

$$\left[\left\|\left\|D^{\alpha}_{e^{x_0}}\left(A_1\circ\log\right)\right\|_p\right\|_{\infty,\left[e^{x_0},e^b\right]}\int_{x_0}^{b}\left\|A_2\left(x\right)\right\|_q\left(e^x-e^{x_0}\right)^{\alpha}dx\right]+$$

$$\left[\left\|\left\|D^{\alpha}_{e^{x_0}-}\left(A_2\circ\log\right)\right\|_q\right\|_{\infty,\left[e^{a},e^{x_0}\right]}\int_{a}^{x_0}\left\|A_1\left(x\right)\right\|_p\left(e^{x_0}-e^{x}\right)^{\alpha}dx\right]+$$

$$\left[\left\|\left\|D^{\alpha}_{e^{x_0}}\left(A_2\circ\log\right)\right\|_q\right\|_{\infty,\left[e^{x_0},e^b\right]}\int_{x_0}^{b}\left\|A_1\left(x\right)\right\|_p\left(e^x-e^{x_0}\right)^{\alpha}dx\right]\Bigg\}. \tag{9.131}$$

We continue with applications on Opial inequalities

Corollary 9.41 (to Theorem 9.19) *All as in Theorem 9.19 with $g(t) = t$. Let $p, q > 1 : \frac{1}{p} + \frac{1}{q} = 1$. Then*

$$\int_{x_0}^{z} \left\| f(w) \left(D_{x_0}^{\nu} f\right)(w) \right\|_1 dw \leq$$

$$\frac{2^{-\frac{1}{q}} (z - x_0)^{\nu + \frac{1}{p} - \frac{1}{q}}}{\Gamma(\nu) \left[(p(\nu - 1) + 1)(p(\nu - 1) + 2)\right]^{\frac{1}{p}}} \left(\int_{x_0}^{z} \left\| \left(D_{x_0}^{\nu} f\right)(w) \right\|_2^q dw \right)^{\frac{2}{q}}, \quad (9.132)$$

for all $x_0 \leq z \leq b$.

It follows:

Corollary 9.42 (to Theorem 9.20) *All as in Theorem 9.20, $\gamma \geq 1$, with $g(t) = e^t$. Let also $p, q > 1 : \frac{1}{p} + \frac{1}{q} = 1$. Then*

$$\int_{e^{x_0}}^{z} \left\| ((f \circ \log)(w)) \left(\left(D_{e^{x_0}}^{\nu} (f \circ \log)\right)(w)\right) \right\|_{\gamma} dw \leq$$

$$\frac{2^{-\frac{1}{q}} (z - e^{x_0})^{\nu + \frac{1}{p} - \frac{1}{q}}}{\Gamma(\nu) \left[(p(\nu - 1) + 1)(p(\nu - 1) + 2)\right]^{\frac{1}{p}}} \left(\int_{e^{x_0}}^{z} \left\| \left(D_{e^{x_0}}^{\nu} (f \circ \log)\right)(w) \right\|_{\gamma}^q dw \right)^{\frac{2}{q}}, \quad (9.133)$$

for all $e^{x_0} \leq z \leq e^b$.

We finish with applications on Grüss inequalities:

Corollary 9.43 (to Theorem 9.35) *All as in Theorem 9.35 with $g(t) = t$ $(1 \leq \nu < 2)$. Then*

(i)

$$\|\Delta(f_1, ..., f_r)\|_1 \leq \frac{M_1(f_1, ..., f_r)(b - a)^{\nu + 2}}{\Gamma(\nu + 1)} \sum_{i=1}^{r} \left(\left\| \prod_{\substack{j=1 \\ j \neq i}}^{r} \|f_j(x)\|_2 \right\|_{\infty, [a,b]} \right), \quad (9.134)$$

where $M_1(f_1, ..., f_r)$ is as in (9.92),

(ii)

$$\|\Delta(f_1, ..., f_r)\|_1 \leq \frac{M_2(f_1, ..., f_r)(b - a)^{\nu + 1}}{\Gamma(\nu)} \sum_{i=1}^{r} \left(\left\| \prod_{\substack{j=1 \\ j \neq i}}^{r} \|f_j(x)\|_2 \right\|_{\infty, [a,b]} \right), \quad (9.135)$$

where $M_2(f_1, ..., f_r)$ is as in (9.94),

(iii) when $p, q > 1 : \frac{1}{p} + \frac{1}{q} = 1$, we have

$$\|\Delta(f_1, ..., f_r)\|_1 \leq \frac{M_3(f_1, ..., f_r)(b-a)^{\nu+1+\frac{1}{p}}}{(p(\nu-1)+1)^{\frac{1}{p}}\Gamma(\nu)} \sum_{i=1}^{r} \left(\left\| \prod_{\substack{j=1 \\ j \neq i}}^{r} \|f_j(x)\|_2 \right\|_{\infty,[a,b]} \right), \tag{9.136}$$

where $M_3(f_1, ..., f_r)$ is as in (9.96).

It follows ($r = 2$ case)

Corollary 9.44 (to Theorem 9.37) *All as in Theorem 9.37, with $[a, b] \subset \mathbb{R}_+ - \{0\}$, and $g(t) = \log t$. Then*

(i) for $\gamma, \delta > 1 : \frac{1}{\gamma} + \frac{1}{\delta} = 1$, we have

$$\|\Delta(A_1, A_2)\|_1 \leq \frac{N_1(A_1, A_2)\left(\log \frac{b}{a}\right)^{\alpha-\frac{1}{\delta}}(b-a)}{\Gamma(\alpha)(\gamma(\alpha-1)+1)^{\frac{1}{\gamma}}}$$

$$\left[\int_a^b \|A_1(x)\|_p\, dx + \int_a^b \|A_2(x)\|_q\, dx\right], \tag{9.137}$$

where $N_1(A_1, A_2)$ is as in (9.104),

(ii)

$$\|\Delta(A_1, A_2)\|_1 \leq \frac{N_2(A_1, A_2)\left(\log \frac{b}{a}\right)^{\alpha-1}(b-a)}{\Gamma(\alpha)}$$

$$\left[\int_a^b \|A_1(x)\|_p\, dx + \int_a^b \|A_2(x)\|_q\, dx\right], \tag{9.138}$$

where $N_2(A_1, A_2)$ is as in (9.106),

and

(iii)

$$\|\Delta(A_1, A_2)\|_1 \leq \frac{N_3(A_1, A_2)\left(\log \frac{b}{a}\right)^{\alpha}(b-a)}{\Gamma(\alpha+1)}$$

$$\left[\int_a^b \|A_1(x)\|_p\, dx + \int_a^b \|A_2(x)\|_q\, dx\right], \tag{9.139}$$

where $N_3(A_1, A_2)$ is as in (9.108).

References

1. Anastassiou, G.A.: Fractional Differentiation Inequalities. Research Monograph, Springer, New York (2009)
2. Anastassiou, G.A.: Advances on Fractional Inequalities. Research Monograph, Springer, New York (2011)
3. Anastassiou, G.A.: Intelligent Comparisons: Analytic Inequalities. Springer, Heidelberg, New York (2016)
4. Anastassiou, G.A.: Strong mixed and generalized fractional calculus for Banach space valued functions. Mat. Vesnik **69**(3), 176–191 (2017)
5. Anastassiou, G.A.: Intelligent Computations: Abstract Fractional Calculus, Inequalities, Approximations. Springer, Heidelberg, New York (2018)
6. Anastassiou, G.A.: Generalized Canavati Fractional Ostrowski, Opial and Grüss type inequalities for Banach algebra valued functions, submitted (2021)
7. Anastassiou, G.A.: p-Schatten norm generalized Canavati fractional Ostrowski, Opial and Grüss type inequalities involving several functions. Panamerican Math. journal, accepted for publication (2021)
8. Bellman, R.: Some inequalities for positive definite matrices. In: Beckenbach, E.F. (ed.), General Inequalities 2, Proceedings of the 2nd International Conference on General Inequalities, Birkhauser, Basel, pp. 89–90 (1980)
9. Čebyšev: Sur les expressions approximatives des intégrales définies par les aures prises entre les mêmes limites. Proc. Math. Soc. Charkov **2**, 93–98 (1882)
10. Chang, D.: A matrix trace inequality for products of Hermitian matrices. J. Math. Anal. Appl. **237**, 721–725 (1999)
11. Coop, I.D.: On matrix trace inequalities and related topics for products of Hermitian matrix. J. Math. Anal. Appl. **188**, 999–1001 (1994)
12. Dragomir, S.S.: p-Schatten norm inequalities of Ostrowski's type. RGMIA Res. Rep. Coll. **24**, Art. 108, 19 pp (2021)
13. Dragomir, S.S.: p-Schatten norm inequalities of Grüss type. RGMIA Res. Rep. Coll. **24**, Art. 115, 16 pp (2021)
14. Mikusinski, J.: The Bochner Integral. Academic Press, New York (1978)
15. Neudecker, H.: A matrix trace inequality. J. Math. Anal. Appl. **166**, 302–303 (1992)
16. Opial, Z.: Sur une inegalite. Ann. Polon. Math. **8**, 29–32 (1960)
17. Ostrowski, A.: Über die Absolutabweichung einer differentiabaren Funcktion von ihrem Integralmittelwert. Comment. Math. Helv. **10**, 226–227 (1938)
18. Rudin, W.: Functional Analysis, 2nd edn. McGraw-Hill Inc, New York (1991)
19. Simon, B.: Trace Ideals and Their Applications. Cambridge University Press, Cambridge (1979)
20. Zagrebvov, V.A.: Gibbs semigroups. Operator Theory: Advances and Applications, vol. 273, Birkhauser (2019)

Chapter 10
γ-Schatten Norm Generalized Canavati Fractional Hilbert–Pachpatte Inequalities with von Neumann–Schatten Class $\mathcal{B}_\gamma(H)$ Valued Functions

Employing generalized Canavati fractional left and right vectorial Taylor formulae we prove corresponding left and right fractional Hilbert–Pachpatte inequalities for von Neumann–Schatten class $\mathcal{B}_\gamma(H)$ valued functions. We cover also the sequential fractional case. We finish with applications. It follows [4].

10.1 Introduction

Here all motivation comes from [3].

All fractional terminology comes from Sect. 10.2 next.

We start with a left generalized Canavati fractional Hilbert–Pachpatte type inequality over a Banach algebra.

Theorem 10.1 *Let $p, q > 1$, such that $\frac{1}{p} + \frac{1}{q} = 1$, and $(A, \|\cdot\|)$ is a Banach algebra; and $i = 1, 2$. Let also $x_{0i} \in [a_i, b_i] \subset \mathbb{R}$, $\nu_i \geq 1$, $n_i = [\nu_i]$, $f_i \in C^{n_i}([a_i, b_i], A)$; $g_i \in C^1([a_i, b_i])$, strictly increasing, such that $g_i^{-1} \in C^{n_i}([g_i(a_i), g_i(b_i)])$, with $\left(f_i \circ g_i^{-1}\right)^{(k_i)}(g_i(x_{0i})) = 0$, $k_i = 0, 1, ..., n_i - 1$. Assume further that $f_i \circ g_i^{-1} \in C^{\nu_i}_{g_i(x_{0i})}([g_i(a_i), g_i(b_i)], A)$. Then*

$$\int_{g_1(x_{01})}^{g_1(b_1)} \int_{g_2(x_{02})}^{g_2(b_2)} \frac{\left\|\left(f_1 \circ g_1^{-1}\right)(z_1)\left(f_2 \circ g_2^{-1}\right)(z_2)\right\| dz_1 dz_2}{\left(\frac{(z_1 - g_1(x_{01}))^{p(\nu_1 - 1)+1}}{p(p(\nu_1 - 1)+1)} + \frac{(z_2 - g_2(x_{02}))^{q(\nu_2 - 1)+1}}{q(q(\nu_2 - 1)+1)}\right)} \leq$$

$$\frac{(g_1(b_1) - g_1(x_{01}))(g_2(b_2) - g_2(x_{02}))}{\Gamma(\nu_1)\Gamma(\nu_2)} \tag{10.1}$$

G. A. Anastassiou, *Fractional Inequalities In Banach Algebras*,
Studies in Systems, Decision and Control 441,
https://doi.org/10.1007/978-3-031-05148-7_10

$$\left\|\left\|D^{\nu_1}_{g_1(x_{01})}\left(f_1 \circ g_1^{-1}\right)\right\|\right\|_{L_q([g_1(x_{01}),g_1(b_1)],A)} \left\|\left\|D^{\nu_2}_{g_2(x_{02})}\left(f_2 \circ g_2^{-1}\right)\right\|\right\|_{L_p([g_2(x_{02}),g_2(b_2)],A)}.$$

Integrals in this chapter are of Bochner type [9].

We continue with a right generalized Canavati fractional Hilbert–Pachpatte type inequality over a Banach algebra.

Theorem 10.2 *All as in Theorem 10.1, however now it is* $f_i \circ g_i^{-1} \in C^{\nu_i}_{g_i(x_{0i})-}([g_i(a_i), g_i(b_i)], A)$, *for* $i = 1, 2$. *Then*

$$\int_{g_1(a_1)}^{g_1(x_{01})} \int_{g_2(a_2)}^{g_2(x_{02})} \frac{\left\|\left(f_1 \circ g_1^{-1}\right)(z_1)\left(f_2 \circ g_2^{-1}\right)(z_2)\right\| dz_1 dz_2}{\left(\frac{(g_1(x_{01})-z_1)^{p(\nu_1-1)+1}}{p(p(\nu_1-1)+1)} + \frac{(g_2(x_{02})-z_2)^{q(\nu_2-1)+1}}{q(q(\nu_2-1)+1)}\right)} \leq$$

$$\frac{(g_1(x_{01}) - g_1(a_1))(g_2(x_{02}) - g_2(a_2))}{\Gamma(\nu_1)\Gamma(\nu_2)} \tag{10.2}$$

$$\left\|\left\|D^{\nu_1}_{g_1(x_{01})-}\left(f_1 \circ g_1^{-1}\right)\right\|\right\|_{L_q([g_1(a_1),g_1(x_{01})],A)} \left\|\left\|D^{\nu_2}_{g_2(x_{02})-}\left(f_2 \circ g_2^{-1}\right)\right\|\right\|_{L_p([g_2(a_2),g_2(x_{02})],A)}.$$

Next comes a sequential left generalized Canavati fractional Hilbert–Pachpatte type inequality over a Banach algebra.

Theorem 10.3 *Let* $p, q > 1$, *such that* $\frac{1}{p} + \frac{1}{q} = 1$, *and* $(A, \|\cdot\|)$ *is a Banach algebra; and* $i = 1, 2$. *Let also* $f_i \in C^1([a_i, b_i], A)$; $g_i \in C^1([a_i, b_i])$, *strictly increasing, such that* $g_i^{-1} \in C^1([g_i(a_i), g_i(b_i)])$. *Assume that* $\frac{1}{(m_i+1)q} < \nu_i < 1$, $x_{0i} \in [a_i, b_i]$, *and* $D^{j_i\nu_i}_{g_i(x_{0i})}\left(f_i \circ g_i^{-1}\right) \in C^{\nu_i}_{g_i(x_{0i})}([g_i(a_i), g_i(b_i)], A)$, *for* $j_i = 0, 1, ..., m_i \in \mathbb{N}$. *Then*

$$\int_{g_1(x_{01})}^{g_1(b_1)} \int_{g_2(x_{02})}^{g_2(b_2)} \frac{\left\|\left(f_1 \circ g_1^{-1}\right)(z_1)\left(f_2 \circ g_2^{-1}\right)(z_2)\right\| dz_1 dz_2}{\left(\frac{(z_1-g_1(x_{01}))^{p((m_1+1)\nu_1-1)+1}}{p(p((m_1+1)\nu_1-1)+1)} + \frac{(z_2-g_2(x_{02}))^{q((m_2+1)\nu_2-1)+1}}{q(q((m_2+1)\nu_2-1)+1)}\right)} \leq$$

$$\frac{(g_1(b_1) - g_1(x_{01}))(g_2(b_2) - g_2(x_{02}))}{\Gamma((m_1+1)\nu_1)\Gamma((m_2+1)\nu_2)} \tag{10.3}$$

$$\left\|\left\|D^{(m_1+1)\nu_1}_{g_1(x_{01})}\left(f_1 \circ g_1^{-1}\right)\right\|\right\|_{L_q([g_1(x_{01}),g_1(b_1)],A)} \left\|\left\|D^{(m_2+1)\nu_2}_{g_2(x_{02})}\left(f_2 \circ g_2^{-1}\right)\right\|\right\|_{L_p([g_2(x_{02}),g_2(b_2)],A)}.$$

The right side analog of Theorem 10.3 follows:

Theorem 10.4 *Let* $p, q > 1$, *such that* $\frac{1}{p} + \frac{1}{q} = 1$, *and* $(A, \|\cdot\|)$ *is a Banach algebra; and* $i = 1, 2$. *Let also* $f_i \in C^1([a_i, b_i], A)$; $g_i \in C^1([a_i, b_i])$, *strictly increasing, such that* $g_i^{-1} \in C^1([g_i(a_i), g_i(b_i)])$. *Assume that* $\frac{1}{(m_i+1)q} < \nu_i < 1$, $x_{0i} \in [a_i, b_i]$, *and* $D^{j_i\nu_i}_{g_i(x_{0i})-}\left(f_i \circ g_i^{-1}\right) \in C^{\nu_i}_{g_i(x_{0i})-}([g_i(a_i), g_i(b_i)], A)$, *for* $j_i = 0, 1, ..., m_i \in \mathbb{N}$. *Then*

$$\int_{g_1(a_1)}^{g_1(x_{01})} \int_{g_2(a_2)}^{g_2(x_{02})} \frac{\left\| \left(f_1 \circ g_1^{-1}\right)(z_1) \left(f_2 \circ g_2^{-1}\right)(z_2) \right\| dz_1 dz_2}{\left(\frac{(g_1(x_{01})-z_1)^{p((m_1+1)\nu_1-1)+1}}{p(p((m_1+1)\nu_1-1)+1)} + \frac{(g_2(x_{02})-z_2)^{q((m_2+1)\nu_2-1)+1}}{q(q((m_2+1)\nu_2-1)+1)} \right)} \leq$$

$$\frac{(g_1(x_{01}) - g_1(a_1))(g_2(x_{02}) - g_2(a_2))}{\Gamma((m_1+1)\nu_1)\Gamma((m_2+1)\nu_2)} \tag{10.4}$$

$$\left\| \left\| D_{g_1(x_{01})-}^{(m_1+1)\nu_1} \left(f_1 \circ g_1^{-1}\right) \right\| \right\|_{L_q([g_1(a_1), g_1(x_{01})], A)} \left\| \left\| D_{g_2(x_{02})-}^{(m_2+1)\nu_2} \left(f_2 \circ g_2^{-1}\right) \right\| \right\|_{L_p([g_2(a_2), g_2(x_{02})], A)}.$$

Other related inspiration comes from [2].

Let $\gamma \geq 1$, in this chapter we derive γ-Schatten left and right Hilbert–Pachpatte inequalities for Banach algebra $\mathcal{B}_\gamma(H)$ valued functions with respect to their Canavati type generalized left and right fractional derivatives. We cover also the sequential fractional case. We finish with applications.

For $\mathcal{B}_\gamma(H)$ definition see Sect. 10.4 later.

10.2 Background on Vectorial Generalized Canavati Fractional Calculus

All in this section come from [2], pp. 109–115 and [1].

Let $g : [a, b] \to \mathbb{R}$ be a strictly increasing function. such that $g \in C^1([a, b])$, and $g^{-1} \in C^n([g(a), g(b)])$, $n \in \mathbb{N}$, $(X, \|\cdot\|)$ is a Banach space. Let $f \in C^n([a, b], X)$, and call $l := f \circ g^{-1} : [g(a), g(b)] \to X$. It is clear that $l, l', \ldots, l^{(n)}$ are continuous functions from $[g(a), g(b)]$ into $f([a, b]) \subseteq X$.

Let $\nu \geq 1$ such that $[\nu] = n$, $n \in \mathbb{N}$ as above, where $[\cdot]$ is the integral part of the number.

Clearly when $0 < \nu < 1$, $[\nu] = 0$.

(I) Let $h \in C([g(a), g(b)], X)$, we define the left Riemann-Liouville Bochner fractional integral as

$$\left(J_\nu^{z_0} h\right)(z) := \frac{1}{\Gamma(\nu)} \int_{z_0}^{z} (z-t)^{\nu-1} h(t)\, dt, \tag{10.5}$$

for $g(a) \leq z_0 \leq z \leq g(b)$, where Γ is the gamma function; $\Gamma(\nu) = \int_0^\infty e^{-t} t^{\nu-1} dt$. We set $J_0^{z_0} h = h$.

Let $\alpha := \nu - [\nu]$ $(0 < \alpha < 1)$. We define the subspace $C_{g(x_0)}^{\nu}([g(a), g(b)], X)$ of $C^{[\nu]}([g(a), g(b)], X)$, where $x_0 \in [a, b]$ as:

$$C_{g(x_0)}^{\nu}([g(a), g(b)], X) =$$

$$\left\{ h \in C^{[\nu]}([g(a), g(b)], X) : J_{1-\alpha}^{g(x_0)} h^{([\nu])} \in C^1([g(x_0), g(b)], X) \right\}. \tag{10.6}$$

So let $h \in C^{\nu}_{g(x_0)}([g(a), g(b)], X)$, we define the left g-generalized X-valued fractional derivative of h of order ν, of Canavati type, over $[g(x_0), g(b)]$ as

$$D^{\nu}_{g(x_0)} h := \left(J^{g(x_0)}_{1-\alpha} h^{([\nu])} \right)'. \tag{10.7}$$

Clearly, for $h \in C^{\nu}_{g(x_0)}([g(a), g(b)], X)$, there exists

$$\left(D^{\nu}_{g(x_0)} h \right)(z) = \frac{1}{\Gamma(1-\alpha)} \frac{d}{dz} \int_{g(x_0)}^{z} (z-t)^{-\alpha} h^{([\nu])}(t)\, dt, \tag{10.8}$$

for all $g(x_0) \leq z \leq g(b)$.

In particular, when $f \circ g^{-1} \in C^{\nu}_{g(x_0)}([g(a), g(b)], X)$, we have that

$$\left(D^{\nu}_{g(x_0)} \left(f \circ g^{-1} \right) \right)(z) = \frac{1}{\Gamma(1-\alpha)} \frac{d}{dz} \int_{g(x_0)}^{z} (z-t)^{-\alpha} \left(f \circ g^{-1} \right)^{([\nu])}(t)\, dt, \tag{10.9}$$

for all $g(x_0) \leq z \leq g(b)$. We have that $D^{n}_{g(x_0)} \left(f \circ g^{-1} \right) = \left(f \circ g^{-1} \right)^{(n)}$ and $D^{0}_{g(x_0)} \left(f \circ g^{-1} \right) = f \circ g^{-1}$, see [1].

By [1], we have for $f \circ g^{-1} \in C^{\nu}_{g(x_0)}([g(a), g(b)], X)$, where $x_0 \in [a, b]$ the following left generalized g-fractional, of Canavati type, X-valued Taylor's formula:

Theorem 10.5 *Let $f \circ g^{-1} \in C^{\nu}_{g(x_0)}([g(a), g(b)], X)$, where $x_0 \in [a, b]$ is fixed.*

(i) If $\nu \geq 1$, then

$$f(x) - f(x_0) = \sum_{k=1}^{[\nu]-1} \frac{\left(f \circ g^{-1} \right)^{(k)} (g(x_0))}{k!} (g(x) - g(x_0))^k +$$

$$\frac{1}{\Gamma(\nu)} \int_{g(x_0)}^{g(x)} (g(x) - t)^{\nu-1} \left(D^{\nu}_{g(x_0)} \left(f \circ g^{-1} \right) \right)(t)\, dt, \tag{10.10}$$

for all $x_0 \leq x \leq b$.

(ii) If $0 < \nu < 1$, we get

$$f(x) = \frac{1}{\Gamma(\nu)} \int_{g(x_0)}^{g(x)} (g(x) - t)^{\nu-1} \left(D^{\nu}_{g(x_0)} \left(f \circ g^{-1} \right) \right)(t)\, dt, \tag{10.11}$$

for all $x_0 \leq x \leq b$.

(II) Let $h \in C([g(a), g(b)], X)$, we define the right Riemann-Liouville Bochner fractional integral as

$$\left(J^{\nu}_{z_0 -} h \right)(z) := \frac{1}{\Gamma(\nu)} \int_{z}^{z_0} (t - z)^{\nu-1} h(t)\, dt, \tag{10.12}$$

for $g(a) \leq z \leq z_0 \leq g(b)$. We set $J^0_{z_0-}h = h$.

Let $\alpha := \nu - [\nu]$ $(0 < \alpha < 1)$. We define the subspace $C^{\nu}_{g(x_0)-}([g(a), g(b)], X)$ of $C^{[\nu]}([g(a), g(b)], X)$, where $x_0 \in [a, b]$ as:

$$C^{\nu}_{g(x_0)-}([g(a), g(b)], X) :=$$

$$\left\{h \in C^{[\nu]}([g(a), g(b)], X) : J^{1-\alpha}_{g(x_0)-}h^{([\nu])} \in C^1([g(a), g(x_0)], X)\right\}. \quad (10.13)$$

So let $h \in C^{\nu}_{g(x_0)-}([g(a), g(b)], X)$, we define the right g-generalized X-valued fractional derivative of h of order ν, of Canavati type, over $[g(a), g(x_0)]$ as

$$D^{\nu}_{g(x_0)-}h := (-1)^{n-1}\left(J^{1-\alpha}_{g(x_0)-}h^{([\nu])}\right)'. \quad (10.14)$$

Clearly, for $h \in C^{\nu}_{g(x_0)-}([g(a), g(b)], X)$, there exists

$$\left(D^{\nu}_{g(x_0)-}h\right)(z) = \frac{(-1)^{n-1}}{\Gamma(1-\alpha)}\frac{d}{dz}\int_z^{g(x_0)}(t-z)^{-\alpha}h^{([\nu])}(t)\,dt, \quad (10.15)$$

for all $g(a) \leq z \leq g(x_0) \leq g(b)$.

In particular, when $f \circ g^{-1} \in C^{\nu}_{g(x_0)-}([g(a), g(b)], X)$, we have that

$$\left(D^{\nu}_{g(x_0)-}\left(f \circ g^{-1}\right)\right)(z) = \frac{(-1)^{n-1}}{\Gamma(1-\alpha)}\frac{d}{dz}\int_z^{g(x_0)}(t-z)^{-\alpha}\left(f \circ g^{-1}\right)^{([\nu])}(t)\,dt, \quad (10.16)$$

for all $g(a) \leq z \leq g(x_0) \leq g(b)$.

We get that

$$\left(D^{n}_{g(x_0)-}\left(f \circ g^{-1}\right)\right)(z) = (-1)^n\left(f \circ g^{-1}\right)^{(n)}(z) \quad (10.17)$$

and $\left(D^{0}_{g(x_0)-}\left(f \circ g^{-1}\right)\right)(z) = \left(f \circ g^{-1}\right)(z)$, all $z \in [g(a), g(b)]$, see [1].

By [1], we have for $f \circ g^{-1} \in C^{\nu}_{g(x_0)-}([g(a), g(b)], X)$, where $x_0 \in [a, b]$ is fixed, the following right generalized g-fractional, of Canavati type, X-valued Taylor's formula:

Theorem 10.6 *Let $f \circ g^{-1} \in C^{\nu}_{g(x_0)-}([g(a), g(b)], X)$, where $x_0 \in [a, b]$ is fixed.*

(i) If $\nu \geq 1$, then

$$f(x) - f(x_0) = \sum_{k=1}^{[\nu]-1}\frac{\left(f \circ g^{-1}\right)^{(k)}(g(x_0))}{k!}(g(x) - g(x_0))^k +$$

$$\frac{1}{\Gamma(\nu)}\int_{g(x)}^{g(x_0)}(t - g(x))^{\nu-1}\left(D^{\nu}_{g(x_0)-}\left(f \circ g^{-1}\right)\right)(t)\,dt, \quad (10.18)$$

for all $a \leq x \leq x_0$,

(ii) If $0 < \nu < 1$, *we get*

$$f(x) = \frac{1}{\Gamma(\nu)} \int_{g(x)}^{g(x_0)} (t - g(x))^{\nu-1} \left(D_{g(x_0)-}^{\nu} \left(f \circ g^{-1} \right) \right)(t)\, dt, \tag{10.19}$$

all $a \leq x \leq x_0$.

(III) Denote by

$$D_{g(x_0)}^{m\nu} = D_{g(x_0)}^{\nu} D_{g(x_0)}^{\nu} \ldots D_{g(x_0)}^{\nu} \ (m\text{-times}),\ m \in \mathbb{N}. \tag{10.20}$$

We mention the following modified and generalized left X-valued fractional Taylor's formula of Canavati type:

Theorem 10.7 *Let* $f \in C^1([a,b], X)$, $g \in C^1([a,b])$, *strictly increasing:* $g^{-1} \in C^1([g(a), g(b)])$. *Assume that* $\left(D_{g(x_0)}^{i\nu} \left(f \circ g^{-1} \right) \right) \in C_{g(x_0)}^{\nu}([g(a), g(b)], X)$, $0 < \nu < 1$, $x_0 \in [a,b]$, *for* $i = 0, 1, ..., m$. *Then*

$$f(x) = \frac{1}{\Gamma((m+1)\nu)} \int_{g(x_0)}^{g(x)} (g(x) - z)^{(m+1)\nu-1} \left(D_{g(x_0)}^{(m+1)\nu} \left(f \circ g^{-1} \right) \right)(z)\, dz, \tag{10.21}$$

all $x_0 \leq x \leq b$.

(IV) Denote by

$$D_{g(x_0)-}^{m\nu} = D_{g(x_0)-}^{\nu} D_{g(x_0)-}^{\nu} \ldots D_{g(x_0)-}^{\nu} \ (m \text{ times}),\ m \in \mathbb{N}. \tag{10.22}$$

We mention the following modified and generalized right X-valued fractional Taylor's formula of Canavati type:

Theorem 10.8 *Let* $f \in C^1([a,b], X)$, $g \in C^1([a,b])$, *strictly increasing:* $g^{-1} \in C^1([g(a), g(b)])$. *Assume that* $\left(D_{g(x_0)-}^{i\nu} \left(f \circ g^{-1} \right) \right) \in C_{g(x_0)-}^{\nu}([g(a), g(b)], X)$, $0 < \nu < 1$, $x_0 \in [a,b]$, *for all* $i = 0, 1, ..., m$. *Then*

$$f(x) = \frac{1}{\Gamma((m+1)\nu)} \int_{g(x)}^{g(x_0)} (z - g(x))^{(m+1)\nu-1} \left(D_{g(x_0)-}^{(m+1)\nu} \left(f \circ g^{-1} \right) \right)(z)\, dz, \tag{10.23}$$

all $a \leq x \leq x_0 \leq b$.

10.3 Basic Banach Algebras Background

All here come from [11].

We need

Definition 10.9 ([11], *p. 245*) A complex algebra is a vector space A over the complex filed $\mathbb{C}$ in which a multiplication is defined that satisfies

$$x(yz) = (xy)z, \tag{10.24}$$

$$(x+y)z = xz + yz, \ x(y+z) = xy + xz, \tag{10.25}$$

and

$$\alpha(xy) = (\alpha x)y = x(\alpha y), \tag{10.26}$$

for all x, y and z in A and for all scalars α.

Additionally if A is a Banach space with respect to a norm that satisfies the multiplicative inequality

$$\|xy\| \leq \|x\| \|y\| \ (x \in A, y \in A) \tag{10.27}$$

and if A contains a unit element e such that

$$xe = ex = x \ (x \in A) \tag{10.28}$$

and

$$\|e\| = 1, \tag{10.29}$$

then A is called a Banach algebra.

A is commutative iff $xy = yx$ for all $x, y \in A$.

We make

Remark 10.10 Commutativity of A will be explicated stated when needed.

There exists at most one $e \in A$ that satisfies (10.28).

Inequality (10.27) makes multiplication to be continuous, more precisely left and right continuous, see [11], p. 246.

Multiplication in A is not necessarily the numerical multiplication, it is something more general and it is defined abstractly, that is for $x, y \in A$ we have $xy \in A$, e.g. composition or convolution, etc.

For nice examples about Banach algebras see [11], pp. 247–248, Sect. 10.3.

We also make

Remark 10.11 Next we mention about integration of A-valued functions, see [11], p. 259, Sect. 10.22:

If A is a Banach algebra and f is a continuous A-valued function on some compact Hausdorff space Q on which a complex Borel measure μ is defined, then $\int f d\mu$ exists and has all the properties that were discussed in Chap. 3 of [11], simply because A is a Banach space. However, an additional property can be added to these, namely: If $x \in A$, then

$$x \int_Q f \, d\mu = \int_Q x f(p) \, d\mu(p) \tag{10.30}$$

and

$$\left(\int_Q f \, d\mu \right) x = \int_Q f(p) x \, d\mu(p). \tag{10.31}$$

The Bochner integrals we will involve in our chapter follow (10.30) and (10.31). Also, let $f \in C([a,b], X)$, where $[a,b] \subset \mathbb{R}$, $(X, \|\cdot\|)$ is a Banach space. By [2], p. 3, f is Bochner integrable.

10.4 *p*-Schatten Norms Background

In this advanced section all come from [8].

Let $(H, \langle \cdot, \cdot \rangle)$ be a complex Hilbert space and $\mathcal{B}(H)$ the Banach algebra of all bounded linear operators on H. If $\{e_i\}_{i \in I}$ an orthonormal basis of H, we say that $A \in \mathcal{B}(H)$ is of trace class if

$$\|A\|_1 := \sum_{i \in I} \langle |A| e_i, e_i \rangle < \infty. \tag{10.32}$$

The definition of $\|A\|_1$ does not depend on the choice of the orthornormal basis $\{e_i\}_{i \in I}$. We denote by $\mathcal{B}_1(H)$ the set of trace class operators in $\mathcal{B}(H)$.

We define the trace of a trace class operator $A \in \mathcal{B}_1(H)$ to be

$$tr(A) := \sum_{i \in I} \langle A e_i, e_i \rangle, \tag{10.33}$$

where $\{e_i\}_{i \in I}$ an orthonormal basis of H. Note that this coincides with the usual definition of the trace if H is finite-dimensional. We observe that the series (10.33) converges absolutely and it is independent from the choice of basis.

The following result collects some properties of the trace:

Theorem 10.12 *We have:*

(i) If $A \in \mathcal{B}_1(H)$ then $A^* \in \mathcal{B}_1(H)$ and

$$tr(A^*) = \overline{tr(A)}; \tag{10.34}$$

(ii) If $A \in \mathcal{B}_1(H)$ and $T \in \mathcal{B}(H)$, then $AT, TA \in \mathcal{B}_1(H)$ and

$$tr(AT) = tr(TA) \text{ and } |tr(AT)| \leq \|A\|_1 \|T\|; \tag{10.35}$$

(iii) $tr(\cdot)$ is a bounded linear functional on $\mathcal{B}_1(H)$ with $\|tr\| = 1$;
(iv) If $A, B \in \mathcal{B}_2(H)$ then $AB, BA \in \mathcal{B}_1(H)$ and $tr(AB) = tr(BA)$;

(v) $\mathcal{B}_{fin}(H)$ *(finite rank operators) is a dense subspace of* $\mathcal{B}_1(H)$.

An operator $A \in \mathcal{B}(H)$ is said to belong to the von Neumann–Schatten class $\mathcal{B}_p(H)$, $1 \le p < \infty$ if the p-Schatten norm is finite [13, pp. 60–64]

$$\|A\|_p := \left[tr\left(|A|^p\right)\right]^{\frac{1}{p}} < \infty,$$

$|A|^p$ is an operator notation and not a power.

For $1 < p < q < \infty$ we have that

$$\mathcal{B}_1(H) \subset \mathcal{B}_p(H) \subset \mathcal{B}_q(H) \subset \mathcal{B}(H) \tag{10.36}$$

and

$$\|A\|_1 \ge \|A\|_p \ge \|A\|_q \ge \|A\|. \tag{10.37}$$

For $p \ge 1$ the functional $\|\cdot\|_p$ is a norm on the $*$-ideal $\mathcal{B}_p(H)$, which is a Banach algebra, and $\left(\mathcal{B}_p(H), \|\cdot\|_p\right)$ is a Banach space.

Also, see for instance [13, pp. 60–64], for $p \ge 1$,

$$\|A\|_p = \left\|A^*\right\|_p, \quad A \in \mathcal{B}_p(H) \tag{10.38}$$

$$\|AB\|_p \le \|A\|_p \|B\|_p, \quad A, B \in \mathcal{B}_p(H) \tag{10.39}$$

and

$$\|AB\|_p \le \|A\|_p \|B\|, \quad \|BA\|_p \le \|B\| \|A\|_p, \quad A \in \mathcal{B}_p(H), B \in \mathcal{B}(H). \tag{10.40}$$

This implies that

$$\|CAB\|_p \le \|C\| \|A\|_p \|B\|, \quad A \in \mathcal{B}_p(H), B, C \in \mathcal{B}(H). \tag{10.41}$$

In terms of p-Schatten norm we have the Hölder inequality for $p, q > 1$ with $\frac{1}{p} + \frac{1}{q} = 1$:

$$(|tr(AB)| \le) \|AB\|_1 \le \|A\|_p \|B\|_q, \quad A \in \mathcal{B}_p(H), B \in \mathcal{B}_q(H). \tag{10.42}$$

For the theory of trace functionals and their applications the interested reader is referred to [12, 13].

For some classical trace inequalities see [6, 7, 10], which are continuations of the work of Bellman [5].

10.5 Main Results

We start with a 1-2-Schatten norms left generalized Canavati fractional Hilbert–Pachpatte type inequality over $\mathcal{B}_2(H)$.

Theorem 10.13 *Let $p, q > 1$, such that $\frac{1}{p} + \frac{1}{q} = 1$, and $(\mathcal{B}_2(H), \|\cdot\|_2)$ is the $*$-ideal; $i = 1, 2$. Let also $x_{0i} \in [a_i, b_i] \subset \mathbb{R}$, $\nu_i \geq 1$, $n_i = [\nu_i]$, $f_i \in C^{n_i}([a_i, b_i], \mathcal{B}_2(H))$; $g_i \in C^1([a_i, b_i])$, strictly increasing, such that $g_i^{-1} \in C^{n_i}([g_i(a_i), g_i(b_i)])$, with $\left(f_i \circ g_i^{-1}\right)^{(k_i)}(g_i(x_{0i})) = 0$, $k_i = 0, 1, ..., n_i - 1$. Assume further that $f_i \circ g_i^{-1} \in C^{\nu_i}_{g_i(x_{0i})}([g_i(a_i), g_i(b_i)], \mathcal{B}_2(H))$. Then*

$$\int_{g_1(x_{01})}^{g_1(b_1)} \int_{g_2(x_{02})}^{g_2(b_2)} \frac{\left\| \left(f_1 \circ g_1^{-1}\right)(z_1) \left(f_2 \circ g_2^{-1}\right)(z_2) \right\|_1 dz_1 dz_2}{\left(\frac{(z_1 - g_1(x_{01}))^{p(\nu_1 - 1)+1}}{p(p(\nu_1 - 1)+1)} + \frac{(z_2 - g_2(x_{02}))^{q(\nu_2 - 1)+1}}{q(q(\nu_2 - 1)+1)} \right)} \leq$$

$$\frac{(g_1(b_1) - g_1(x_{01}))(g_2(b_2) - g_2(x_{02}))}{\Gamma(\nu_1)\Gamma(\nu_2)} \tag{10.43}$$

$$\left\| \left\| D^{\nu_1}_{g_1(x_{01})} \left(f_1 \circ g_1^{-1}\right) \right\|_2 \right\|_{L_q([g_1(x_{01}), g_1(b_1)], \mathcal{B}_2(H))} \left\| \left\| D^{\nu_2}_{g_2(x_{02})} \left(f_2 \circ g_2^{-1}\right) \right\|_2 \right\|_{L_p([g_2(x_{02}), g_2(b_2)], \mathcal{B}_2(H))}.$$

Proof By (10.10) and assumptions we get that

$$\left(f_i \circ g_i^{-1}\right)(z_i) = \frac{1}{\Gamma(\nu_i)} \int_{g_i(x_{0i})}^{z_i} (z_i - t_i)^{\nu_i - 1} \left(D^{\nu_i}_{g_i(x_{0i})} \left(f_i \circ g_i^{-1}\right) \right)(t_i)\, dt_i, \tag{10.44}$$

for all $g_i(x_{0i}) \leq z_i \leq g_i(b_i)$; $i = 1, 2$.

By Hölder's inequality we obtain

$$\left\| \left(f_1 \circ g_1^{-1}\right)(z_1) \right\|_2 \leq \frac{1}{\Gamma(\nu_1)} \int_{g_1(x_{01})}^{z_1} (z_1 - t_1)^{\nu_1 - 1} \left\| \left(D^{\nu_1}_{g_1(x_{01})} \left(f_1 \circ g_1^{-1}\right) \right)(t_1) \right\|_2 dt_1 \leq$$

$$\frac{1}{\Gamma(\nu_1)} \left(\int_{g_1(x_{01})}^{z_1} (z_1 - t_1)^{p(\nu_1 - 1)} dt_1 \right)^{\frac{1}{p}} \left(\int_{g_1(x_{01})}^{z_1} \left\| \left(D^{\nu_1}_{g_1(x_{01})} \left(f_1 \circ g_1^{-1}\right) \right)(t_1) \right\|_2^q dt_1 \right)^{\frac{1}{q}} =$$

$$\frac{1}{\Gamma(\nu_1)} \frac{(z_1 - g_1(x_{01}))^{\frac{p(\nu_1 - 1)+1}{p}}}{(p(\nu_1 - 1)+1)^{\frac{1}{p}}} \left(\int_{g_1(x_{01})}^{z_1} \left\| \left(D^{\nu_1}_{g_1(x_{01})} \left(f_1 \circ g_1^{-1}\right) \right)(t_1) \right\|_2^q dt_1 \right)^{\frac{1}{q}}. \tag{10.45}$$

That is

$$\left\| \left(f_1 \circ g_1^{-1}\right)(z_1) \right\|_2 \leq \frac{1}{\Gamma(\nu_1)} \frac{(z_1 - g_1(x_{01}))^{\frac{p(\nu_1 - 1)+1}{p}}}{(p(\nu_1 - 1)+1)^{\frac{1}{p}}}$$

$$\left(\int_{g_1(x_{01})}^{z_1} \left\| \left(D^{\nu_1}_{g_1(x_{01})} \left(f_1 \circ g_1^{-1}\right) \right)(t_1) \right\|_2^q dt_1 \right)^{\frac{1}{q}}, \tag{10.46}$$

for all $g_1(x_{01}) \leq z_1 \leq g_1(b_1)$.

Similarly, we prove that

$$\left\| \left(f_2 \circ g_2^{-1} \right)(z_2) \right\|_2 \leq \frac{1}{\Gamma(\nu_2)} \frac{(z_2 - g_2(x_{02}))^{\frac{q(\nu_2-1)+1}{q}}}{(q(\nu_2-1)+1)^{\frac{1}{q}}}$$

$$\left(\int_{g_2(x_{02})}^{z_2} \left\| \left(D_{g_2(x_{02})}^{\nu_2} \left(f_2 \circ g_2^{-1} \right) \right)(t_2) \right\|_2^p dt_2 \right)^{\frac{1}{p}}, \tag{10.47}$$

for all $g_2(x_{02}) \leq z_2 \leq g_2(b_2)$.

Therefore we have

$$\left\| \left(f_1 \circ g_1^{-1} \right)(z_1) \right\|_2 \leq \frac{1}{\Gamma(\nu_1)} \frac{(z_1 - g_1(x_{01}))^{\frac{p(\nu_1-1)+1}{p}}}{(p(\nu_1-1)+1)^{\frac{1}{p}}}$$

$$\left\| \left\| \left(D_{g_1(x_{01})}^{\nu_1} \left(f_1 \circ g_1^{-1} \right) \right) \right\|_2 \right\|_{q,[g_1(x_{01}),g_1(b_1)]}, \tag{10.48}$$

for all $g_1(x_{01}) \leq z_1 \leq g_1(b_1)$;

and

$$\left\| \left(f_2 \circ g_2^{-1} \right)(z_2) \right\|_2 \leq \frac{1}{\Gamma(\nu_2)} \frac{(z_2 - g_2(x_{02}))^{\frac{q(\nu_2-1)+1}{q}}}{(q(\nu_2-1)+1)^{\frac{1}{q}}}$$

$$\left\| \left\| \left(D_{g_2(x_{02})}^{\nu_2} \left(f_2 \circ g_2^{-1} \right) \right) \right\|_2 \right\|_{p,[g_2(x_{02}),g_2(b_2)]}, \tag{10.49}$$

for all $g_2(x_{02}) \leq z_2 \leq g_2(b_2)$.

Hence we get that

$$\left\| \left(f_1 \circ g_1^{-1} \right)(z_1) \right\|_2 \left\| \left(f_2 \circ g_2^{-1} \right)(z_2) \right\|_2 \leq \frac{1}{\Gamma(\nu_1)\Gamma(\nu_2)(p(\nu_1-1)+1)^{\frac{1}{p}}(q(\nu_2-1)+1)^{\frac{1}{q}}}$$

$$(z_1 - g_1(x_{01}))^{\frac{p(\nu_1-1)+1}{p}} (z_2 - g_2(x_{02}))^{\frac{q(\nu_2-1)+1}{q}} \tag{10.50}$$

$$\left\| \left\| \left(D_{g_1(x_{01})}^{\nu_1} \left(f_1 \circ g_1^{-1} \right) \right) \right\|_2 \right\|_{q,[g_1(x_{01}),g_1(b_1)]} \left\| \left\| \left(D_{g_2(x_{02})}^{\nu_2} \left(f_2 \circ g_2^{-1} \right) \right) \right\|_2 \right\|_{p,[g_2(x_{02}),g_2(b_2)]} \leq$$

(using Young's inequality for $a, b \geq 0$, $a^{\frac{1}{p}} b^{\frac{1}{q}} \leq \frac{a}{p} + \frac{b}{q}$)

$$\frac{1}{\Gamma(\nu_1)\Gamma(\nu_2)} \left(\frac{(z_1 - g_1(x_{01}))^{p(\nu_1-1)+1}}{p(p(\nu_1-1)+1)} + \frac{(z_2 - g_2(x_{02}))^{q(\nu_2-1)+1}}{q(q(\nu_2-1)+1)} \right)$$

$$\left\|\left\|\left(D^{\nu_1}_{g_1(x_{01})}\left(f_1 \circ g_1^{-1}\right)\right)\right\|_2\right\|_{L_q([g_1(x_{01}),g_1(b_1)],\mathcal{B}_2(H))} \left\|\left\|\left(D^{\nu_2}_{g_2(x_{02})}\left(f_2 \circ g_2^{-1}\right)\right)\right\|_2\right\|_{L_p([g_2(x_{02}),g_2(b_2)],\mathcal{B}_2(H))},$$

$\forall\, (z_1, z_2) \in [g_1(x_{01}), g_1(b_1)] \times [g_2(x_{02}), g_2(b_2)]$.

So far we have

$$\frac{\left\|\left(f_1 \circ g_1^{-1}\right)(z_1)\left(f_2 \circ g_2^{-1}\right)(z_2)\right\|_1}{\left(\frac{(z_1-g_1(x_{01}))^{p(\nu_1-1)+1}}{p(p(\nu_1-1)+1)} + \frac{(z_2-g_2(x_{02}))^{q(\nu_2-1)+1}}{q(q(\nu_2-1)+1)}\right)} \overset{(10.42)}{\leq} \tag{10.51}$$

$$\frac{\left\|\left(f_1 \circ g_1^{-1}\right)(z_1)\right\|_2\left\|\left(f_2 \circ g_2^{-1}\right)(z_2)\right\|_2}{\left(\frac{(z_1-g_1(x_{01}))^{p(\nu_1-1)+1}}{p(p(\nu_1-1)+1)} + \frac{(z_2-g_2(x_{02}))^{q(\nu_2-1)+1}}{q(q(\nu_2-1)+1)}\right)} \leq \tag{10.52}$$

$$\frac{1}{\Gamma(\nu_1)\Gamma(\nu_2)}\left\|\left\|\left(D^{\nu_1}_{g_1(x_{01})}\left(f_1 \circ g_1^{-1}\right)\right)\right\|_2\right\|_{L_q([g_1(x_{01}),g_1(b_1)],\mathcal{B}_2(H))}$$

$$\left\|\left\|\left(D^{\nu_2}_{g_2(x_{02})}\left(f_2 \circ g_2^{-1}\right)\right)\right\|_2\right\|_{L_p([g_2(x_{02}),g_2(b_2)],\mathcal{B}_2(H))},$$

$\forall\, (z_1, z_2) \in [g_1(x_{01}), g_1(b_1)] \times [g_2(x_{02}), g_2(b_2)]$.

The denominators in (10.51), (10.52) can be zero only when both $z_1 = g_1(x_{01})$ and $z_2 = g_2(x_{02})$.

Therefore we obtain (10.43), by integrating (10.51), (10.52) over $[g_1(x_{01}), g_1(b_1)] \times [g_2(x_{02}), g_2(b_2)]$. ∎

We continue with the corresponding right generalized Canavati fractional Hilbert–Pachpatte type inequality over $\mathcal{B}_2(H)$.

Theorem 10.14 *All as in Theorem 10.13, however now it is* $f_i \circ g_i^{-1} \in C^{\nu_i}_{g_i(x_{0i})-}([g_i(a_i), g_i(b_i)], \mathcal{B}_2(H))$, *for* $i = 1, 2$. *Then*

$$\int_{g_1(a_1)}^{g_1(x_{01})}\int_{g_2(a_2)}^{g_2(x_{02})} \frac{\left\|\left(f_1 \circ g_1^{-1}\right)(z_1)\left(f_2 \circ g_2^{-1}\right)(z_2)\right\|_1 dz_1 dz_2}{\left(\frac{(g_1(x_{01})-z_1)^{p(\nu_1-1)+1}}{p(p(\nu_1-1)+1)} + \frac{(g_2(x_{02})-z_2)^{q(\nu_2-1)+1}}{q(q(\nu_2-1)+1)}\right)} \leq$$

$$\frac{(g_1(x_{01}) - g_1(a_1))(g_2(x_{02}) - g_2(a_2))}{\Gamma(\nu_1)\Gamma(\nu_2)} \tag{10.53}$$

$$\left\|\left\|D^{\nu_1}_{g_1(x_{01})-}\left(f_1 \circ g_1^{-1}\right)\right\|_2\right\|_{L_q([g_1(a_1),g_1(x_{01})],\mathcal{B}_2(H))}\left\|\left\|D^{\nu_2}_{g_2(x_{02})-}\left(f_2 \circ g_2^{-1}\right)\right\|_2\right\|_{L_p([g_2(a_2),g_2(x_{02})],\mathcal{B}_2(H))}.$$

Proof Similar to Theorem 10.13, by using now (10.18). ∎

Next comes a sequential analogous left generalized Canavati fractional Hilbert–Pachpatte type inequality over $\mathcal{B}_2(H)$.

Theorem 10.15 *Let $p, q > 1$, such that $\frac{1}{p} + \frac{1}{q} = 1$, and $(\mathcal{B}_2(H), \|\cdot\|_2)$ is the $*$-ideal; $i = 1, 2$. Let also $f_i \in C^1([a_i, b_i], \mathcal{B}_2(H))$; $g_i \in C^1([a_i, b_i])$, strictly increasing, such that $g_i^{-1} \in C^1([g_i(a_i), g_i(b_i)])$. Assume that $\frac{1}{(m_i+1)q} < \nu_i < 1$, $x_{0i} \in [a_i, b_i]$, and $D_{g_i(x_{0i})}^{j_i \nu_i}(f_i \circ g_i^{-1}) \in C_{g_i(x_{0i})}^{\nu_i}([g_i(a_i), g_i(b_i)], \mathcal{B}_2(H))$, for $j_i = 0, 1, ..., m_i \in \mathbb{N}$. Then*

$$\int_{g_1(x_{01})}^{g_1(b_1)} \int_{g_2(x_{02})}^{g_2(b_2)} \frac{\left\| (f_1 \circ g_1^{-1})(z_1)(f_2 \circ g_2^{-1})(z_2) \right\|_1 dz_1 dz_2}{\left(\frac{(z_1 - g_1(x_{01}))^{p((m_1+1)\nu_1 - 1)+1}}{p(p((m_1+1)\nu_1-1)+1)} + \frac{(z_2 - g_2(x_{02}))^{q((m_2+1)\nu_2-1)+1}}{q(q((m_2+1)\nu_2-1)+1)} \right)} \leq$$

$$\frac{(g_1(b_1) - g_1(x_{01}))(g_2(b_2) - g_2(x_{02}))}{\Gamma((m_1+1)\nu_1)\Gamma((m_2+1)\nu_2)} \tag{10.54}$$

$$\left\| \left\| D_{g_1(x_{01})}^{(m_1+1)\nu_1}(f_1 \circ g_1^{-1}) \right\|_2 \right\|_{L_q([g_1(x_{01}), g_1(b_1)], \mathcal{B}_2(H))} \left\| \left\| D_{g_2(x_{02})}^{(m_2+1)\nu_2}(f_2 \circ g_2^{-1}) \right\|_2 \right\|_{L_p([g_2(x_{02}), g_2(b_2)], \mathcal{B}_2(H))}.$$

Proof Using (10.21), as similar to Theorem 10.13 the proof is omitted. ■

The right side analog of Theorem 10.15 follows:

Theorem 10.16 *Let $p, q > 1$, such that $\frac{1}{p} + \frac{1}{q} = 1$, and $(\mathcal{B}_2(H), \|\cdot\|_2)$ is the $*$-ideal; $i = 1, 2$. Let also $f_i \in C^1([a_i, b_i], \mathcal{B}_2(H))$; $g_i \in C^1([a_i, b_i])$, strictly increasing, such that $g_i^{-1} \in C^1([g_i(a_i), g_i(b_i)])$. Assume that $\frac{1}{(m_i+1)q} < \nu_i < 1$, $x_{0i} \in [a_i, b_i]$, and $D_{g_i(x_{0i})-}^{j_i \nu_i}(f_i \circ g_i^{-1}) \in C_{g_i(x_{0i})-}^{\nu_i}([g_i(a_i), g_i(b_i)], \mathcal{B}_2(H))$, for $j_i = 0, 1, ..., m_i \in \mathbb{N}$. Then*

$$\int_{g_1(a_1)}^{g_1(x_{01})} \int_{g_2(a_2)}^{g_2(x_{02})} \frac{\left\| (f_1 \circ g_1^{-1})(z_1)(f_2 \circ g_2^{-1})(z_2) \right\|_1 dz_1 dz_2}{\left(\frac{(g_1(x_{01}) - z_1)^{p((m_1+1)\nu_1 - 1)+1}}{p(p((m_1+1)\nu_1-1)+1)} + \frac{(g_2(x_{02}) - z_2)^{q((m_2+1)\nu_2-1)+1}}{q(q((m_2+1)\nu_2-1)+1)} \right)} \leq$$

$$\frac{(g_1(x_{01}) - g_1(a_1))(g_2(x_{02}) - g_2(a_2))}{\Gamma((m_1+1)\nu_1)\Gamma((m_2+1)\nu_2)} \tag{10.55}$$

$$\left\| \left\| D_{g_1(x_{01})-}^{(m_1+1)\nu_1}(f_1 \circ g_1^{-1}) \right\|_2 \right\|_{L_q([g_1(a_1), g_1(x_{01})], \mathcal{B}_2(H))} \left\| \left\| D_{g_2(x_{02})-}^{(m_2+1)\nu_2}(f_2 \circ g_2^{-1}) \right\|_2 \right\|_{L_p([g_2(a_2), g_2(x_{02})], \mathcal{B}_2(H))}.$$

Proof Using (10.23), as similar to Theorem 10.13 is omitted. ■

We continue with a γ-Schatten norm left generalized Canavati fractional Hilbert–Pachpatte type inequality over $\mathcal{B}_\gamma(H)$, $\gamma \geq 1$.

Theorem 10.17 *Let $\gamma \geq 1$, $p, q > 1$, such that $\frac{1}{p} + \frac{1}{q} = 1$, and $(\mathcal{B}_\gamma(H), \|\cdot\|_\gamma)$ is the $*$-ideal; $i = 1, 2$. Let also $x_{0i} \in [a_i, b_i] \subset \mathbb{R}$, $\nu_i \geq 1$, $n_i = [\nu_i]$, $f_i \in C^{n_i}([a_i, b_i], \mathcal{B}_\gamma(H))$; $g_i \in C^1([a_i, b_i])$, strictly increasing, such that $g_i^{-1} \in C^{n_i}([g_i(a_i), g_i(b_i)])$, with $(f_i \circ g_i^{-1})^{(k_i)}(g_i(x_{0i})) = 0$, $k_i = 0, 1, ..., n_i - 1$. Assume further that $f_i \circ g_i^{-1} \in C_{g_i(x_{0i})}^{\nu_i}([g_i(a_i), g_i(b_i)], \mathcal{B}_\gamma(H))$. Then*

$$\int_{g_1(x_{01})}^{g_1(b_1)} \int_{g_2(x_{02})}^{g_2(b_2)} \frac{\left\| \left(f_1 \circ g_1^{-1}\right)(z_1) \left(f_2 \circ g_2^{-1}\right)(z_2) \right\|_\gamma dz_1 dz_2}{\left(\frac{(z_1 - g_1(x_{01}))^{p(\nu_1 - 1)+1}}{p(p(\nu_1-1)+1)} + \frac{(z_2 - g_2(x_{02}))^{q(\nu_2-1)+1}}{q(q(\nu_2-1)+1)} \right)} \leq$$

$$\frac{(g_1(b_1) - g_1(x_{01}))(g_2(b_2) - g_2(x_{02}))}{\Gamma(\nu_1)\Gamma(\nu_2)} \tag{10.56}$$

$$\left\| \left\| D_{g_1(x_{01})}^{\nu_1} \left(f_1 \circ g_1^{-1}\right) \right\|_\gamma \right\|_{L_q([g_1(x_{01}), g_1(b_1)], \mathcal{B}_\gamma(H))} \left\| \left\| D_{g_2(x_{02})}^{\nu_2} \left(f_2 \circ g_2^{-1}\right) \right\|_\gamma \right\|_{L_p([g_2(x_{02}), g_2(b_2)], \mathcal{B}_\gamma(H))}.$$

Proof Similar to Theorem 10.13, by using norm (10.39). ■

We continue with the corresponding right generalized Canavati fractional Hilbert–Pachpatte type inequality over $\mathcal{B}_\gamma(H)$.

Theorem 10.18 *All as in Theorem 10.17, however now it is $f_i \circ g_i^{-1} \in C_{g_i(x_{0i})-}^{\nu_i}\left([g_i(a_i), g_i(b_i)], \mathcal{B}_\gamma(H)\right)$, for $i = 1, 2$. Then*

$$\int_{g_1(a_1)}^{g_1(x_{01})} \int_{g_2(a_2)}^{g_2(x_{02})} \frac{\left\| \left(f_1 \circ g_1^{-1}\right)(z_1) \left(f_2 \circ g_2^{-1}\right)(z_2) \right\|_\gamma dz_1 dz_2}{\left(\frac{(g_1(x_{01}) - z_1)^{p(\nu_1 - 1)+1}}{p(p(\nu_1-1)+1)} + \frac{(g_2(x_{02}) - z_2)^{q(\nu_2-1)+1}}{q(q(\nu_2-1)+1)} \right)} \leq$$

$$\frac{(g_1(x_{01}) - g_1(a_1))(g_2(x_{02}) - g_2(a_2))}{\Gamma(\nu_1)\Gamma(\nu_2)} \tag{10.57}$$

$$\left\| \left\| D_{g_1(x_{01})-}^{\nu_1} \left(f_1 \circ g_1^{-1}\right) \right\|_\gamma \right\|_{L_q([g_1(a_1), g_1(x_{01})], \mathcal{B}_\gamma(H))} \left\| \left\| D_{g_2(x_{02})-}^{\nu_2} \left(f_2 \circ g_2^{-1}\right) \right\|_\gamma \right\|_{L_p([g_2(a_2), g_2(x_{02})], \mathcal{B}_\gamma(H))}.$$

Proof Similar to Theorem 10.13, by using now (10.39). ■

Next comes a sequential analogous left generalized Canavati fractional Hilbert–Pachpatte type inequality over $\mathcal{B}_\gamma(H)$.

Theorem 10.19 *Let $\gamma \geq 1$, $p, q > 1$, such that $\frac{1}{p} + \frac{1}{q} = 1$, and $\left(\mathcal{B}_\gamma(H), \|\cdot\|_\gamma\right)$ is the $*$-ideal; $i = 1, 2$. Let also $f_i \in C^1\left([a_i, b_i], \mathcal{B}_\gamma(H)\right)$; $g_i \in C^1([a_i, b_i])$, strictly increasing, such that $g_i^{-1} \in C^1([g_i(a_i), g_i(b_i)])$. Assume that $\frac{1}{(m_i+1)q} < \nu_i < 1$, $x_{0i} \in [a_i, b_i]$, and $D_{g_i(x_{0i})}^{j_i \nu_i} \left(f_i \circ g_i^{-1}\right) \in C_{g_i(x_{0i})}^{\nu_i}\left([g_i(a_i), g_i(b_i)], \mathcal{B}_\gamma(H)\right)$, for $j_i = 0, 1, ..., m_i \in \mathbb{N}$. Then*

$$\int_{g_1(x_{01})}^{g_1(b_1)} \int_{g_2(x_{02})}^{g_2(b_2)} \frac{\left\| \left(f_1 \circ g_1^{-1}\right)(z_1) \left(f_2 \circ g_2^{-1}\right)(z_2) \right\|_\gamma dz_1 dz_2}{\left(\frac{(z_1 - g_1(x_{01}))^{p((m_1+1)\nu_1 - 1)+1}}{p(p((m_1+1)\nu_1-1)+1)} + \frac{(z_2 - g_2(x_{02}))^{q((m_2+1)\nu_2-1)+1}}{q(q((m_2+1)\nu_2-1)+1)} \right)} \leq$$

$$\frac{(g_1(b_1) - g_1(x_{01}))(g_2(b_2) - g_2(x_{02}))}{\Gamma((m_1+1)\nu_1)\Gamma((m_2+1)\nu_2)} \tag{10.58}$$

$$\left\| \left\| D_{g_1(x_{01})}^{(m_1+1)\nu_1} \left(f_1 \circ g_1^{-1} \right) \right\|_\gamma \right\|_{L_q([g_1(x_{01}), g_1(b_1)], \mathcal{B}_\gamma(H))} \left\| \left\| D_{g_2(x_{02})}^{(m_2+1)\nu_2} \left(f_2 \circ g_2^{-1} \right) \right\|_\gamma \right\|_{L_p([g_2(x_{02}), g_2(b_2)], \mathcal{B}_\gamma(H))}.$$

Proof Using (10.21), as similar to Theorem 10.13 the proof is omitted. ∎

The right side analog of Theorem 10.19 follows:

Theorem 10.20 *Let $\gamma \geq 1$, $p, q > 1$, such that $\frac{1}{p} + \frac{1}{q} = 1$, and $\left(\mathcal{B}_\gamma(H), \|\cdot\|_\gamma\right)$ is the $*$-ideal; $i = 1, 2$. Let also $f_i \in C^1\left([a_i, b_i], \mathcal{B}_\gamma(H)\right)$; $g_i \in C^1([a_i, b_i])$, strictly increasing, such that $g_i^{-1} \in C^1([g_i(a_i), g_i(b_i)])$. Assume that $\frac{1}{(m_i+1)q} < \nu_i < 1$, $x_{0i} \in [a_i, b_i]$, and $D_{g_i(x_{0i})-}^{j_i \nu_i}\left(f_i \circ g_i^{-1}\right) \in C_{g_i(x_{0i})-}^{\nu_i}\left([g_i(a_i), g_i(b_i)], \mathcal{B}_\gamma(H)\right)$, for $j_i = 0, 1, ..., m_i \in \mathbb{N}$. Then*

$$\int_{g_1(a_1)}^{g_1(x_{01})} \int_{g_2(a_2)}^{g_2(x_{02})} \frac{\left\| \left(f_1 \circ g_1^{-1}\right)(z_1) \left(f_2 \circ g_2^{-1}\right)(z_2) \right\|_\gamma dz_1 dz_2}{\left(\frac{(g_1(x_{01})-z_1)^{p((m_1+1)\nu_1-1)+1}}{p(p((m_1+1)\nu_1-1)+1)} + \frac{(g_2(x_{02})-z_2)^{q((m_2+1)\nu_2-1)+1}}{q(q((m_2+1)\nu_2-1)+1)} \right)} \leq$$

$$\frac{(g_1(x_{01}) - g_1(a_1))(g_2(x_{02}) - g_2(a_2))}{\Gamma((m_1+1)\nu_1)\Gamma((m_2+1)\nu_2)} \tag{10.59}$$

$$\left\| \left\| D_{g_1(x_{01})-}^{(m_1+1)\nu_1} \left(f_1 \circ g_1^{-1} \right) \right\|_\gamma \right\|_{L_q([g_1(a_1), g_1(x_{01})], \mathcal{B}_\gamma(H))} \left\| \left\| D_{g_2(x_{02})-}^{(m_2+1)\nu_2} \left(f_2 \circ g_2^{-1} \right) \right\|_\gamma \right\|_{L_p([g_2(a_2), g_2(x_{02})], \mathcal{B}_\gamma(H))}.$$

Proof Using (10.23), as similar to Theorem 10.13 is omitted. ∎

10.6 Applications

We give the following γ-Schatten Hilbert–Pachpatte fractional inequalities:

Corollary 10.21 (to Theorem 10.13) *All as in Theorem 10.13 for $g_1(t) = g_2(t) = t$. Then*

$$\int_{x_{01}}^{b_1} \int_{x_{02}}^{b_2} \frac{\|f_1(z_1) f_2(z_2)\|_1 dz_1 dz_2}{\left(\frac{(z_1-x_{01})^{p(\nu_1-1)+1}}{p(p(\nu_1-1)+1)} + \frac{(z_2-x_{02})^{q(\nu_2-1)+1}}{q(q(\nu_2-1)+1)} \right)} \leq \tag{10.60}$$

$$\frac{(b_1 - x_{01})(b_2 - x_{02})}{\Gamma(\nu_1)\Gamma(\nu_2)} \left\| \left\| D_{x_{01}}^{\nu_1} f_1 \right\|_2 \right\|_{L_q([x_{01}, b_1], \mathcal{B}_2(H))} \left\| \left\| D_{x_{02}}^{\nu_2} f_2 \right\|_2 \right\|_{L_p([x_{02}, b_2], \mathcal{B}_2(H))}.$$

We continue with

Corollary 10.22 (to Theorem 10.15) *All as in Theorem 10.15 for $g_1(t) = g_2(t) = t$. Then*

$$\int_{x_{01}}^{b_1} \int_{x_{02}}^{b_2} \frac{\|f_1(z_1) f_2(z_2)\|_1 dz_1 dz_2}{\left(\frac{(z_1-x_{01})^{p((m_1+1)\nu_1-1)+1}}{p(p((m_1+1)\nu_1-1)+1)} + \frac{(z_2-x_{02})^{q((m_2+1)\nu_2-1)+1}}{q(q((m_2+1)\nu_2-1)+1)} \right)} \leq \tag{10.61}$$

$$\frac{(b_1 - x_{01})(b_2 - x_{02})}{\Gamma((m_1+1)\nu_1)\Gamma((m_2+1)\nu_2)}$$

$$\left\| \left\| D_{x_{01}}^{(m_1+1)\nu_1} f_1 \right\|_2 \right\|_{L_q([x_{01},b_1],\mathcal{B}_2(H))} \left\| \left\| D_{x_{02}}^{(m_2+1)\nu_2} f_2 \right\|_2 \right\|_{L_p([x_{02},b_2],\mathcal{B}_2(H))}.$$

Next we present

Corollary 10.23 (to Theorem 10.18) *All as in Theorem 10.18 for $g_1(t) = g_2(t) = e^t$. Then*

$$\int_{e^{a_1}}^{e^{x_{01}}} \int_{e^{a_2}}^{e^{x_{02}}} \frac{\|(f_1 \circ \log)(z_1)(f_2 \circ \log)(z_2)\|_\gamma dz_1 dz_2}{\left(\frac{(e^{x_{01}}-z_1)^{p(\nu_1-1)+1}}{p(p(\nu_1-1)+1)} + \frac{(e^{x_{02}}-z_2)^{q(\nu_2-1)+1}}{q(q(\nu_2-1)+1)} \right)} \leq$$

$$\frac{(e^{x_{01}} - e^{a_1})(e^{x_{02}} - e^{a_2})}{\Gamma(\nu_1)\Gamma(\nu_2)} \tag{10.62}$$

$$\left\| \left\| D_{e^{x_{01}}-}^{\nu_1} (f_1 \circ \log) \right\|_\gamma \right\|_{L_q([e^{a_1},e^{x_{01}}],\mathcal{B}_\gamma(H))} \left\| \left\| D_{e^{x_{02}}-}^{\nu_2} (f_2 \circ \log) \right\|_\gamma \right\|_{L_p([e^{a_2},e^{x_{02}}],\mathcal{B}_\gamma(H))}.$$

We finish with

Corollary 10.24 (to Theorem 10.20) *All as in Theorem 10.20, with $g_1(t) = g_2(t) = \log t$, where $[a_i, b_i] \subset \mathbb{R}_+ - \{0\}$, $i = 1, 2$. Then*

$$\int_{\log a_1}^{\log x_{01}} \int_{\log a_2}^{\log x_{02}} \frac{\left\|(f_1 \circ e^t)(z_1)(f_2 \circ e^t)(z_2)\right\|_\gamma dz_1 dz_2}{\left(\frac{(\log x_{01}-z_1)^{p((m_1+1)\nu_1-1)+1}}{p(p((m_1+1)\nu_1-1)+1)} + \frac{(\log x_{02}-z_2)^{q((m_2+1)\nu_2-1)+1}}{q(q((m_2+1)\nu_2-1)+1)} \right)} \leq$$

$$\frac{\left(\log \frac{x_{01}}{a_1}\right)\left(\log \frac{x_{02}}{a_2}\right)}{\Gamma((m_1+1)\nu_1)\Gamma((m_2+1)\nu_2)} \tag{10.63}$$

$$\left\| \left\| D_{\log x_{01}-}^{(m_1+1)\nu_1} (f_1 \circ e^t) \right\|_\gamma \right\|_{L_q([\log a_1,\log x_{01}],\mathcal{B}_\gamma(H))} \left\| \left\| D_{\log x_{02}-}^{(m_2+1)\nu_2} (f_2 \circ e^t) \right\|_\gamma \right\|_{L_p([\log a_2,\log x_{02}],\mathcal{B}_\gamma(H))}.$$

References

1. Anastassiou, G.A.: Strong mixed and generalized fractional calculus for Banach space valued functions. Mat. Vesnik **69**(3), 176–191 (2017)
2. Anastassiou, G.A.: Intelligent Computations: Abstract Fractional Calculus, Inequalities, Approximations. Springer, Heidelberg (2018)
3. Anastassiou, G.A.: Generalized Canavati Fractional Hilbert-Pachpatte type inequalities for Banach algebra valued functions. J. Comput. Anal. Appl. **30**(1), 66–77 (2022)
4. Anastassiou, G.A.: γ- Schatten norm Generalized Canavati Fractional Hilbert-Pachpatte type inequalities for von Neumann-Schatten class $B_\gamma(H)$ valued functions. Analele Universitatii Oradea, Fasc. Matematica, accepted (2021)

5. Bellman, R.: Some inequalities for positive definite matrices. In: Beckenbach, E.F. (ed.), General Inequalities 2, Proceedings of the 2nd International Conference on General Inequalities, pp. 89–90. Birkhauser, Basel (1980)
6. Chang, D.: A matrix trace inequality for products of Hermitian matrices. J. Math. Anal. Appl. **237**, 721–725 (1999)
7. Coop, I.D.: On matrix trace inequalities and related topics for products of Hermitian matrix. J. Math. Anal. Appl. **188**, 999–1001 (1994)
8. Dragomir, S.S.: p-Schatten norm inequalities of Ostrowski's type. RGMIA Res. Rep. Coll. **24**, Art. 108, 19 pp (2021)
9. Mikusinski, J.: The Bochner Integral. Academic, New York (1978)
10. Neudecker, H.: A matrix trace inequality. J. Math. Anal. Appl. **166**, 302–303 (1992)
11. Rudin, W.: Functional Analysis, 2nd edn. McGraw-Hill Inc, New York (1991)
12. Simon, B.: Trace ideals and Their Applications. Cambridge University Press, Cambridge (1979)
13. Zagrebvov, V.A.: Gibbs Semigroups. Operator Theory: Advances and Applications, vol. 273. Birkhauser (2019)

Chapter 11
γ-Schatten Norm Generalized Ostrowski, Opial and Hilbert–Pachpatte Inequalities with von Neumann–Schatten Class $\mathcal{B}_\gamma(H)$ Valued Functions Using Ordinary Vectorial Derivatives

Using a generalized vectorial Taylor formula involving ordinary vector derivatives we establish mixed Ostrowski, Opial and Hilbert–Pachpatte type inequalities for several von Neumann–Schatten class $\mathcal{B}_\gamma(H)$ valued functions. The estimates are with respect to all norms $\|\cdot\|_p$, $1 \leq p \leq \infty$. We finish with applications. It follows [4].

11.1 Introduction

Our main motivation is [3].

We mention a uniform mixed generalized Ostrowski type inequality for several functions that are Banach algebra valued.

Theorem 11.1 ([3]) *Let* $n \in \mathbb{N}$ *and* $f_i \in C^n([a,b], A)$, $i = 1, ..., r \in \mathbb{N} - \{1\}$; *where* $[a,b] \subset \mathbb{R}$ *and* $(A, \|\cdot\|)$ *is a Banach algebra. Let* $g \in C^1([a,b])$, *strictly increasing, such that* $g^{-1} \in C^n([g(a), g(b)])$. *We assume that* $(f_i \circ g^{-1})^{(j)}(g(x_0)) = 0$, $j = 1, ..., n-1$; $i = 1, ..., r$; *where* $x_0 \in [a,b]$ *be fixed. Denote by*

$$E(f_1, ..., f_r)(x_0) :=$$

$$\sum_{i=1}^{r}\left[\int_a^b\left(\prod_{\substack{j=1\\ j\neq i}}^{r} f_j(x)\right) f_i(x)\,dx - \left(\int_a^b\left(\prod_{\substack{j=1\\ j\neq i}}^{r} f_j(x)\right)dx\right) f_i(x_0)\right]. \quad (11.1)$$

G. A. Anastassiou, *Fractional Inequalities In Banach Algebras*,
Studies in Systems, Decision and Control 441,
https://doi.org/10.1007/978-3-031-05148-7_11

Then

(1)

$$E\left(f_1,...,f_r\right)\left(x_0\right)=\frac{1}{(n-1)!}$$

$$\sum_{i=1}^{r}\left[(-1)^n\left[\int_a^{x_0}\left(\prod_{\substack{j=1\\j\neq i}}^{r}f_j\left(x\right)\right)\left(\int_{g(x)}^{g(x_0)}\left(z-g\left(x\right)\right)^{n-1}\left(f_i\circ g^{-1}\right)^{(n)}\left(z\right)dz\right)dx\right]+\right.$$

$$\left.\left[\int_{x_0}^{b}\left(\prod_{\substack{j=1\\j\neq i}}^{r}f_j\left(x\right)\right)\left(\int_{g(x_0)}^{g(x)}\left(g\left(x\right)-z\right)^{n-1}\left(f_i\circ g^{-1}\right)^{(n)}\left(z\right)dz\right)dx\right]\right], \tag{11.2}$$

and

(2)

$$\left\|E\left(f_1,...,f_r\right)\left(x_0\right)\right\|\leq\frac{1}{n!}$$

$$\left\{\sum_{i=1}^{r}\left[\left[\left\|\left(f_i\circ g^{-1}\right)^{(n)}\right\|_{\infty,[g(a),g(x_0)]}\left(g\left(x_0\right)-g\left(a\right)\right)^n\left(\int_a^{x_0}\left(\prod_{\substack{j=1\\j\neq i}}^{r}\left\|f_j\left(x\right)\right\|\right)dx\right)\right]+\right.\right.$$

$$\left.\left.\left[\left\|\left(f_i\circ g^{-1}\right)^{(n)}\right\|_{\infty,[g(x_0),g(b)]}\left(g\left(b\right)-g\left(x_0\right)\right)^n\left(\int_{x_0}^{b}\left(\prod_{\substack{j=1\\j\neq i}}^{r}\left\|f_j\left(x\right)\right\|\right)dx\right)\right]\right]\right\}. \tag{11.3}$$

We also mention a left generalized Opial type inequality for ordinary vector valued derivatives:

Theorem 11.2 ([3]) *Let* $p,q>1:\frac{1}{p}+\frac{1}{q}=1,$ *and* $n\in\mathbb{N},\ f\in C^n\left([a,b],A\right);$ *where* $[a,b]\subset\mathbb{R}$ *and* $(A,\|\cdot\|)$ *is a Banach algebra. Let* $g\in C^1\left([a,b]\right),$ *strictly increasing, such that* $g^{-1}\in C^n\left([g(a),g(b)]\right).$ *We assume that* $\left(f\circ g^{-1}\right)^{(j)}\left(g\left(x_0\right)\right)=0,\ j=0,1,...,n-1;$ *where* $x_0\in[a,b]$ *be fixed. Then*

$$\int_{g(x_0)}^{g(x)}\left\|\left(f\circ g^{-1}\right)\left(z\right)\left(f\circ g^{-1}\right)^{(n)}\left(z\right)\right\|dz\leq$$

$$\frac{\left(g\left(x\right)-g\left(x_0\right)\right)^{n+\frac{1}{p}-\frac{1}{q}}}{2^{\frac{1}{q}}\left(n-1\right)!\left[\left(p\left(n-1\right)+1\right)\left(p\left(n-1\right)+2\right)\right]^{\frac{1}{p}}}\left(\int_{g(x_0)}^{g(x)}\left\|\left(f\circ g^{-1}\right)^{(n)}\left(z\right)\right\|^q dz\right)^{\frac{2}{q}}, \tag{11.4}$$

for all $x_0\leq x\leq b.$

We also mention a left generalized Hilbert–Pachpatte inequality for ordinary vector valued derivatives.

Theorem 11.3 ([3]) *Let* $i = 1, 2;$ $p, q > 1 : \frac{1}{p} + \frac{1}{q} = 1,$ *and* $n_i \in \mathbb{N},$ $f_i \in C^{n_i}([a_i, b_i], A)$*; where* $[a_i, b_i] \subset \mathbb{R}$ *and* $(A, \|\cdot\|)$ *is a Banach algebra. Let* $g_i \in C^1([a_i, b_i])$*, strictly increasing, such that* $g_i^{-1} \in C^{n_i}([g_i(a_i), g_i(b_i)])$*. We assume that* $\left(f_i \circ g_i^{-1}\right)^{(j_i)}(g_i(x_{0i})) = 0,$ $j_i = 0, 1, ..., n_i - 1;$ *where* $x_{0i} \in [a_i, b_i]$ *be fixed. Then*

$$\int_{g_1(x_{01})}^{g_1(b_1)} \int_{g_2(x_{02})}^{g_2(b_2)} \frac{\left\|\left(f_1 \circ g_1^{-1}\right)(z_1)\left(f_2 \circ g_2^{-1}\right)(z_2)\right\| dz_1 dz_2}{\left(\frac{(z_1 - g_1(x_{01}))^{p(n_1-1)+1}}{p(p(n_1-1)+1)} + \frac{(z_2 - g_2(x_{02}))^{q(n_2-1)+1}}{q(q(n_2-1)+1)}\right)} \leq$$

$$\frac{(g_1(b_1) - g_1(x_{01}))(g_2(b_2) - g_2(x_{02}))}{(n_1-1)!(n_2-1)!} \tag{11.5}$$

$$\left\|\left\|\left(f_1 \circ g_1^{-1}\right)^{(n_1)}\right\|\right\|_{L_q([g_1(x_{01}), g_1(b_1)], A)} \left\|\left\|\left(f_2 \circ g_2^{-1}\right)^{(n_2)}\right\|\right\|_{L_p([g_2(x_{02}), g_2(b_2)], A)}.$$

We are motivated also by Dragomir [8] recent work:

Let $(H, \langle\cdot,\cdot\rangle)$ be a complex Hilbert space and $\mathcal{B}(H)$ the Banach algebra of all bounded linear operators on H.

An operator $A \in \mathcal{B}(H)$ is said to belong to the von Neumann–Schatten class $\mathcal{B}_p(H)$, $1 \leq p < \infty$ if the p-Schatten norm is finite

$$\|A\|_p := \left[tr\left(|A|^p\right)\right]^{\frac{1}{p}} < \infty.$$

Assume that $A : [a, b] \to \mathcal{B}_p(H)$, $B : [a, b] \to \mathcal{B}_q(H)$, $p, q > 1$ with $\frac{1}{p} + \frac{1}{q} = 1$, are continuous and B is strongly differentiable on (a, b), then

$$\left\|\int_a^b A(t) B(t) dt - \left(\int_a^b A(s) ds\right) B(u)\right\|_1 \leq$$

$$\sup_{t\in[a,b]} \left\|B'(t)\right\|_q \times \begin{cases} \left[\frac{1}{2}(b-a) + \left|u - \frac{a+b}{2}\right|\right] \int_a^b \|A(t)\|_p dt, \\ \left[\frac{(u-a)^{\beta+1} + (b-u)^{\beta+1}}{\beta+1}\right]^{\frac{1}{\beta}} \left(\int_a^b \|A(t)\|_p^{\alpha}\right)^{\frac{1}{\alpha}}, \\ \text{for } \alpha, \beta > 1 \text{ with } \frac{1}{\alpha} + \frac{1}{\beta} = 1 \\ \left[\frac{1}{4}(b-a)^2 + \left(u - \frac{a+b}{2}\right)^2\right] \sup_{t\in[a,b]} \|A(t)\|_p, \end{cases} \tag{11.6}$$

for all $u \in [a, b]$, an Ostrowski type inequality.

Ostrowski type inequalities have great applications to integral approximations in Numerical Analysis.

We presented ([1], Chaps. 8, 9) mixed fractional Ostrowski inequalities for several functions for various norms.

In this chapter we generalize [1], Chaps. 8, 9 for several $\mathcal{B}_p(H)$ valued functions by using ordinary vector valued derivatives and our integrals here are of Bochner type [9].

Opial-type inequalities are used a lot in proving uniqueness of solutions to differential equations and also to give upper bounds to their solutions.

In this chapter we also derive Opial type inequalities for $\mathcal{B}_p(H)$ valued functions with respect to ordinary vector valued derivatives.

Additionally we include in this article related to $\mathcal{B}_p(H)$ Hilbert–Pachpatte type inequalities, [11]. We finish the chapter with selective applications to Ostrowski, Opial and Hilbert–Pachpatte inequalities.

11.2 Background

We use the following generalized vector Taylor's formula:

Theorem 11.4 ([2], p. 97) *Let $n \in \mathbb{N}$ and $f \in C^n([a,b], X)$, where $[a,b] \subset \mathbb{R}$ and $(X, \|\cdot\|)$ is a Banach space. Let $g \in C^1([a,b])$, strictly increasing, such that $g^{-1} \in C^n([g(a), g(b)])$. Let any $x, y \in [a,b]$. Then*

$$f(x) = f(y) + \sum_{i=1}^{n-1} \frac{(g(x) - g(y))^i}{i!} \left(f \circ g^{-1}\right)^{(i)} (g(y)) \tag{11.7}$$

$$+ \frac{1}{(n-1)!} \int_{g(y)}^{g(x)} (g(x) - z)^{n-1} \left(f \circ g^{-1}\right)^{(n)} (z)\, dz.$$

The derivatives here are defined similarly to the numerical ones, see [13], pp. 83–86.

The above integral is of Bochner type [9], and so are the integrals in this work. By [2], p. 3, if $f \in C([a,b], X)$ then f is Bochner integrable.

11.3 About Basic Banach Algebras

All here come from [12].

We need

Definition 11.5 ([12], *p. 245*) A complex algebra is a vector space A over the complex field $\mathbb{C}$ in which a multiplication is defined that satisfies

$$x(yz) = (xy)z, \tag{11.8}$$

$$(x + y) z = xz + yz, x (y + z) = xy + xz, \tag{11.9}$$

and

$$\alpha (xy) = (\alpha x) y = x (\alpha y), \tag{11.10}$$

for all x, y and z in A and for all scalars α.

Additionally if A is a Banach space with respect to a norm that satisfies the multiplicative inequality

$$\|xy\| \leq \|x\| \|y\| \quad (x \in A, y \in A) \tag{11.11}$$

and if A contains a unit element e such that

$$xe = ex = x \quad (x \in A) \tag{11.12}$$

and

$$\|e\| = 1, \tag{11.13}$$

then A is called a Banach algebra.

A is commutative iff $xy = yx$ for all $x, y \in A$.

We make

Remark 11.6 Commutativity of A will be explicated stated when needed.

There exists at most one $e \in A$ that satisfies (11.12).

Inequality (11.11) makes multiplication to be continuous, more precisely left and right continuous, see [12], p. 246.

Multiplication in A is not necessarily the numerical multiplication, it is something more general and it is defined abstractly, that is for $x, y \in A$ we have $xy \in A$, e.g. composition or convolution, etc.

For nice examples about Banach algebras see [12], pp. 247–248, Sect. 10.3.

We also make

Remark 11.7 Next we mention about integration of A-valued functions, see [12], p. 259, Sect. 10.22:

If A is a Banach algebra and f is a continuous A-valued function on some compact Hausdorff space Q on which a complex Borel measure μ is defined, then $\int f d\mu$ exists and has all the properties that were discussed in Chap. 3 of [12], simply because A is a Banach space. However, an additional property can be added to these, namely: If $x \in A$, then

$$x \int_Q f \, d\mu = \int_Q xf(p) \, d\mu(p) \tag{11.14}$$

and

$$\left(\int_{Q} f \, d\mu\right) x = \int_{Q} f(p) \, x \, d\mu(p). \tag{11.15}$$

The Bochner integrals we will involve in our chapter follow (11.14) and (11.15).

11.4 *p*-Schatten Norms Background

In this advanced section all come from [8].

Let $(H, \langle \cdot, \cdot \rangle)$ be a complex Hilbert space and $\mathcal{B}(H)$ the Banach algebra of all bounded linear operators on H. If $\{e_i\}_{i \in I}$ an orthonormal basis of H, we say that $A \in \mathcal{B}(H)$ is of trace class if

$$\|A\|_1 := \sum_{i \in I} \langle |A| e_i, e_i \rangle < \infty. \tag{11.16}$$

The definition of $\|A\|_1$ does not depend on the choice of the orthornormal basis $\{e_i\}_{i \in I}$. We denote by $\mathcal{B}_1(H)$ the set of trace class operators in $\mathcal{B}(H)$.

We define the trace of a trace class operator $A \in \mathcal{B}_1(H)$ to be

$$tr(A) := \sum_{i \in I} \langle Ae_i, e_i \rangle, \tag{11.17}$$

where $\{e_i\}_{i \in I}$ an orthonormal basis of H. Note that this coincides with the usual definition of the trace if H is finite-dimensional. We observe that the series (11.17) converges absolutely and it is independent from the choice of basis.

The following result collects some properties of the trace:

Theorem 11.8 *We have:*

(i) If $A \in \mathcal{B}_1(H)$ then $A^* \in \mathcal{B}_1(H)$ and

$$tr(A^*) = \overline{tr(A)}; \tag{11.18}$$

(ii) If $A \in \mathcal{B}_1(H)$ and $T \in \mathcal{B}(H)$, then $AT, TA \in \mathcal{B}_1(H)$ and

$$tr(AT) = tr(TA) \text{ and } |tr(AT)| \le \|A\|_1 \|T\|; \tag{11.19}$$

(iii) $tr(\cdot)$ is a bounded linear functional on $\mathcal{B}_1(H)$ with $\|tr\| = 1$;
(iv) If $A, B \in \mathcal{B}_2(H)$ then $AB, BA \in \mathcal{B}_1(H)$ and $tr(AB) = tr(BA)$;
(v) $\mathcal{B}_{fin}(H)$ (finite rank operators) is a dense subspace of $\mathcal{B}_1(H)$.

An operator $A \in \mathcal{B}(H)$ is said to belong to the von Neumann–Schatten class $\mathcal{B}_p(H)$, $1 \le p < \infty$ if the p-Schatten norm is finite [15, pp. 60–64]

$$\|A\|_p := \left[tr\left(|A|^p\right)\right]^{\frac{1}{p}} < \infty,$$

$|A|^p$ is an operator notation and not a power.

For $1 < p < q < \infty$ we have that

$$\mathcal{B}_1(H) \subset \mathcal{B}_p(H) \subset \mathcal{B}_q(H) \subset \mathcal{B}(H) \tag{11.20}$$

and

$$\|A\|_1 \geq \|A\|_p \geq \|A\|_q \geq \|A\|. \tag{11.21}$$

For $p \geq 1$ the functional $\|\cdot\|_p$ is a norm on the $*$-ideal $\mathcal{B}_p(H)$, which is a Banach algebra, and $\left(\mathcal{B}_p(H), \|\cdot\|_p\right)$ is a Banach space.

Also, see for instance [15, pp. 60–64], for $p \geq 1$,

$$\|A\|_p = \left\|A^*\right\|_p, A \in \mathcal{B}_p(H) \tag{11.22}$$

$$\|AB\|_p \leq \|A\|_p \|B\|_p, A, B \in \mathcal{B}_p(H) \tag{11.23}$$

and

$$\|AB\|_p \leq \|A\|_p \|B\|, \|BA\|_p \leq \|B\| \|A\|_p, A \in \mathcal{B}_p(H), B \in \mathcal{B}(H). \tag{11.24}$$

This implies that

$$\|CAB\|_p \leq \|C\| \|A\|_p \|B\|, A \in \mathcal{B}_p(H), B, C \in \mathcal{B}(H). \tag{11.25}$$

In terms of p-Schatten norm we have the Hölder inequality for $p, q > 1$ with $\frac{1}{p} + \frac{1}{q} = 1$:

$$(|tr(AB)| \leq) \|AB\|_1 \leq \|A\|_p \|B\|_q, \ A \in \mathcal{B}_p(H), B \in \mathcal{B}_q(H). \tag{11.26}$$

For the theory of trace functionals and their applications the interested reader is referred to [14, 15].

For some classical trace inequalities see [6, 7, 10], which are continuations of the work of Bellman [5].

11.5 Main Results

We start with 1-2-Schatten norms mixed generalized Ostrowski type inequalities for several functions that are Banach algebra $\mathcal{B}_2(H) \subset \mathcal{B}(H)$ valued. A uniform estimate follows.

Theorem 11.9 *Let $n \in \mathbb{N}$ and $f_i \in C^n([a,b], \mathcal{B}_2(H))$, $i = 1, ..., r \in \mathbb{N} - \{1\}$; where $[a,b] \subset \mathbb{R}$ and $\mathcal{B}_2(H)$ is a $*$-ideal, which $(\mathcal{B}_2(H), \|\cdot\|_2)$ is a Banach algebra. Let $g \in C^1([a,b])$, strictly increasing, such that $g^{-1} \in C^n([g(a), g(b)])$. We assume that $\left(f_i \circ g^{-1}\right)^{(j)}(g(x_0)) = 0$, $j = 1, ..., n-1$; $i = 1, ..., r$; where $x_0 \in [a,b]$ be fixed. Denote by*

$$E(f_1, ..., f_r)(x_0) :=$$

$$\sum_{i=1}^{r}\left[\int_a^b \left(\prod_{\substack{j=1 \\ j\neq i}}^{r} f_j(x)\right) f_i(x)\,dx - \left(\int_a^b \left(\prod_{\substack{j=1 \\ j\neq i}}^{r} f_j(x)\right) dx\right) f_i(x_0)\right]. \tag{11.27}$$

Then

$$\|E(f_1, ..., f_r)(x_0)\|_1 \leq \frac{1}{n!}$$

$$\left\{\sum_{i=1}^{r}\left[\left[\left\|\left\|\left(f_i \circ g^{-1}\right)^{(n)}\right\|_2\right\|_{\infty,[g(a),g(x_0)]} (g(x_0) - g(a))^n \left(\int_a^{x_0}\left(\prod_{\substack{j=1 \\ j\neq i}}^{r} \|f_j(x)\|_2\right) dx\right)\right]+\right.\right.$$

$$\left.\left.\left[\left\|\left\|\left(f_i \circ g^{-1}\right)^{(n)}\right\|_2\right\|_{\infty,[g(x_0),g(b)]} (g(b) - g(x_0))^n \left(\int_{x_0}^{b}\left(\prod_{\substack{j=1 \\ j\neq i}}^{r} \|f_j(x)\|_2\right) dx\right)\right]\right]\right\}. \tag{11.28}$$

Proof Let $x_0 \in [a,b]$ such that $\left(f_i \circ g^{-1}\right)^{(j)}(g(x_0)) = 0$, $j = 1, ..., n-1$; $i = 1, ..., r$. Let $x \in [a, x_0]$, then by Theorem 11.4 we have

$$f_i(x) - f_i(x_0) = \frac{1}{(n-1)!}\int_{g(x_0)}^{g(x)} (g(x) - z)^{n-1}\left(f_i \circ g^{-1}\right)^{(n)}(z)\,dz \tag{11.29}$$

$$= \frac{(-1)^n}{(n-1)!}\int_{g(x)}^{g(x_0)} (z - g(x))^{n-1}\left(f_i \circ g^{-1}\right)^{(n)}(z)\,dz,$$

for $i = 1, ..., r$.

And for $x \in [x_0, b]$, then again by Theorem 11.4 we get

$$f_i(x) - f_i(x_0) = \frac{1}{(n-1)!}\int_{g(x_0)}^{g(x)} (g(x) - z)^{n-1}\left(f_i \circ g^{-1}\right)^{(n)}(z)\,dz, \tag{11.30}$$

for $i = 1, ..., r$.

We multiply (11.29) by $\left(\prod_{\substack{j=1 \\ j\neq i}}^{r} f_j(x)\right)$ to get:

$$\left(\prod_{\substack{j=1\\ j\neq i}}^{r} f_j(x)\right) f_i(x) - \left(\prod_{\substack{j=1\\ j\neq i}}^{r} f_j(x)\right) f_i(x_0) =$$

$$\frac{\left(\prod_{\substack{j=1\\ j\neq i}}^{r} f_j(x)\right)(-1)^n}{(n-1)!} \int_{g(x)}^{g(x_0)} (z - g(x))^{n-1} \left(f_i \circ g^{-1}\right)^{(n)}(z)\, dz, \tag{11.31}$$

$\forall\, x \in [a, x_0]$; for $i = 1, ..., r$.

Similarly, we get (by (11.30))

$$\left(\prod_{\substack{j=1\\ j\neq i}}^{r} f_j(x)\right) f_i(x) - \left(\prod_{\substack{j=1\\ j\neq i}}^{r} f_j(x)\right) f_i(x_0) =$$

$$\frac{\left(\prod_{\substack{j=1\\ j\neq i}}^{r} f_j(x)\right)}{(n-1)!} \int_{g(x_0)}^{g(x)} (g(x) - z)^{n-1} \left(f_i \circ g^{-1}\right)^{(n)}(z)\, dz, \tag{11.32}$$

$\forall\, x \in [x_0, b]$; for $i = 1, ..., r$.

Adding (11.31) and (11.32) as separate groups, we obtain

$$\sum_{i=1}^{r}\left(\prod_{\substack{j=1\\ j\neq i}}^{r} f_j(x)\right) f_i(x) - \sum_{i=1}^{r}\left(\prod_{\substack{j=1\\ j\neq i}}^{r} f_j(x)\right) f_i(x_0) =$$

$$\frac{(-1)^n}{(n-1)!}\sum_{i=1}^{r}\left(\prod_{\substack{j=1\\ j\neq i}}^{r} f_j(x)\right) \int_{g(x)}^{g(x_0)} (z - g(x))^{n-1} \left(f_i \circ g^{-1}\right)^{(n)}(z)\, dz, \tag{11.33}$$

$\forall\, x \in [a, x_0]$,

and

$$\sum_{i=1}^{r}\left(\prod_{\substack{j=1\\ j\neq i}}^{r} f_j(x)\right) f_i(x) - \sum_{i=1}^{r}\left(\prod_{\substack{j=1\\ j\neq i}}^{r} f_j(x)\right) f_i(x_0) =$$

$$\frac{1}{(n-1)!}\sum_{i=1}^{r}\left(\prod_{\substack{j=1\\ j\neq i}}^{r} f_j(x)\right)\int_{g(x_0)}^{g(x)} (g(x)-z)^{n-1}\left(f_i\circ g^{-1}\right)^{(n)}(z)\,dz, \tag{11.34}$$

$\forall\, x \in [x_0, b]$.

Next, we integrate (11.33) and (11.34) with respect to $x \in [a, b]$. We have

$$\sum_{i=1}^{r}\int_{a}^{x_0}\left(\prod_{\substack{j=1\\ j\neq i}}^{r} f_j(x)\right) f_i(x)\,dx - \sum_{i=1}^{r}\left(\int_{a}^{x_0}\left(\prod_{\substack{j=1\\ j\neq i}}^{r} f_j(x)\right)dx\right) f_i(x_0) = \tag{11.35}$$

$$\frac{(-1)^n}{(n-1)!}\sum_{i=1}^{r}\left[\int_{a}^{x_0}\left(\prod_{\substack{j=1\\ j\neq i}}^{r} f_j(x)\right)\left(\int_{g(x)}^{g(x_0)} (z-g(x))^{n-1}\left(f_i\circ g^{-1}\right)^{(n)}(z)\,dz\right)dx\right],$$

and

$$\sum_{i=1}^{r}\int_{x_0}^{b}\left(\prod_{\substack{j=1\\ j\neq i}}^{r} f_j(x)\right) f_i(x)\,dx - \sum_{i=1}^{r}\left(\int_{x_0}^{b}\left(\prod_{\substack{j=1\\ j\neq i}}^{r} f_j(x)\right)dx\right) f_i(x_0) = \tag{11.36}$$

$$\frac{1}{(n-1)!}\sum_{i=1}^{r}\left[\int_{x_0}^{b}\left(\prod_{\substack{j=1\\ j\neq i}}^{r} f_j(x)\right)\left(\int_{g(x_0)}^{g(x)} (g(x)-z)^{n-1}\left(f_i\circ g^{-1}\right)^{(n)}(z)\,dz\right)dx\right].$$

Finally, adding (11.35) and (11.36) we obtain the useful identity

$$E(f_1, ..., f_r)(x_0) =$$

$$\sum_{i=1}^{r}\left[\int_{a}^{b}\left(\prod_{\substack{j=1\\ j\neq i}}^{r} f_j(x)\right) f_i(x)\,dx - \left(\int_{a}^{b}\left(\prod_{\substack{j=1\\ j\neq i}}^{r} f_j(x)\right)dx\right) f_i(x_0)\right] = \frac{1}{(n-1)!}$$

$$\sum_{i=1}^{r}\left[(-1)^n\left[\int_{a}^{x_0}\left(\prod_{\substack{j=1\\ j\neq i}}^{r} f_j(x)\right)\left(\int_{g(x)}^{g(x_0)} (z-g(x))^{n-1}\left(f_i\circ g^{-1}\right)^{(n)}(z)\,dz\right)dx\right] +\right.$$

$$\left[\int_{x_0}^{b}\left(\prod_{\substack{j=1\\ j\neq i}}^{r} f_j(x)\right)\left(\int_{g(x_0)}^{g(x)}(g(x)-z)^{n-1}\left(f_i\circ g^{-1}\right)^{(n)}(z)\,dz\right)dx\right]\Bigg], \quad (11.37)$$

see also (11.2).

Therefore, we get that

$$\|E(f_1,...,f_r)(x_0)\|_1 =$$

$$\left\|\sum_{i=1}^{r}\left[\int_a^b\left(\prod_{\substack{j=1\\ j\neq i}}^{r} f_j(x)\right) f_i(x)\,dx-\left(\int_a^b\left(\prod_{\substack{j=1\\ j\neq i}}^{r} f_j(x)\right)dx\right) f_i(x_0)\right]\right\|_1 \leq \frac{1}{(n-1)!}$$

$$\left\{\sum_{i=1}^{r}\left[\left\|\left[\int_a^{x_0}\left(\prod_{\substack{j=1\\ j\neq i}}^{r} f_j(x)\right)\left(\int_{g(x)}^{g(x_0)}(z-g(x))^{n-1}\left(f_i\circ g^{-1}\right)^{(n)}(z)\,dz\right)dx\right]\right\|_1\right.\right.$$

$$\left.\left.+\left\|\left[\int_{x_0}^{b}\left(\prod_{\substack{j=1\\ j\neq i}}^{r} f_j(x)\right)\left(\int_{g(x_0)}^{g(x)}(g(x)-z)^{n-1}\left(f_i\circ g^{-1}\right)^{(n)}(z)\,dz\right)dx\right]\right\|_1\right]\right\}\leq$$

$$\frac{1}{(n-1)!}\left\{\sum_{i=1}^{r}\left[\left[\int_a^{x_0}\left\|\left(\prod_{\substack{j=1\\ j\neq i}}^{r} f_j(x)\right)\left(\int_{g(x)}^{g(x_0)}(z-g(x))^{n-1}\left(f_i\circ g^{-1}\right)^{(n)}(z)\,dz\right)\right\|_1 dx\right]\right.\right.$$

$$\left.\left.+\left[\int_{x_0}^{b}\left\|\left(\prod_{\substack{j=1\\ j\neq i}}^{r} f_j(x)\right)\left(\int_{g(x_0)}^{g(x)}(g(x)-z)^{n-1}\left(f_i\circ g^{-1}\right)^{(n)}(z)\,dz\right)\right\|_1 dx\right]\right]\right\}\overset{(11.20)}{\leq} \quad (11.38)$$

$$\frac{1}{(n-1)!}\left\{\sum_{i=1}^{r}\left[\left[\int_a^{x_0}\left(\prod_{\substack{j=1\\ j\neq i}}^{r}\|f_j(x)\|_2\right)\left(\int_{g(x)}^{g(x_0)}(z-g(x))^{n-1}\left\|\left(f_i\circ g^{-1}\right)^{(n)}(z)\right\|_2 dz\right)dx\right]\right.\right.$$

$$\left.\left.+\left[\int_{x_0}^{b}\left(\prod_{\substack{j=1\\ j\neq i}}^{r}\|f_j(x)\|_2\right)\left(\int_{g(x_0)}^{g(x)}(g(x)-z)^{n-1}\left\|\left(f_i\circ g^{-1}\right)^{(n)}(z)\right\|_2 dz\right)dx\right]\right]\right\}=:(\xi). \quad (11.39)$$

Hence it holds

$$\|E(f_1,...,f_r)(x_0)\|_1 \leq (\xi). \quad (11.40)$$

We have that

$$(\xi) \leq \frac{1}{n!} \left\{ \sum_{i=1}^{r} \left[\left[\left\| \left\| \left(f_i \circ g^{-1} \right)^{(n)} \right\|_2 \right\|_{\infty,[g(a),g(x_0)]} \int_a^{x_0} \left(\prod_{\substack{j=1 \\ j \neq i}}^{r} \| f_j (x) \|_2 \right) (g (x_0) - g (x))^n \, dx \right] \right. \right.$$

$$\left. \left. + \left[\left\| \left\| \left(f_i \circ g^{-1} \right)^{(n)} \right\|_2 \right\|_{\infty,[g(x_0),g(b)]} \int_{x_0}^{b} \left(\prod_{\substack{j=1 \\ j \neq i}}^{r} \| f_j (x) \|_2 \right) (g (x) - g (x_0))^n \, dx \right] \right] \right\} \leq \tag{11.41}$$

$$\frac{1}{n!} \left\{ \sum_{i=1}^{r} \left[\left[\left\| \left\| \left(f_i \circ g^{-1} \right)^{(n)} \right\|_2 \right\|_{\infty,[g(a),g(x_0)]} (g (x_0) - g (a))^n \left(\int_a^{x_0} \left(\prod_{\substack{j=1 \\ j \neq i}}^{r} \| f_j (x) \|_2 \right) dx \right) \right] \right. \right.$$

$$\left. \left. + \left[\left\| \left\| \left(f_i \circ g^{-1} \right)^{(n)} \right\|_2 \right\|_{\infty,[g(x_0),g(b)]} (g (b) - g (x_0))^n \left(\int_{x_0}^{b} \left(\prod_{\substack{j=1 \\ j \neq i}}^{r} \| f_j (x) \|_2 \right) dx \right) \right] \right] \right\}, \tag{11.42}$$

proving (11.28). ∎

Next comes an L_1 estimate.

Theorem 11.10 *All as in Theorem 11.9. Then*

$$\| E (f_1, ..., f_r) (x_0) \|_1 \leq \frac{1}{(n-1)!}$$

$$\left\{ \sum_{i=1}^{r} \left[\left[\left\| \left\| \left(f_i \circ g^{-1} \right)^{(n)} \right\|_2 \right\|_{L_1([g(a),g(x_0)])} \int_a^{x_0} \left(\prod_{\substack{j=1 \\ j \neq i}}^{r} \| f_j (x) \|_2 \right) (g (x_0) - g (x))^{n-1} \, dx \right] \right. \right.$$

$$\left. \left. + \left[\left\| \left\| \left(f_i \circ g^{-1} \right)^{(n)} \right\|_2 \right\|_{L_1([g(x_0),g(b)])} \int_{x_0}^{b} \left(\prod_{\substack{j=1 \\ j \neq i}}^{r} \| f_j (x) \|_2 \right) (g (x) - g (x_0))^{n-1} \, dx \right] \right] \right\}. \tag{11.43}$$

Proof By (11.39), (11.40), we get that

$$\| E (f_1, ..., f_r) (x_0) \|_1 \leq (\xi) \leq \frac{1}{(n-1)!}$$

$$\left\{ \sum_{i=1}^{r} \left[\left[\left\| \left\| \left(f_i \circ g^{-1} \right)^{(n)} \right\|_2 \right\|_{L_1([g(a),g(x_0)])} \int_a^{x_0} \left(\prod_{\substack{j=1 \\ j \neq i}}^{r} \| f_j (x) \|_2 \right) (g (x_0) - g (x))^{n-1} \, dx \right] \right. \right.$$

$$+\left[\left\|\left\|\left(f_i\circ g^{-1}\right)^{(n)}\right\|_2\right\|_{L_1([g(x_0),g(b)])}\int_{x_0}^{b}\left(\prod_{\substack{j=1\\j\neq i}}^{r}\|f_j(x)\|_2\right)(g(x)-g(x_0))^{n-1}dx\right]\Bigg]\Bigg\},\tag{11.44}$$

proving (11.43). ■

An L_p estimate follows.

Theorem 11.11 *All as in Theorem 11.9, and let $p, q > 1 : \frac{1}{p}+\frac{1}{q}=1$. Then*

$$\|E(f_1,...,f_r)(x_0)\|_1\leq\frac{1}{(n-1)!\,(p(n-1)+1)^{\frac{1}{p}}}$$

$$\sum_{i=1}^{r}\left[\left\|\left\|\left(f_i\circ g^{-1}\right)^{(n)}\right\|_2\right\|_{L_q([g(a),g(x_0)])}\left(\int_a^{x_0}(g(x_0)-g(x))^{n-\frac{1}{q}}\left(\prod_{\substack{j=1\\j\neq i}}^{r}\|f_j(x)\|_2\right)dx\right)\right.$$

$$\left.+\left\|\left\|\left(f_i\circ g^{-1}\right)^{(n)}\right\|_2\right\|_{L_q([g(x_0),g(b)])}\left(\int_{x_0}^{b}(g(x)-g(x_0))^{n-\frac{1}{q}}\left(\prod_{\substack{j=1\\j\neq i}}^{r}\|f_j(x)\|_2\right)dx\right)\right].\tag{11.45}$$

Proof By (11.39), (11.40), we get that

$$\|E(f_1,...,f_r)(x_0)\|_1\leq(\xi)\leq\frac{1}{(n-1)!}$$

$$\left\{\sum_{i=1}^{r}\left[\left[\int_a^{x_0}\left(\prod_{\substack{j=1\\j\neq i}}^{r}\|f_j(x)\|_2\right)\left(\int_{g(x)}^{g(x_0)}(z-g(x))^{p(n-1)}dz\right)^{\frac{1}{p}}\right.\right.\right.$$

$$\left.\left(\int_{g(x)}^{g(x_0)}\left\|\left(f_i\circ g^{-1}\right)^{(n)}(z)\right\|_2^q dz\right)^{\frac{1}{q}}dx\right]+$$

$$\left[\int_{x_0}^{b}\left(\prod_{\substack{j=1\\j\neq i}}^{r}\|f_j(x)\|_2\right)\left(\int_{g(x_0)}^{g(x)}(g(x)-z)^{p(n-1)}dz\right)^{\frac{1}{p}}\right.\tag{11.46}$$

$$\left.\left.\left.\left(\int_{g(x_0)}^{g(x)}\left\|\left(f_i\circ g^{-1}\right)^{(n)}(z)\right\|_2^q dz\right)^{\frac{1}{q}}dx\right]\right]\right\}=\frac{1}{(n-1)!}$$

$$\left\{\sum_{i=1}^{r}\left[\left[\int_a^{x_0}\left(\prod_{\substack{j=1\\j\neq i}}^{r}\|f_j(x)\|_2\right)\frac{(g(x_0)-g(x))^{\frac{p(n-1)|+1}{p}}}{(p(n-1)+1)^{\frac{1}{p}}}\left\|\left\|\left(f_i\circ g^{-1}\right)^{(n)}\right\|_2\right\|_{L_q([g(a),g(x_0)])}dx\right]\right.\right.$$

$$\left.\left.+\left[\int_{x_0}^{b}\left(\prod_{\substack{j=1\\j\neq i}}^{r}\|f_j(x)\|_2\right)\frac{(g(x)-g(x_0))^{\frac{p(n-1)|+1}{p}}}{(p(n-1)+1)^{\frac{1}{p}}}\left\|\left\|\left(f_i\circ g^{-1}\right)^{(n)}\right\|_2\right\|_{L_q([g(x_0),g(b)])}dx\right]\right]\right\}$$

$$=\frac{1}{(n-1)!\,(p(n-1)+1)^{\frac{1}{p}}}$$

$$\left\{\sum_{i=1}^{r}\left[\left\|\left\|\left(f_i\circ g^{-1}\right)^{(n)}\right\|_2\right\|_{L_q([g(a),g(x_0)])}\left(\int_a^{x_0}(g(x_0)-g(x))^{n-\frac{1}{q}}\left(\prod_{\substack{j=1\\j\neq i}}^{r}\|f_j(x)\|_2\right)dx\right)\right.\right.$$

$$\left.\left.+\left\|\left\|\left(f_i\circ g^{-1}\right)^{(n)}\right\|_2\right\|_{L_q([g(x_0),g(b)])}\left(\int_{x_0}^{b}(g(x)-g(x_0))^{n-\frac{1}{q}}\left(\prod_{\substack{j=1\\j\neq i}}^{r}\|f_j(x)\|_2\right)dx\right)\right]\right\}, \tag{11.47}$$

proving (11.45). ■

We continue with mixed generalized Ostrowski type inequalities for several functions that are Banach algebra $\mathcal{B}_\gamma(H)\subset\mathcal{B}(H)$, $\gamma\geq 1$, valued. A uniform estimate follows.

Theorem 11.12 *Let $\gamma\geq 1$, $n\in\mathbb{N}$ and $f_i\in C^n\left([a,b],\mathcal{B}_\gamma(H)\right)$, $i=1,\dots,r\in\mathbb{N}-\{1\}$; where $[a,b]\subset\mathbb{R}$ and $\mathcal{B}_\gamma(H)$ is a $*$-ideal, which $\left(\mathcal{B}_\gamma(H),\|\cdot\|_\gamma\right)$ is a Banach algebra. Let $g\in C^1([a,b])$, strictly increasing, such that $g^{-1}\in C^n([g(a),g(b)])$. We assume that $\left(f_i\circ g^{-1}\right)^{(j)}(g(x_0))=0$, $j=1,\dots,n-1$; $i=1,\dots,r$; where $x_0\in[a,b]$ be fixed. Denote again by*

$$E(f_1,\dots,f_r)(x_0):=$$

$$\sum_{i=1}^{r}\left[\int_a^b\left(\prod_{\substack{j=1\\j\neq i}}^{r}f_j(x)\right)f_i(x)\,dx-\left(\int_a^b\left(\prod_{\substack{j=1\\j\neq i}}^{r}f_j(x)\right)dx\right)f_i(x_0)\right]. \tag{11.48}$$

Then

$$\|E(f_1,\dots,f_r)(x_0)\|_\gamma\leq\frac{1}{n!}$$

$$\left\{\sum_{i=1}^{r}\left[\left[\left\|\left\|\left(f_i\circ g^{-1}\right)^{(n)}\right\|_{\gamma}\right\|_{\infty,[g(a),g(x_0)]}(g(x_0)-g(a))^n\left(\int_a^{x_0}\left(\prod_{\substack{j=1\\j\neq i}}^{r}\|f_j(x)\|_{\gamma}\right)dx\right)\right]+\right.\right.$$

$$\left.\left.\left[\left\|\left\|\left(f_i\circ g^{-1}\right)^{(n)}\right\|_{\gamma}\right\|_{\infty,[g(x_0),g(b)]}(g(b)-g(x_0))^n\left(\int_{x_0}^{b}\left(\prod_{\substack{j=1\\j\neq i}}^{r}\|f_j(x)\|_{\gamma}\right)dx\right)\right]\right]\right\}. \tag{11.49}$$

Proof As similar to Theorem 11.9 is omitted, use of (11.23). ∎

An L_1 estimate follows:

Theorem 11.13 *All as in Theorem 11.12. Then*

$$\|E(f_1,...,f_r)(x_0)\|_{\gamma}\leq\frac{1}{(n-1)!}$$

$$\left\{\sum_{i=1}^{r}\left[\left[\left\|\left\|\left(f_i\circ g^{-1}\right)^{(n)}\right\|_{\gamma}\right\|_{L_1([g(a),g(x_0)])}\int_a^{x_0}\left(\prod_{\substack{j=1\\j\neq i}}^{r}\|f_j(x)\|_{\gamma}\right)(g(x_0)-g(x))^{n-1}dx\right]\right.\right.$$

$$\left.\left.+\left[\left\|\left\|\left(f_i\circ g^{-1}\right)^{(n)}\right\|_{\gamma}\right\|_{L_1([g(x_0),g(b)])}\int_{x_0}^{b}\left(\prod_{\substack{j=1\\j\neq i}}^{r}\|f_j(x)\|_{\gamma}\right)(g(x)-g(x_0))^{n-1}dx\right]\right]\right\}. \tag{11.50}$$

Proof As similar to Theorem 11.10 is omitted. ∎

An L_p estimate follows.

Theorem 11.14 *All as in Theorem 11.12, and let* $p,q>1:\frac{1}{p}+\frac{1}{q}=1$. *Then*

$$\|E(f_1,...,f_r)(x_0)\|_{\gamma}\leq\frac{1}{(n-1)!\,(p(n-1)+1)^{\frac{1}{p}}}$$

$$\sum_{i=1}^{r}\left[\left\|\left\|\left(f_i\circ g^{-1}\right)^{(n)}\right\|_{\gamma}\right\|_{L_q([g(a),g(x_0)])}\left(\int_a^{x_0}(g(x_0)-g(x))^{n-\frac{1}{q}}\left(\prod_{\substack{j=1\\j\neq i}}^{r}\|f_j(x)\|_{\gamma}\right)dx\right)\right.$$

$$\left.+\left\|\left\|\left(f_i\circ g^{-1}\right)^{(n)}\right\|_{\gamma}\right\|_{L_q([g(x_0),g(b)])}\left(\int_{x_0}^{b}(g(x)-g(x_0))^{n-\frac{1}{q}}\left(\prod_{\substack{j=1\\j\neq i}}^{r}\|f_j(x)\|_{\gamma}\right)dx\right)\right]. \tag{11.51}$$

Proof As similar to Theorem 11.11 is omitted. ■

When $r = 2$ we derive the following p-Schatten norm operator related Ostrowski type inequalities.

Theorem 11.15 *Let $p, q > 1 : \frac{1}{p} + \frac{1}{q} = 1$, and let the $*$-ideals $\mathcal{B}_p(H)$, $\mathcal{B}_q(H)$, for which $\left(\mathcal{B}_p(H), \|\cdot\|_p\right)$, $\left(\mathcal{B}_q(H), \|\cdot\|_q\right)$ are Banach algebras; $n \in \mathbb{N}$, $x_0 \in [a,b] \subset \mathbb{R}$, $A_1 \in C^n\left([a,b], \mathcal{B}_p(H)\right)$, $A_2 \in C^n\left([a,b], \mathcal{B}_q(H)\right)$; $g \in C^1([a,b])$, strictly increasing, such that $g^{-1} \in C^n([g(a), g(b)])$, with $\left(A_i \circ g^{-1}\right)^{(k)}(g(x_0)) = 0$, $k = 1, ..., n-1$; $i = 1, 2$. Then*

(1)

$$\Phi(A_1, A_2)(x_0) := \int_a^b A_2(x) A_1(x)\,dx + \int_a^b A_1(x) A_2(x)\,dx -$$

$$\left(\int_a^b A_2(x)\,dx\right) A_1(x_0) - \left(\int_a^b A_1(x)\,dx\right) A_2(x_0) =$$

$$\frac{1}{(n-1)!}\left\{(-1)^n\left[\int_a^{x_0} A_2(x)\left(\int_{g(x)}^{g(x_0)} (z-g(x))^{n-1}\left(A_1 \circ g^{-1}\right)^{(n)}(z)\,dz\right)dx\right] +\right.$$

$$\left[\int_{x_0}^b A_2(x)\left(\int_{g(x_0)}^{g(x)} (g(x)-z)^{n-1}\left(A_1 \circ g^{-1}\right)^{(n)}(z)\,dz\right)dx\right] +$$

$$(-1)^n\left[\int_a^{x_0} A_1(x)\left(\int_{g(x)}^{g(x_0)} (z-g(x))^{n-1}\left(A_2 \circ g^{-1}\right)^{(n)}(z)\,dz\right)dx\right] + \quad (11.52)$$

$$\left.\left[\int_{x_0}^b A_1(x)\left(\int_{g(x_0)}^{g(x)} (g(x)-z)^{n-1}\left(A_2 \circ g^{-1}\right)^{(n)}(z)\,dz\right)dx\right]\right\}.$$

(2) for $\gamma, \delta > 1 : \frac{1}{\gamma} + \frac{1}{\delta} = 1$, we obtain

$$\|\Phi(A_1, A_2)(x_0)\|_1 \leq \frac{1}{(n-1)!\,(\gamma(n-1)+1)^{\frac{1}{\gamma}}}$$

$$\left\{\left[\left\|\left\|\left(A_1 \circ g^{-1}\right)^{(n)}\right\|_p\right\|_{\delta,[g(a),g(x_0)]} \int_a^{x_0} \|A_2(x)\|_q (g(x_0)-g(x))^{n-\frac{1}{\delta}}\,dx\right] +\right.$$

$$\left[\left\|\left\|\left(A_1 \circ g^{-1}\right)^{(n)}\right\|_p\right\|_{\delta,[g(x_0),g(b)]} \int_{x_0}^b \|A_2(x)\|_q (g(x)-g(x_0))^{n-\frac{1}{\delta}}\,dx\right] + \quad (11.53)$$

$$\left[\left\|\left\|\left(A_2 \circ g^{-1}\right)^{(n)}\right\|_q\right\|_{\delta,[g(a),g(x_0)]} \int_a^{x_0} \|A_1(x)\|_p (g(x_0) - g(x))^{n-\frac{1}{\delta}} dx\right] +$$

$$\left[\left\|\left\|\left(A_2 \circ g^{-1}\right)^{(n)}\right\|_q\right\|_{\delta,[g(x_0),g(b)]} \int_{x_0}^{b} \|A_1(x)\|_p (g(x) - g(x_0))^{n-\frac{1}{\delta}} dx\right]\Bigg\},$$

(3)

$$\|\Phi(A_1, A_2)(x_0)\|_1 \leq \frac{1}{(n-1)!}$$

$$\Bigg\{\left[\left\|\left\|\left(A_1 \circ g^{-1}\right)^{(n)}\right\|_p\right\|_{L_1([g(a),g(x_0)])} \int_a^{x_0} \|A_2(x)\|_q (g(x_0) - g(x))^{n-1} dx\right] +$$

$$\left[\left\|\left\|\left(A_1 \circ g^{-1}\right)^{(n)}\right\|_p\right\|_{L_1([g(x_0),g(b)])} \int_{x_0}^{b} \|A_2(x)\|_q (g(x) - g(x_0))^{n-1} dx\right] + \tag{11.54}$$

$$\left[\left\|\left\|\left(A_2 \circ g^{-1}\right)^{(n)}\right\|_q\right\|_{L_1([g(a),g(x_0)])} \int_a^{x_0} \|A_1(x)\|_p (g(x_0) - g(x))^{n-1} dx\right] +$$

$$\left[\left\|\left\|\left(A_2 \circ g^{-1}\right)^{(n)}\right\|_q\right\|_{L_1([g(x_0),g(b)])} \int_{x_0}^{b} \|A_1(x)\|_p (g(x) - g(x_0))^{n-1} dx\right]\Bigg\},$$

and

(4)

$$\|\Phi(A_1, A_2)(x_0)\|_1 \leq \frac{1}{n!}$$

$$\Bigg\{\left[\left\|\left\|\left(A_1 \circ g^{-1}\right)^{(n)}\right\|_p\right\|_{\infty,[g(a),g(x_0)]} \int_a^{x_0} \|A_2(x)\|_q (g(x_0) - g(x))^{n} dx\right] +$$

$$\left[\left\|\left\|\left(A_1 \circ g^{-1}\right)^{(n)}\right\|_p\right\|_{\infty,[g(x_0),g(b)]} \int_{x_0}^{b} \|A_2(x)\|_q (g(x) - g(x_0))^{n} dx\right] + \tag{11.55}$$

$$\left[\left\|\left\|\left(A_2 \circ g^{-1}\right)^{(n)}\right\|_q\right\|_{\infty,[g(a),g(x_0)]} \int_a^{x_0} \|A_1(x)\|_p (g(x_0) - g(x))^{n} dx\right] +$$

$$\left[\left\|\left\|\left(A_2 \circ g^{-1}\right)^{(n)}\right\|_q\right\|_{\infty,[g(x_0),g(b)]} \int_{x_0}^{b} \|A_1(x)\|_p (g(x) - g(x_0))^{n} dx\right]\Bigg\}.$$

Proof Writing (11.37) for $r = 2$ and for A_1, A_2, we derive

$$\Phi(A_1, A_2)(x_0) = \int_a^b A_2(x) A_1(x)\,dx + \int_a^b A_1(x) A_2(x)\,dx -$$

$$\left(\int_a^b A_2(x)\,dx\right) A_1(x_0) - \left(\int_a^b A_1(x)\,dx\right) A_2(x_0) =$$

$$\frac{1}{(n-1)!}\left\{(-1)^n \left[\int_a^{x_0} A_2(x)\left(\int_{g(x)}^{g(x_0)} (z-g(x))^{n-1}\left(A_1 \circ g^{-1}\right)^{(n)}(z)\,dz\right)dx\right] +\right.$$

$$\left[\int_{x_0}^b A_2(x)\left(\int_{g(x_0)}^{g(x)} (g(x)-z)^{n-1}\left(A_1 \circ g^{-1}\right)^{(n)}(z)\,dz\right)dx\right] +$$

$$(-1)^n \left[\int_a^{x_0} A_1(x)\left(\int_{g(x)}^{g(x_0)} (z-g(x))^{n-1}\left(A_2 \circ g^{-1}\right)^{(n)}(z)\,dz\right)dx\right] + \tag{11.56}$$

$$\left.\left[\int_{x_0}^b A_1(x)\left(\int_{g(x_0)}^{g(x)} (g(x)-z)^{n-1}\left(A_2 \circ g^{-1}\right)^{(n)}(z)\,dz\right)dx\right]\right\},$$

proving (11.52).

Hence it holds

$$\|\Phi(A_1, A_2)(x_0)\|_1 = \left\|\int_a^b A_2(x) A_1(x)\,dx + \int_a^b A_1(x) A_2(x)\,dx -\right.$$

$$\left.\left(\int_a^b A_2(x)\,dx\right) A_1(x_0) - \left(\int_a^b A_1(x)\,dx\right) A_2(x_0)\right\|_1 =$$

$$\frac{1}{(n-1)!}\left\|\left\{(-1)^n \left[\int_a^{x_0} A_2(x)\left(\int_{g(x)}^{g(x_0)} (z-g(x))^{n-1}\left(A_1 \circ g^{-1}\right)^{(n)}(z)\,dz\right)dx\right] +\right.\right.$$

$$\left[\int_{x_0}^b A_2(x)\left(\int_{g(x_0)}^{g(x)} (g(x)-z)^{n-1}\left(A_1 \circ g^{-1}\right)^{(n)}(z)\,dz\right)dx\right] +$$

$$(-1)^n \left[\int_a^{x_0} A_1(x)\left(\int_{g(x)}^{g(x_0)} (z-g(x))^{n-1}\left(A_2 \circ g^{-1}\right)^{(n)}(z)\,dz\right)dx\right] + \tag{11.57}$$

$$\left.\left.\left[\int_{x_0}^b A_1(x)\left(\int_{g(x_0)}^{g(x)} (g(x)-z)^{n-1}\left(A_2 \circ g^{-1}\right)^{(n)}(z)\,dz\right)dx\right]\right\}\right\|_1 \leq$$

$$\frac{1}{(n-1)!}\left\{\left\|\left[\int_a^{x_0} A_2(x)\left(\int_{g(x)}^{g(x_0)} (z-g(x))^{n-1}\left(A_1 \circ g^{-1}\right)^{(n)}(z)\,dz\right)dx\right]\right\|_1 +\right.$$

$$\left\| \left[\int_{x_0}^{b} A_2(x) \left(\int_{g(x_0)}^{g(x)} (g(x) - z)^{n-1} \left(A_1 \circ g^{-1} \right)^{(n)}(z)\, dz \right) dx \right] \right\|_1 +$$

$$\left\| \left[\int_{a}^{x_0} A_1(x) \left(\int_{g(x)}^{g(x_0)} (z - g(x))^{n-1} \left(A_2 \circ g^{-1} \right)^{(n)}(z)\, dz \right) dx \right] \right\|_1 + \tag{11.58}$$

$$\left. \left\| \left[\int_{x_0}^{b} A_1(x) \left(\int_{g(x_0)}^{g(x)} (g(x) - z)^{n-1} \left(A_2 \circ g^{-1} \right)^{(n)}(z)\, dz \right) dx \right] \right\|_1 \right\} \leq$$

$$\frac{1}{(n-1)!} \left\{ \left[\int_{a}^{x_0} \left\| A_2(x) \left(\int_{g(x)}^{g(x_0)} (z - g(x))^{n-1} \left(A_1 \circ g^{-1} \right)^{(n)}(z)\, dz \right) \right\|_1 dx \right] + \right.$$

$$\left[\int_{x_0}^{b} \left\| A_2(x) \left(\int_{g(x_0)}^{g(x)} (g(x) - z)^{n-1} \left(A_1 \circ g^{-1} \right)^{(n)}(z)\, dz \right) \right\|_1 dx \right] +$$

$$\left[\int_{a}^{x_0} \left\| A_1(x) \left(\int_{g(x)}^{g(x_0)} (z - g(x))^{n-1} \left(A_2 \circ g^{-1} \right)^{(n)}(z)\, dz \right) \right\|_1 dx \right] + \tag{11.59}$$

$$\left. \left[\int_{x_0}^{b} \left\| A_1(x) \left(\int_{g(x_0)}^{g(x)} (g(x) - z)^{n-1} \left(A_2 \circ g^{-1} \right)^{(n)}(z)\, dz \right) \right\|_1 dx \right] \right\} \leq$$

(by using the p-Schatten norm and Hölder's type inequality (11.26) for $p, q > 1$: $\frac{1}{p} + \frac{1}{q} = 1$)

$$\frac{1}{(n-1)!} \left\{ \left[\int_{a}^{x_0} \| A_2(x) \|_q \left\| \left(\int_{g(x)}^{g(x_0)} (z - g(x))^{n-1} \left(A_1 \circ g^{-1} \right)^{(n)}(z)\, dz \right) \right\|_p dx \right] + \right.$$

$$\left[\int_{x_0}^{b} \| A_2(x) \|_q \left\| \left(\int_{g(x_0)}^{g(x)} (g(x) - z)^{n-1} \left(A_1 \circ g^{-1} \right)^{(n)}(z)\, dz \right) \right\|_p dx \right] +$$

$$\left[\int_{a}^{x_0} \| A_1(x) \|_p \left\| \left(\int_{g(x)}^{g(x_0)} (z - g(x))^{n-1} \left(A_2 \circ g^{-1} \right)^{(n)}(z)\, dz \right) \right\|_q dx \right] + \tag{11.60}$$

$$\left. \left[\int_{x_0}^{b} \| A_1(x) \|_p \left\| \left(\int_{g(x_0)}^{g(x)} (g(x) - z)^{n-1} \left(A_2 \circ g^{-1} \right)^{(n)}(z)\, dz \right) \right\|_q dx \right] \right\} \leq$$

$$\frac{1}{(n-1)!} \left\{ \left[\int_{a}^{x_0} \| A_2(x) \|_q \left(\int_{g(x)}^{g(x_0)} (z - g(x))^{n-1} \left\| \left(A_1 \circ g^{-1} \right)^{(n)}(z) \right\|_p dz \right) dx \right] + \right.$$

$$\left[\int_{x_0}^{b} \|A_2(x)\|_q \left(\int_{g(x_0)}^{g(x)} (g(x)-z)^{n-1} \left\|\left(A_1 \circ g^{-1}\right)^{(n)}(z)\right\|_p dz\right) dx\right] +$$

$$\left[\int_{a}^{x_0} \|A_1(x)\|_p \left(\int_{g(x)}^{g(x_0)} (z-g(x))^{n-1} \left\|\left(A_2 \circ g^{-1}\right)^{(n)}(z)\right\|_q dz\right) dx\right] + \quad (11.61)$$

$$\left[\int_{x_0}^{b} \|A_1(x)\|_p \left(\int_{g(x_0)}^{g(x)} (g(x)-z)^{n-1} \left\|\left(A_2 \circ g^{-1}\right)^{(n)}(z)\right\|_q dz\right) dx\right]\Bigg\}.$$

So far we have proved that

$$\|\Phi(A_1, A_2)(x_0)\|_1 \leq$$

$$\frac{1}{(n-1)!}\Bigg\{\left[\int_{a}^{x_0} \|A_2(x)\|_q \left(\int_{g(x)}^{g(x_0)} (z-g(x))^{n-1} \left\|\left(A_1 \circ g^{-1}\right)^{(n)}(z)\right\|_p dz\right) dx\right] +$$

$$\left[\int_{x_0}^{b} \|A_2(x)\|_q \left(\int_{g(x_0)}^{g(x)} (g(x)-z)^{n-1} \left\|\left(A_1 \circ g^{-1}\right)^{(n)}(z)\right\|_p dz\right) dx\right] +$$

$$\left[\int_{a}^{x_0} \|A_1(x)\|_p \left(\int_{g(x)}^{g(x_0)} (z-g(x))^{n-1} \left\|\left(A_2 \circ g^{-1}\right)^{(n)}(z)\right\|_q dz\right) dx\right] + \quad (11.62)$$

$$\left[\int_{x_0}^{b} \|A_1(x)\|_p \left(\int_{g(x_0)}^{g(x)} (g(x)-z)^{n-1} \left\|\left(A_2 \circ g^{-1}\right)^{(n)}(z)\right\|_q dz\right) dx\right]\Bigg\} =: (\lambda).$$

Let now $\gamma, \delta > 1$ such that $\frac{1}{\gamma} + \frac{1}{\delta} = 1$, and we apply the usual Hölder's inequality in (11.62). Then, we have that

$$\|\Phi(A_1, A_2)(x_0)\|_1 \leq (\lambda) \leq \frac{1}{(n-1)!\,(\gamma(n-1)+1)^{\frac{1}{\gamma}}}$$

$$\Bigg\{\left[\int_{a}^{x_0} \|A_2(x)\|_q (g(x_0)-g(x))^{\frac{\gamma(n-1)+1}{\gamma}} \left(\int_{g(x)}^{g(x_0)} \left\|\left(A_1 \circ g^{-1}\right)^{(n)}(z)\right\|_p^{\delta} dz\right)^{\frac{1}{\delta}} dx\right] +$$

$$\left[\int_{x_0}^{b} \|A_2(x)\|_q (g(x)-g(x_0))^{\frac{\gamma(n-1)+1}{\gamma}} \left(\int_{g(x_0)}^{g(x)} \left\|\left(A_1 \circ g^{-1}\right)^{(n)}(z)\right\|_p^{\delta} dz\right)^{\frac{1}{\delta}} dx\right] +$$

$$(11.63)$$

$$\left[\int_{a}^{x_0} \|A_1(x)\|_p (g(x_0)-g(x))^{\frac{\gamma(n-1)+1}{\gamma}} \left(\int_{g(x)}^{g(x_0)} \left\|\left(A_2 \circ g^{-1}\right)^{(n)}(z)\right\|_q^{\delta} dz\right)^{\frac{1}{\delta}} dx\right] +$$

$$\left[\int_{x_0}^{b} \|A_1(x)\|_p (g(x)-g(x_0))^{\frac{\gamma(n-1)+1}{\gamma}} \left(\int_{g(x_0)}^{g(x)} \left\|\left(A_2 \circ g^{-1}\right)^{(n)}(z)\right\|_q^{\delta} dz\right)^{\frac{1}{\delta}} dx\right]\Bigg\}$$

$$\leq \frac{1}{(n-1)!\,(\gamma(n-1)+1)^{\frac{1}{\gamma}}}$$

$$\Bigg\{\left[\left\|\left\|\left(A_1 \circ g^{-1}\right)^{(n)}\right\|_p\right\|_{\delta,[g(a),g(x_0)]} \int_a^{x_0} \|A_2(x)\|_q (g(x_0)-g(x))^{n-\frac{1}{\delta}} dx\right]+$$

$$\left[\left\|\left\|\left(A_1 \circ g^{-1}\right)^{(n)}\right\|_p\right\|_{\delta,[g(x_0),g(b)]} \int_{x_0}^{b} \|A_2(x)\|_q (g(x)-g(x_0))^{n-\frac{1}{\delta}} dx\right]+ \tag{11.64}$$

$$\left[\left\|\left\|\left(A_2 \circ g^{-1}\right)^{(n)}\right\|_q\right\|_{\delta,[g(a),g(x_0)]} \int_a^{x_0} \|A_1(x)\|_p (g(x_0)-g(x))^{n-\frac{1}{\delta}} dx\right]+$$

$$\left[\left\|\left\|\left(A_2 \circ g^{-1}\right)^{(n)}\right\|_q\right\|_{\delta,[g(x_0),g(b)]} \int_{x_0}^{b} \|A_1(x)\|_p (g(x)-g(x_0))^{n-\frac{1}{\delta}} dx\right]\Bigg\},$$

proving (11.53).

We also obtain

$$\|\Phi(A_1,A_2)(x_0)\|_1 \leq (\lambda) \leq \frac{1}{(n-1)!}$$

$$\Bigg\{\left[\left\|\left\|\left(A_1 \circ g^{-1}\right)^{(n)}\right\|_p\right\|_{\infty,[g(a),g(x_0)]} \int_a^{x_0} \|A_2(x)\|_q (g(x_0)-g(x))^{n-1} dx\right]+$$

$$\left[\left\|\left\|\left(A_1 \circ g^{-1}\right)^{(n)}\right\|_p\right\|_{L_1([g(x_0),g(b)])} \int_{x_0}^{b} \|A_2(x)\|_q (g(x)-g(x_0))^{n-1} dx\right]+ \tag{11.65}$$

$$\left[\left\|\left\|\left(A_2 \circ g^{-1}\right)^{(n)}\right\|_q\right\|_{L_1([g(a),g(x_0)])} \int_a^{x_0} \|A_1(x)\|_p (g(x_0)-g(x))^{n-1} dx\right]+$$

$$\left[\left\|\left\|\left(A_2 \circ g^{-1}\right)^{(n)}\right\|_q\right\|_{L_1([g(x_0),g(b)])} \int_{x_0}^{b} \|A_1(x)\|_p (g(x)-g(x_0))^{n-1} dx\right]\Bigg\},$$

proving (11.54).

At last we derive

$$\|\Phi(A_1,A_2)(x_0)\|_1 \leq (\lambda) \leq \frac{1}{n!}$$

$$\left\{\left[\left\|\left\|\left(A_1 \circ g^{-1}\right)^{(n)}\right\|_p\right\|_{\infty,[g(a),g(x_0)]} \int_a^{x_0} \|A_2(x)\|_q (g(x_0) - g(x))^n \, dx\right] +\right.$$

$$\left[\left\|\left\|\left(A_1 \circ g^{-1}\right)^{(n)}\right\|_p\right\|_{\infty,[g(x_0),g(b)]} \int_{x_0}^{b} \|A_2(x)\|_q (g(x) - g(x_0))^n \, dx\right] + \quad (11.66)$$

$$\left[\left\|\left\|\left(A_2 \circ g^{-1}\right)^{(n)}\right\|_q\right\|_{\infty,[g(a),g(x_0)]} \int_a^{x_0} \|A_1(x)\|_p (g(x_0) - g(x))^n \, dx\right] +$$

$$\left.\left[\left\|\left\|\left(A_2 \circ g^{-1}\right)^{(n)}\right\|_q\right\|_{\infty,[g(x_0),g(b)]} \int_{x_0}^{b} \|A_1(x)\|_p (g(x) - g(x_0))^n \, dx\right]\right\},$$

proving (11.55).

The theorem is proved. ■

Next we present a left generalized Opial type inequality involving 1-2-Schatten norms.

Theorem 11.16 *Let $p, q > 1 : \frac{1}{p} + \frac{1}{q} = 1$, and $n \in \mathbb{N}$, $f \in C^n([a,b], \mathcal{B}_2(H))$; where $[a,b] \subset \mathbb{R}$ and $\mathcal{B}_2(H)$ is the $*$-ideal. Let $g \in C^1([a,b])$, strictly increasing, such that $g^{-1} \in C^n([g(a), g(b)])$. We assume that $\left(f \circ g^{-1}\right)^{(j)}(g(x_0)) = 0$, $j = 0, 1, ..., n-1$; where $x_0 \in [a,b]$ be fixed. Then*

$$\int_{g(x_0)}^{g(x)} \left\|\left(f \circ g^{-1}\right)(z) \left(f \circ g^{-1}\right)^{(n)}(z)\right\|_1 dz \leq$$

$$\frac{(g(x) - g(x_0))^{n+\frac{1}{p}-\frac{1}{q}}}{2^{\frac{1}{q}} (n-1)! \left[(p(n-1)+1)(p(n-1)+2)\right]^{\frac{1}{p}}} \left(\int_{g(x_0)}^{g(x)} \left\|\left(f \circ g^{-1}\right)^{(n)}(z)\right\|_2^q dz\right)^{\frac{2}{q}}, \quad (11.67)$$

for all $x_0 \leq x \leq b$.

Proof Let $x_0 \in [a,b]$ such that $\left(f \circ g^{-1}\right)^{(j)}(g(x_0)) = 0$, $j = 0, 1, ..., n-1$.

For $x \in [x_0, b]$ by Theorem 11.4 we have

$$\left(f \circ g^{-1}\right)(g(z)) = \frac{1}{(n-1)!} \int_{g(x_0)}^{g(x)} (g(x) - z)^{n-1} \left(f \circ g^{-1}\right)^{(n)}(z) \, dz. \quad (11.68)$$

By Hölder's inequality we obtain

$$\left\|\left(f \circ g^{-1}\right)(g(x))\right\|_2 \leq \frac{1}{(n-1)!} \int_{g(x_0)}^{g(x)} (g(x) - z)^{n-1} \left\|\left(f \circ g^{-1}\right)^{(n)}(z)\right\|_2 dz \leq \quad (11.69)$$

$$\frac{1}{(n-1)!}\left(\int_{g(x_0)}^{g(x)} (g(x)-z)^{p(n-1)}\,dt\right)^{\frac{1}{p}}\left(\int_{g(x_0)}^{g(x)} \left\|\left(f\circ g^{-1}\right)^{(n)}(z)\right\|_2^q dz\right)^{\frac{1}{q}} =$$

$$\frac{1}{(n-1)!}\frac{(g(x)-g(x_0))^{\frac{p(n-1)+1}{p}}}{(p(n-1)+1)^{\frac{1}{p}}}\left(\int_{g(x_0)}^{g(x)} \left\|\left(f\circ g^{-1}\right)^{(n)}(z)\right\|_2^q dz\right)^{\frac{1}{q}}.$$

Call

$$\varphi(g(x)) := \int_{g(x_0)}^{g(x)} \left\|\left(f\circ g^{-1}\right)^{(n)}(z)\right\|_2^q dz, \tag{11.70}$$

$\varphi(g(x_0)) = 0$.

Thus

$$\frac{d\varphi(g(x))}{dg(x)} = \left\|\left(f\circ g^{-1}\right)^{(n)}(g(x))\right\|_2^q \geq 0, \tag{11.71}$$

and

$$\left(\frac{d\varphi(g(x))}{dg(x)}\right)^{\frac{1}{q}} = \left\|\left(f\circ g^{-1}\right)^{(n)}(g(x))\right\|_2 \geq 0, \tag{11.72}$$

$\forall\, g(x) \in [g(x_0), g(b)]$.

Consequently, we get

$$\left\|\left(f\circ g^{-1}\right)(g(w))\right\|_2 \left\|\left(f\circ g^{-1}\right)^{(n)}(g(w))\right\|_2 \leq$$

$$\frac{(g(w)-g(x_0))^{\frac{p(n-1)+1}{p}}}{(n-1)!\,(p(n-1)+1)^{\frac{1}{p}}}\left(\varphi(g(w))\,\frac{d\varphi(g(w))}{dg(w)}\right)^{\frac{1}{q}}, \tag{11.73}$$

$\forall\, g(w) \in [g(x_0), g(b)]$.

Then we observe that

$$\int_{g(x_0)}^{g(x)} \left\|\left(f\circ g^{-1}\right)(g(w))\left(f\circ g^{-1}\right)^{(n)}(g(w))\right\|_1 dg(w) \overset{(11.26)}{\leq}$$

$$\int_{g(x_0)}^{g(x)} \left\|\left(f\circ g^{-1}\right)(g(w))\right\|_2 \left\|\left(f\circ g^{-1}\right)^{(n)}(g(w))\right\|_2 dg(w) \leq$$

$$\frac{1}{(n-1)!\,(p(n-1)+1)^{\frac{1}{p}}}$$

$$\int_{g(x_0)}^{g(x)} (g(w)-g(x_0))^{\frac{p(n-1)+1}{p}}\left(\varphi(g(w))\,\frac{d\varphi(g(w))}{dg(w)}\right)^{\frac{1}{q}} dg(w) \leq \tag{11.74}$$

$$\frac{1}{(n-1)!\,(p(n-1)+1)^{\frac{1}{p}}}$$

$$\left(\int_{g(x_0)}^{g(x)} (g(w)-g(x_0))^{p(n-1)+1}\, dg(w)\right)^{\frac{1}{p}} \left(\int_{g(x_0)}^{g(x)} \varphi(g(w)) \frac{d\varphi(g(w))}{dg(w)} dg(w)\right)^{\frac{1}{q}} =$$

$$\frac{1}{(n-1)!\,(p(n-1)+1)^{\frac{1}{p}}\,(p(n-1)+2)^{\frac{1}{p}}}$$

$$(g(x)-g(x_0))^{\frac{p(n-1)+2}{p}} \left(\int_{g(x_0)}^{g(x)} \varphi(g(w))\, d\varphi(g(w))\right)^{\frac{1}{q}} = \tag{11.75}$$

$$\frac{(g(x)-g(x_0))^{n+\frac{1}{p}-\frac{1}{q}}}{(n-1)!\,(p(n-1)+1)^{\frac{1}{p}}\,(p(n-1)+2)^{\frac{1}{p}}} \left(\frac{\varphi^2(g(x))}{2}\right)^{\frac{1}{q}} =$$

$$\frac{(g(x)-g(x_0))^{n+\frac{1}{p}-\frac{1}{q}}}{2^{\frac{1}{q}}\,(n-1)!\,((p(n-1)+1)(p(n-1)+2))^{\frac{1}{p}}} \left(\int_{g(x_0)}^{g(x)} \left\| \left(f\circ g^{-1}\right)^{(n)}(z)\right\|_2^q dz\right)^{\frac{2}{q}}, \tag{11.76}$$

for all $g(x_0) \le g(x) \le g(b)$, proving (11.67). ■

The corresponding $\mathcal{B}_2(H)$ right generalized Opial type inequality follows:

Theorem 11.17 *All as in Theorem 11.16. Then*

$$\int_{g(x)}^{g(x_0)} \left\| \left(f\circ g^{-1}\right)(z)\left(f\circ g^{-1}\right)^{(n)}(z)\right\|_1 dz \le$$

$$\frac{(g(x_0)-g(x))^{n+\frac{1}{p}-\frac{1}{q}}}{2^{\frac{1}{q}}\,(n-1)!\,((p(n-1)+1)(p(n-1)+2))^{\frac{1}{p}}} \left(\int_{g(x)}^{g(x_0)} \left\| \left(f\circ g^{-1}\right)^{(n)}(z)\right\|_2^q dz\right)^{\frac{2}{q}}, \tag{11.77}$$

for all $a \le x \le x_0$.

Proof As similar to Theorem 11.16 is omitted. ■

A $\mathcal{B}_\gamma(H)$, $\gamma \ge 1$, left Opial inequality follows:

Theorem 11.18 *Let* $\gamma \ge 1$, $p, q > 1 : \frac{1}{p}+\frac{1}{q}=1$, *and* $n \in \mathbb{N}$, $f \in C^n\left([a,b], \mathcal{B}_\gamma(H)\right)$; *where* $[a,b] \subset \mathbb{R}$ *and* $\mathcal{B}_\gamma(H)$ *is the* $*$*-ideal. Let* $g \in C^1([a,b])$, *strictly increasing, such that* $g^{-1} \in C^n([g(a), g(b)])$. *We assume that* $\left(f\circ g^{-1}\right)^{(j)}(g(x_0)) = 0$, $j = 0, 1, ..., n-1$; *where* $x_0 \in [a,b]$ *be fixed. Then*

$$\int_{g(x_0)}^{g(x)} \left\| \left(f\circ g^{-1}\right)(z)\left(f\circ g^{-1}\right)^{(n)}(z)\right\|_\gamma dz \le$$

$$\frac{\left(g\left(x\right)-g\left(x_0\right)\right)^{n+\frac{1}{p}-\frac{1}{q}}}{2^{\frac{1}{q}}\left(n-1\right)!\left[\left(p\left(n-1\right)+1\right)\left(p\left(n-1\right)+2\right)\right]^{\frac{1}{p}}}\left(\int_{g(x_0)}^{g(x)}\left\|\left(f\circ g^{-1}\right)^{(n)}(z)\right\|_{\gamma}^{q}dz\right)^{\frac{2}{q}}, \tag{11.78}$$

for all $x_0 \leq x \leq b$.

Proof As similar to Theorem 11.16 is omitted. Use of (11.23). ■

A $\mathcal{B}_{\gamma}(H)$, $\gamma \geq 1$, right Opial inequality follows:

Theorem 11.19 *All as in Theorem 11.18. Then*

$$\int_{g(x)}^{g(x_0)}\left\|\left(f\circ g^{-1}\right)(z)\left(f\circ g^{-1}\right)^{(n)}(z)\right\|_{\gamma}dz \leq$$

$$\frac{\left(g\left(x_0\right)-g\left(x\right)\right)^{n+\frac{1}{p}-\frac{1}{q}}}{2^{\frac{1}{q}}\left(n-1\right)!\left(\left(p\left(n-1\right)+1\right)\left(p\left(n-1\right)+2\right)\right)^{\frac{1}{p}}}\left(\int_{g(x)}^{g(x_0)}\left\|\left(f\circ g^{-1}\right)^{(n)}(z)\right\|_{\gamma}^{q}dz\right)^{\frac{2}{q}}, \tag{11.79}$$

for all $a \leq x \leq x_0$.

Proof As similar to Theorem 11.16 is omitted. ■

Next we present a $\mathcal{B}_2(H)$ left generalized Hilbert–Pachpatte inequality for ordinary derivatives.

Theorem 11.20 *Let* $i = 1, 2$; $p, q > 1 : \frac{1}{p} + \frac{1}{q} = 1$, *and* $n_i \in \mathbb{N}$, $f_i \in C^{n_i}([a_i, b_i], \mathcal{B}_2(H))$; *where* $[a_i, b_i] \subset \mathbb{R}$ *and* $\mathcal{B}_2(H)$ *is the* $*$-*ideal. Let* $g_i \in C^1([a_i, b_i])$, *strictly increasing, such that* $g_i^{-1} \in C^{n_i}([g_i(a_i), g_i(b_i)])$. *We assume that* $\left(f_i \circ g_i^{-1}\right)^{(j_i)}(g_i(x_{0i})) = 0$, $j_i = 0, 1, ..., n_i - 1$; *where* $x_{0i} \in [a_i, b_i]$ *be fixed. Then*

$$\int_{g_1(x_{01})}^{g_1(b_1)}\int_{g_2(x_{02})}^{g_2(b_2)}\frac{\left\|\left(f_1\circ g_1^{-1}\right)(z_1)\left(f_2\circ g_2^{-1}\right)(z_2)\right\|_1 dz_1dz_2}{\left(\frac{(z_1-g_1(x_{01}))^{p(n_1-1)+1}}{p(p(n_1-1)+1)}+\frac{(z_2-g_2(x_{02}))^{q(n_2-1)+1}}{q(q(n_2-1)+1)}\right)} \leq$$

$$\frac{\left(g_1\left(b_1\right)-g_1\left(x_{01}\right)\right)\left(g_2\left(b_2\right)-g_2\left(x_{02}\right)\right)}{\left(n_1-1\right)!\left(n_2-1\right)!} \tag{11.80}$$

$$\left\|\left\|\left(f_1\circ g_1^{-1}\right)^{(n_1)}\right\|_2\right\|_{L_q([g_1(x_{01}),g_1(b_1)],\mathcal{B}_2(H))}\left\|\left\|\left(f_2\circ g_2^{-1}\right)^{(n_2)}\right\|_2\right\|_{L_p([g_2(x_{02}),g_2(b_2)],\mathcal{B}_2(H))}.$$

Proof Let $i = 1, 2$; $x_0 \in [a_i, b_i]$, such that $\left(f_i \circ g_i^{-1}\right)^{(j_i)}(g_i(x_{0i})) = 0$, $j_i = 0, 1, ..., n_i - 1$.

For $x_i \in [x_{0i}, b_i]$ by Theorem 11.4 we have

$$\left(f_i \circ g_i^{-1}\right)\left(g_i\left(x_i\right)\right) = \frac{1}{\left(n_i - 1\right)!} \int_{g_i\left(x_{0i}\right)}^{g_i\left(x_i\right)} \left(g_i\left(x_i\right) - z_i\right)^{n_i - 1} \left(f_i \circ g_i^{-1}\right)^{\left(n_i\right)} \left(z_i\right) dz_i. \tag{11.81}$$

As in (11.69) we have

$$\left\|\left(f_1 \circ g_1^{-1}\right)\left(g_1\left(x_1\right)\right)\right\|_2 \leq \frac{1}{\left(n_1 - 1\right)!} \frac{\left(g_1\left(x_1\right) - g_1\left(x_{01}\right)\right)^{\frac{p\left(n_1 - 1\right) + 1}{p}}}{\left(p\left(n_1 - 1\right) + 1\right)^{\frac{1}{p}}}$$

$$\left(\int_{g_1\left(x_{01}\right)}^{g_1\left(x_1\right)} \left\|\left(f_1 \circ g_1^{-1}\right)^{\left(n_1\right)}(z)\right\|_2^q dz\right)^{\frac{1}{q}} \leq$$

$$\frac{1}{\left(n_1 - 1\right)!} \frac{\left(g_1\left(x_1\right) - g_1\left(x_{01}\right)\right)^{\frac{p\left(n_1 - 1\right) + 1}{p}}}{\left(p\left(n_1 - 1\right) + 1\right)^{\frac{1}{p}}} \left\|\left\|\left(f_1 \circ g_1^{-1}\right)^{\left(n_1\right)}\right\|_2\right\|_{L_q\left(\left[g_1\left(x_{01}\right), g_1\left(b_1\right)\right]\right)}, \tag{11.82}$$

for all $x_1 \in \left[x_{01}, b_1\right]$.

Similarly, we obtain that

$$\left\|\left(f_2 \circ g_2^{-1}\right)\left(g_2\left(x_2\right)\right)\right\|_2 \leq \frac{1}{\left(n_2 - 1\right)!} \frac{\left(g_2\left(x_2\right) - g_2\left(x_{02}\right)\right)^{\frac{q\left(n_2 - 1\right) + 1}{q}}}{\left(q\left(n_2 - 1\right) + 1\right)^{\frac{1}{q}}}$$

$$\left\|\left\|\left(f_2 \circ g_2^{-1}\right)^{\left(n_2\right)}\right\|_2\right\|_{L_p\left(\left[g_2\left(x_{02}\right), g_2\left(b_2\right)\right]\right)}, \tag{11.83}$$

for all $x_2 \in \left[x_{02}, b_2\right]$.

By (11.82) and (11.83) we get

$$\left\|\left(f_1 \circ g_1^{-1}\right)\left(g_1\left(x_1\right)\right)\left(f_2 \circ g_2^{-1}\right)\left(g_2\left(x_2\right)\right)\right\|_1 \overset{(11.26)}{\leq}$$

$$\left\|\left(f_1 \circ g_1^{-1}\right)\left(g_1\left(x_1\right)\right)\right\|_2 \left\|\left(f_2 \circ g_2^{-1}\right)\left(g_2\left(x_2\right)\right)\right\|_2 \leq \frac{1}{\left(n_1 - 1\right)!\left(n_2 - 1\right)!}$$

$$\frac{\left(g_1\left(x_1\right) - g_1\left(x_{01}\right)\right)^{\frac{p\left(n_1 - 1\right) + 1}{p}}}{\left(p\left(n_1 - 1\right) + 1\right)^{\frac{1}{p}}} \frac{\left(g_2\left(x_2\right) - g_2\left(x_{02}\right)\right)^{\frac{q\left(n_2 - 1\right) + 1}{q}}}{\left(q\left(n_2 - 1\right) + 1\right)^{\frac{1}{q}}} \tag{11.84}$$

$$\left\|\left\|\left(f_1 \circ g_1^{-1}\right)^{\left(n_1\right)}\right\|_2\right\|_{L_q\left(\left[g_1\left(x_{01}\right), g_1\left(b_1\right)\right]\right)} \left\|\left\|\left(f_2 \circ g_2^{-1}\right)^{\left(n_2\right)}\right\|_2\right\|_{L_p\left(\left[g_2\left(x_{02}\right), g_2\left(b_2\right)\right]\right)} \leq$$

(using Young's inequality for $a, b \geq 0$, $a^{\frac{1}{p}} b^{\frac{1}{q}} \leq \frac{a}{p} + \frac{b}{q}$)

$$\frac{1}{(n_1-1)!\,(n_2-1)!}\left(\frac{(g_1(x_1)-g_1(x_{01}))^{p(n_1-1)+1}}{p(p(n_1-1)+1)}+\frac{(g_2(x_2)-g_2(x_{02}))^{q(n_2-1)+1}}{q(q(n_2-1)+1)}\right) \tag{11.85}$$

$$\left\|\left\|\left(f_1\circ g_1^{-1}\right)^{(n_1)}\right\|_2\right\|_{L_q([g_1(x_{01}),g_1(b_1)])}\left\|\left\|\left(f_2\circ g_2^{-1}\right)^{(n_2)}\right\|_2\right\|_{L_p([g_2(x_{02}),g_2(b_2)])},$$

$\forall\,(x_1,x_2)\in[x_{01},b_1]\times[x_{02},b_2]$.

So far we have

$$\frac{\left\|\left(f_1\circ g_1^{-1}\right)(g_1(x_1))\left(f_2\circ g_2^{-1}\right)(g_2(x_2))\right\|_1}{\left(\frac{(g_1(x_1)-g_1(x_{01}))^{p(n_1-1)+1}}{p(p(n_1-1)+1)}+\frac{(g_2(x_2)-g_2(x_{02}))^{q(n_2-1)+1}}{q(q(n_2-1)+1)}\right)}\le \tag{11.86}$$

$$\frac{1}{(n_1-1)!\,(n_2-1)!}\left\|\left\|\left(f_1\circ g_1^{-1}\right)^{(n_1)}\right\|_2\right\|_{L_q([g_1(x_{01}),g_1(b_1)],\mathcal{B}_2(H))}$$

$$\left\|\left\|\left(f_2\circ g_2^{-1}\right)^{(n_2)}\right\|_2\right\|_{L_p([g_2(x_{02}),g_2(b_2)],\mathcal{B}_2(H))},$$

$\forall\,(x_1,x_2)\in[x_{01},b_1]\times[x_{02},b_2]$.

The denominator in (11.86) can be zero, only when both $g_1(x_1)=g_1(x_{01})$ and $g_2(x_2)=g_2(x_{02})$.

Therefore we obtain (11.80), by integrating (11.86) over $[g_1(x_{01}),g_1(b_1)]\times[g_2(x_{02}),g_2(b_2)]$. ■

It follows the $\mathcal{B}_2(H)$ right generalized Hilbert–Pachpate inequality for ordinary derivatives.

Theorem 11.21 *All as in Theorem 11.20. Then*

$$\int_{g_1(a_1)}^{g_1(x_{01})}\int_{g_2(a_2)}^{g_2(x_{02})}\frac{\left\|\left(f_1\circ g_1^{-1}\right)(z_1)\left(f_2\circ g_2^{-1}\right)(z_2)\right\|_1dz_1dz_2}{\left(\frac{(g_1(x_{01})-z_1)^{p(n_1-1)+1}}{p(p(n_1-1)+1)}+\frac{(g_2(x_{02})-z_2)^{q(n_2-1)+1}}{q(q(n_2-1)+1)}\right)}\le$$

$$\frac{(g_1(x_{01})-g_1(a_1))(g_2(x_{02})-g_2(a_2))}{(n_1-1)!\,(n_2-1)!} \tag{11.87}$$

$$\left\|\left\|\left(f_1\circ g_1^{-1}\right)^{(n_1)}\right\|_2\right\|_{L_q([g_1(a_1),g_1(x_{01})],\mathcal{B}_2(H))}\left\|\left\|\left(f_2\circ g_2^{-1}\right)^{(n_2)}\right\|_2\right\|_{L_p([g_2(a_2),g_2(x_{02})],\mathcal{B}_2(H))}.$$

Proof As similar to Theorem 11.20 is omitted. ■

Next we present a $\mathcal{B}_\gamma(H)$, $\gamma\ge 1$, left generalized Hilbert–Pachpatte inequality for ordinary derivatives.

Theorem 11.22 *Let* $\gamma\ge 1$, $i=1,2$; $p,q>1:\frac{1}{p}+\frac{1}{q}=1$, *and* $n_i\in\mathbb{N}$, $f_i\in C^{n_i}([a_i,b_i],\mathcal{B}_\gamma(H))$; *where* $[a_i,b_i]\subset\mathbb{R}$ *and* $\mathcal{B}_\gamma(H)$ *is the* $*$*-ideal. Let* $g_i\in$

$C^1([a_i, b_i])$, *strictly increasing, such that* $g_i^{-1} \in C^{n_i}([g_i(a_i), g_i(b_i)])$. *We assume that* $\left(f_i \circ g_i^{-1}\right)^{(j_i)}(g_i(x_{0i})) = 0$, $j_i = 0, 1, ..., n_i - 1$; *where* $x_{0i} \in [a_i, b_i]$ *be fixed. Then*

$$\int_{g_1(x_{01})}^{g_1(b_1)} \int_{g_2(x_{02})}^{g_2(b_2)} \frac{\left\| \left(f_1 \circ g_1^{-1}\right)(z_1) \left(f_2 \circ g_2^{-1}\right)(z_2) \right\|_\gamma dz_1 dz_2}{\left(\frac{(z_1 - g_1(x_{01}))^{p(n_1-1)+1}}{p(p(n_1-1)+1)} + \frac{(z_2 - g_2(x_{02}))^{q(n_2-1)+1}}{q(q(n_2-1)+1)} \right)} \leq$$

$$\frac{(g_1(b_1) - g_1(x_{01}))(g_2(b_2) - g_2(x_{02}))}{(n_1 - 1)!(n_2 - 1)!} \tag{11.88}$$

$$\left\| \left\| \left(f_1 \circ g_1^{-1}\right)^{(n_1)} \right\|_\gamma \right\|_{L_q([g_1(x_{01}), g_1(b_1)], \mathcal{B}_\gamma(H))} \left\| \left\| \left(f_2 \circ g_2^{-1}\right)^{(n_2)} \right\|_\gamma \right\|_{L_p([g_2(x_{02}), g_2(b_2)], \mathcal{B}_\gamma(H))}.$$

Proof As similar to Theorem 11.20 is omitted. Use of (11.23). ∎

It follows the $\mathcal{B}_\gamma(H)$, $\gamma \geq 1$, right generalized Hilbert–Pachpate inequality for ordinary derivatives.

Theorem 11.23 *All as in Theorem 11.22. Then*

$$\int_{g_1(a_1)}^{g_1(x_{01})} \int_{g_2(a_2)}^{g_2(x_{02})} \frac{\left\| \left(f_1 \circ g_1^{-1}\right)(z_1) \left(f_2 \circ g_2^{-1}\right)(z_2) \right\|_\gamma dz_1 dz_2}{\left(\frac{(g_1(x_{01}) - z_1)^{p(n_1-1)+1}}{p(p(n_1-1)+1)} + \frac{(g_2(x_{02}) - z_2)^{q(n_2-1)+1}}{q(q(n_2-1)+1)} \right)} \leq$$

$$\frac{(g_1(x_{01}) - g_1(a_1))(g_2(x_{02}) - g_2(a_2))}{(n_1 - 1)!(n_2 - 1)!} \tag{11.89}$$

$$\left\| \left\| \left(f_1 \circ g_1^{-1}\right)^{(n_1)} \right\|_\gamma \right\|_{L_q([g_1(a_1), g_1(x_{01})], \mathcal{B}_\gamma(H))} \left\| \left\| \left(f_2 \circ g_2^{-1}\right)^{(n_2)} \right\|_\gamma \right\|_{L_p([g_2(a_2), g_2(x_{02})], \mathcal{B}_\gamma(H))}.$$

Proof As similar to Theorem 11.20 is omitted. ∎

11.6 Applications

We start with $\mathcal{B}_2(H)$ Ostrowski type inequalities.

Corollary 11.24 (to Theorem 11.15) *All as in Theorem 11.15, with* $g(t) = t$. *Then*
(1) for $\gamma, \delta > 1 : \frac{1}{\gamma} + \frac{1}{\delta} = 1$, *we have*

$$\|\Phi(A_1, A_2)(x_0)\|_1 \leq \frac{1}{(n-1)!(\gamma(n-1)+1)^{\frac{1}{\gamma}}}$$

$$\left\{\left[\left\|\left\|A_1^{(n)}\right\|_p\right\|_{\delta,[a,x_0]} \int_a^{x_0} \|A_2(x)\|_q (x_0 - x)^{n-\frac{1}{\delta}} dx\right] + \right. \tag{11.90}$$

$$\left[\left\|\left\|A_1^{(n)}\right\|_p\right\|_{\delta,[x_0,b]} \int_{x_0}^{b} \|A_2(x)\|_q (x - x_0)^{n-\frac{1}{\delta}} dx\right] +$$

$$\left[\left\|\left\|A_2^{(n)}\right\|_q\right\|_{\delta,[a,x_0]} \int_a^{x_0} \|A_1(x)\|_p (x_0 - x)^{n-\frac{1}{\delta}} dx\right] +$$

$$\left.\left[\left\|\left\|A_2^{(n)}\right\|_q\right\|_{\delta,[x_0,b]} \int_{x_0}^{b} \|A_1(x)\|_p (x - x_0)^{n-\frac{1}{\delta}} dx\right]\right\},$$

(2)

$$\|\Phi(A_1, A_2)(x_0)\|_1 \leq \frac{1}{(n-1)!}$$

$$\left\{\left[\left\|\left\|A_1^{(n)}\right\|_p\right\|_{L_1([a,x_0])} \int_a^{x_0} \|A_2(x)\|_q (x_0 - x)^{n-1} dx\right] + \right.$$

$$\left[\left\|\left\|A_1^{(n)}\right\|_p\right\|_{L_1([x_0,b])} \int_{x_0}^{b} \|A_2(x)\|_q (x - x_0)^{n-1} dx\right] + \tag{11.91}$$

$$\left[\left\|\left\|A_2^{(n)}\right\|_q\right\|_{L_1([a,x_0])} \int_a^{x_0} \|A_1(x)\|_p (x_0 - x)^{n-1} dx\right] +$$

$$\left.\left[\left\|\left\|A_2^{(n)})\right\|_q\right\|_{L_1([x_0,b])} \int_{x_0}^{b} \|A_1(x)\|_p (x - x_0)^{n-1} dx\right]\right\},$$

and

(3)

$$\|\Phi(A_1, A_2)(x_0)\|_1 \leq \frac{1}{n!}$$

$$\left\{\left[\left\|\left\|A_1^{(n)}\right\|_p\right\|_{\infty,[a,x_0]} \int_a^{x_0} \|A_2(x)\|_q (x_0 - x)^{n} dx\right] + \right.$$

$$\left[\left\|\left\|A_1^{(n)}\right\|_p\right\|_{\infty,[x_0,b]} \int_{x_0}^{b} \|A_2(x)\|_q (x - x_0)^{n} dx\right] + \tag{11.92}$$

$$\left[\left\| \left\| A_2^{(n)} \right\|_q \right\|_{\infty,[a,x_0]} \int_a^{x_0} \| A_1(x) \|_p (x_0 - x)^n \, dx \right] +$$

$$\left[\left\| \left\| A_2^{(n)}) \right\|_q \right\|_{\infty,[x_0,b]} \int_{x_0}^{b} \| A_1(x) \|_p (x - x_0)^n \, dx \right] \Bigg\} .$$

We continue with $\mathcal{B}_\gamma(H)$, $\gamma \geq 1$, Ostrowski type inequalities.

Corollary 11.25 (to Theorems 11.12–11.14) *All as in Theorem 11.12, with $g(t) = e^t$. Then*

(1)

$$\| E(f_1, ..., f_r)(x_0) \|_\gamma \leq \frac{1}{n!} \Bigg\{ \sum_{i=1}^{r} \Bigg[\Bigg[\left\| \left\| (f_i \circ \log)^{(n)} \right\|_\gamma \right\|_{\infty,[e^a,e^{x_0}]}$$

$$\left(e^{x_0} - e^a \right)^n \left(\int_a^{x_0} \left(\prod_{\substack{j=1 \\ j \neq i}}^{r} \| f_j(x) \|_\gamma \right) dx \right) \Bigg] +$$

$$\left[\left\| \left\| (f_i \circ \log)^{(n)} \right\|_\gamma \right\|_{\infty,[e^{x_0},e^b]} \left(e^b - e^{x_0} \right)^n \left(\int_{x_0}^{b} \left(\prod_{\substack{j=1 \\ j \neq i}}^{r} \| f_j(x) \|_\gamma \right) dx \right) \right] \Bigg] \Bigg\} , \tag{11.93}$$

(2)

$$\| E(f_1, ..., f_r)(x_0) \|_\gamma \leq \frac{1}{(n-1)!} \Bigg\{ \sum_{i=1}^{r} \Bigg[\Bigg[\left\| \left\| (f_i \circ \log)^{(n)} \right\|_\gamma \right\|_{L_1([e^a,e^{x_0}])}$$

$$\int_a^{x_0} \left(\prod_{\substack{j=1 \\ j \neq i}}^{r} \| f_j(x) \|_\gamma \right) \left(e^{x_0} - e^x \right)^{n-1} dx \Bigg] +$$

$$\left[\left\| \left\| (f_i \circ \log)^{(n)} \right\|_\gamma \right\|_{L_1([e^{x_0},e^b])} \int_{x_0}^{b} \left(\prod_{\substack{j=1 \\ j \neq i}}^{r} \| f_j(x) \|_\gamma \right) \left(e^x - e^{x_0} \right)^{n-1} dx \right] \Bigg] \Bigg\} , \tag{11.94}$$

and

(3) if $p, q > 1 : \frac{1}{p} + \frac{1}{q} = 1$, we have

$$\|E(f_1, ..., f_r)(x_0)\|_\gamma \leq \frac{1}{(n-1)!\,(p(n-1)+1)^{\frac{1}{p}}}$$

$$\sum_{i=1}^{r}\left[\left\|\|(f_i \circ \log)^{(n)}\|_\gamma\right\|_{L_q([e^a, e^{x_0}])}\left(\int_a^{x_0}\left(e^{x_0}-e^x\right)^{n-\frac{1}{q}}\left(\prod_{\substack{j=1\\j\neq i}}^{r}\|f_j(x)\|_\gamma\right)dx\right)+\right.$$

$$\left.\left\|\|(f_i \circ \log)^{(n)}\|_\gamma\right\|_{L_q([e^{x_0}, e^b])}\left(\int_{x_0}^{b}\left(e^x-e^{x_0}\right)^{n-\frac{1}{q}}\left(\prod_{\substack{j=1\\j\neq i}}^{r}\|f_j(x)\|_\gamma\right)dx\right)\right]. \tag{11.95}$$

We continue with a $\mathcal{B}_2(H)$ Opial type inequality.

Corollary 11.26 (to Theorem 11.16) *All as in Theorem 11.16 with $g(t) = t$. Then*

$$\int_{x_0}^{x}\|f(z)\,f^{(n)}(z)\|_1\,dz \leq$$

$$\frac{(x-x_0)^{n+\frac{1}{p}-\frac{1}{q}}}{2^{\frac{1}{q}}(n-1)!\,[(p(n-1)+1)(p(n-1)+2)]^{\frac{1}{p}}}\left(\int_{x_0}^{z}\|f^{(n)}(z)\|_2^q\,dz\right)^{\frac{2}{q}}, \tag{11.96}$$

for all $x_0 \leq x \leq b$.

Next comes a $\mathcal{B}_\gamma(H)$, $\gamma \geq 1$, Opial type inequality.

Corollary 11.27 (to Theorem 11.18) *All as in Theorem 11.18 with $g(t) = e^t$. Then*

$$\int_{e^{x_0}}^{e^x}\|((f \circ \log)(z))\,(f \circ \log)^{(n)}(z)\|_\gamma\,dz \leq$$

$$\frac{(e^x-e^{x_0})^{n+\frac{1}{p}-\frac{1}{q}}}{2^{\frac{1}{q}}(n-1)!\,[(p(n-1)+1)(p(n-1)+2)]^{\frac{1}{p}}}\left(\int_{e^{x_0}}^{e^x}\|(f \circ \log)^{(n)}(z)\|_\gamma^q\,dz\right)^{\frac{2}{q}}, \tag{11.97}$$

for all $x_0 \leq x \leq b$.

A $\mathcal{B}_2(H)$ Hilbert–Pachpatte type inequality follows.

Corollary 11.28 (to Theorem 11.20) *All as in Theorem 11.20 for $g_1(t) = g_2(t) = t$. Then*

$$\int_{x_{01}}^{b_1}\int_{x_{02}}^{b_2}\frac{\|f_1(z_1)\,f_2(z_2)\|_1\,dz_1dz_2}{\left(\frac{(z_1-x_{01})^{p(n_1-1)+1}}{p(p(n_1-1)+1)}+\frac{(z_2-x_{02})^{q(n_2-1)+1}}{q(q(n_2-1)+1)}\right)} \leq$$

$$\frac{(b_1 - x_{01})(b_2 - x_{02})}{(n_1 - 1)!(n_2 - 1)!} \left\| \left\| f_1^{(n_1)} \right\|_2 \right\|_{L_q([x_{01}, b_1], \mathcal{B}_2(H))} \left\| \left\| f_2^{(n_2)} \right\|_2 \right\|_{L_p([x_{02}, b_2], \mathcal{B}_2(H))}. \tag{11.98}$$

We finish with a $\mathcal{B}_\gamma(H)$, $\gamma \geq 1$, Hilbert–Pachpatte type inequality.

Corollary 11.29 (to Theorem 11.22) *All as in Theorem 11.22 for $g_1(t) = g_2(t) = \log t$, and $[a_i, b_i] \subset \mathbb{R}_+ - \{0\}$, $i = 1, 2$. Then*

$$\int_{\log x_{01}}^{\log b_1} \int_{\log x_{02}}^{\log b_2} \frac{\left\| \left(f_1 \circ e^t\right)(z_1) \left(f_2 \circ e^t\right)(z_2) \right\|_\gamma dz_1 dz_2}{\left(\frac{(z_1 - \log x_{01})^{p(n_1-1)+1}}{p(p(n_1-1)+1)} + \frac{(z_2 - \log x_{02})^{q(n_2-1)+1}}{q(q(n_2-1)+1)} \right)} \leq$$

$$\frac{\left(\log \frac{b_1}{x_{01}}\right)\left(\log \frac{b_2}{x_{02}}\right)}{(n_1 - 1)!(n_2 - 1)!} \tag{11.99}$$

$$\left\| \left\| \left(f_1 \circ e^t\right)^{(n_1)} \right\|_\gamma \right\|_{L_q([\log x_{01}, \log b_1], \mathcal{B}_\gamma(H))} \left\| \left\| \left(f_2 \circ e^t\right)^{(n_2)} \right\|_\gamma \right\|_{L_p([\log x_{02}, \log b_2], \mathcal{B}_\gamma(H))}.$$

References

1. Anastassiou, G.A.: Intelligent Comparisons: Analytic Inequalities. Springer, Heidelberg (2016)
2. Anastassiou, G.A.: Intelligent Computations: Abstract Fractional Calculus, Inequalities, Approximations, Springer, Heidelberg (2018)
3. Anastassiou, G.A.: Generalized Ostrowski, Opial and Hilbert-Pachpatte type inequalities fot Banach algebra valued functions involving integer vectorial derivatives. J. Comput. Anal. Appl. **30**(1), 78–94 (2022)
4. Anastassiou, G.A.: γ-Schatten norm Generalized Ostrowski, Opial and Hilbert-Pachpatte type inequalities for von Neumann-Schatten class $B_\gamma(H)$ valued functions with integer vectorial derivatives. J. Appl. Math. Comput. **2**(1), 32–57 (2021)
5. Bellman, R.: Some inequalities for positive definite matrices. In: Beckenbach, E.F. (ed.), General Inequalities 2, Proceedings of the 2nd International Conference on General Inequalities, pp. 89-90. Birkhauser, Basel (1980)
6. Chang, D.: A matrix trace inequality for products of Hermitian matrices. J. Math. Anal. Appl. **237**, 721–725 (1999)
7. Coop, I.D.: On matrix trace inequalities and related topics for products of Hermitian matrix. J. Math. Anal. Appl. **188**, 999–1001 (1994)
8. Dragomir, S.S.: p-Schatten norm inequalities of Ostrowski's type. RGMIA Res. Rep. Coll. **24**, Art. 108, 19 pp (2021)
9. Mikusinski, J.: The Bochner Integral. Academic, New York (1978)
10. Neudecker, H.: A matrix trace inequality. J. Math. Anal. Appl. **166**, 302–303 (1992)
11. Pachpatte, B.G.: Inequalities similar to the integral analogue of Hilbert's inequalities. Tamkang J. Math. **30**(1), 139–146 (1999)
12. Rudin, W.: Functional Analysis, 2nd edn. McGraw-Hill Inc, New York (1991)
13. Shilov, G.E.: Elementary Functional Analysis. Dover Publications Inc., New York (1996)
14. Simon, B.: Trace ideals and Their Applications. Cambridge University Press, Cambridge (1979)
15. Zagrebvov, V.A.: Gibbs Semigroups. Operator Theory: Advances and Applications, vol 273. Birkhauser (2019)

Chapter 12
γ-Schatten Norm Multivariate Ostrowski Inequalities for Multiple Neumann–Schatten Class $\mathcal{B}_\gamma(H)$ Valued Functions

Here we are dealing with several smooth functions from a compact convex set of $\mathbb{R}^k$, $k \geq 2$ to a Neumann–Schatten class $\mathcal{B}_\gamma(H)$, $\gamma \geq 1$, which is a Banach algebra. For these we prove general multivariate Ostrowski inequalities with estimates in norms $\|\cdot\|_p$, for all $1 \leq p \leq \infty$. We provide also interesting applications. It follows [2].

12.1 Introduction

Our main motivation comes from [1].

We mention a general Ostrowski type inequality result regarding several Banach algebra valued functions.

Theorem 12.1 ([1]) *Let $p, q > 1 : \frac{1}{p} + \frac{1}{q} = 1$; $(A, \|\cdot\|)$ a Banach algebra and $f_i \in C^{n+1}(Q, A)$, $i = 1, ..., r$; $r \in \mathbb{N}$, $n \in \mathbb{Z}_+$, and fixed $\overrightarrow{x_0} \in Q \subset \mathbb{R}^k$, $k \geq 2$, where Q is a compact and convex subset. Here all vector partial derivatives $f_{i\alpha} := \frac{\partial^\alpha f_i}{\partial z^\alpha}$, where $\alpha = (\alpha_1, ..., \alpha_k)$, $\alpha_\lambda \in \mathbb{Z}^+$, $\lambda = 1, ..., k$, $|\alpha| = \sum_{\lambda=1}^{k} \alpha_\lambda = j$, $j = 1, ..., n$, fulfill $f_{i\alpha}(\overrightarrow{x_0}) = 0$, $i = 1, ..., r$.*

Denote

$$D_{n+1}(f_i) := \max_{\alpha:|\alpha|=n+1} \| \| f_{i\alpha} \| \|_{\infty,Q}, \tag{12.1}$$

$i = 1, ..., r$, and

$$\left\| \overrightarrow{z} - \overrightarrow{x_0} \right\|_{l_1} := \sum_{\lambda=1}^{k} |z_\lambda - x_{0\lambda}|. \tag{12.2}$$

Then

G. A. Anastassiou, *Fractional Inequalities In Banach Algebras*,
Studies in Systems, Decision and Control 441,
https://doi.org/10.1007/978-3-031-05148-7_12

$$\left\| \sum_{i=1}^{r} \int_{Q} \left(\prod_{\substack{\rho=1 \\ \rho \neq i}}^{r} f_{\rho}(\overrightarrow{z}) \right) f_{i}(\overrightarrow{z})\, d\overrightarrow{z} - \sum_{i=1}^{r} \left(\int_{Q} \left(\prod_{\substack{\rho=1 \\ \rho \neq i}}^{r} f_{\rho}(\overrightarrow{z}) \right) d\overrightarrow{z} \right) f_{i}(\overrightarrow{x_0}) \right\| \leq \tag{12.3}$$

$$\frac{\max_{i \in \{1,\ldots,r\}} D_{n+1}(f_i)}{(n+1)!} \sum_{i=1}^{r} \left(\int_{Q} \left(\prod_{\substack{\rho=1 \\ \rho \neq i}}^{r} \| f_{\rho}(\overrightarrow{z}) \| \right) \| \overrightarrow{z} - \overrightarrow{x_0} \|_{l_1}^{n+1} d\overrightarrow{z} \right) \leq$$

$$\frac{\max_{i \in \{1,\ldots,r\}} D_{n+1}(f_i)}{(n+1)!} \min \left\{ \left(\int_{Q} \| \overrightarrow{z} - \overrightarrow{x_0} \|_{l_1}^{n+1} d\overrightarrow{z} \right) \left[\sum_{i=1}^{r} \left(\prod_{\substack{\rho=1 \\ \rho \neq i}}^{r} \| \| f_{\rho} \| \|_{\infty, Q} \right) \right], \right.$$

$$\left\| \| \cdot - \overrightarrow{x_0} \|_{l_1}^{n+1} \right\|_{\infty, Q} \left[\sum_{i=1}^{r} \left\| \left(\prod_{\substack{\rho=1 \\ \rho \neq i}}^{r} \| f_{\rho} \| \right) \right\|_{L_1(Q,A)} \right],$$

$$\left. \left\| \| \cdot - \overrightarrow{x_0} \|_{l_1}^{n+1} \right\|_{L_p(Q,A)} \left[\sum_{i=1}^{r} \left[\left\| \left(\prod_{\substack{\rho=1 \\ \rho \neq i}}^{r} \| f_{\rho} \| \right) \right\|_{L_q(Q,A)} \right] \right] \right\}. \tag{12.4}$$

In particular we obtained the following Ostrowski type inequality.

Theorem 12.2 ([1]) *Let* $(A, \|\cdot\|)$ *a Banach algebra and* $f_i \in C^{n+1}\left(\prod_{\lambda=1}^{k} [a_\lambda, b_\lambda], A \right)$, $i = 1, \ldots, r$; $r \in \mathbb{N}$, $n \in \mathbb{Z}_+$, *and fixed* $\overrightarrow{x_0} \in \prod_{\lambda=1}^{k} [a_\lambda, b_\lambda] \subset \mathbb{R}^k$, $k \geq 2$. *Here all vector partial derivatives* $f_{i\alpha} := \frac{\partial^\alpha f_i}{\partial z^\alpha}$, *where* $\alpha = (\alpha_1, \ldots, \alpha_k)$, $\alpha_\lambda \in \mathbb{Z}^+$, $\lambda = 1, \ldots, k$, $|\alpha| = \sum_{\lambda=1}^{k} \alpha_\lambda = j$, $j = 1, \ldots, n$, *fulfill* $f_{i\alpha}(\overrightarrow{x_0}) = 0$, $i = 1, \ldots, r$.

Denote

$$D_{n+1}(f_i) := \max_{\alpha : |\alpha| = n+1} \| \| f_{i\alpha} \| \|_{\infty, \prod_{\lambda=1}^{k} [a_\lambda, b_\lambda]}, \tag{12.5}$$

$i = 1, \ldots, r$.

Then

$$\left\| \sum_{i=1}^{r} \int_{\prod_{\lambda=1}^{k} [a_\lambda, b_\lambda]} \left(\prod_{\substack{\rho=1 \\ \rho \neq i}}^{r} f_{\rho}(\overrightarrow{z}) \right) f_{i}(\overrightarrow{z})\, d\overrightarrow{z} - \right.$$

$$\sum_{i=1}^{r}\left(\int_{\prod_{\lambda=1}^{k}[a_\lambda,b_\lambda]}\left(\prod_{\substack{\rho=1\\ \rho\neq i}}^{r} f_\rho\left(\overrightarrow{z}\right)\right)d\overrightarrow{z}\right)f_i\left(\overrightarrow{x_0}\right)\Bigg\| \leq \tag{12.6}$$

$$\left(\max_{i\in\{1,\ldots,r\}} D_{n+1}(f_i)\right)\left[\sum_{i=1}^{r}\left(\prod_{\substack{\rho=1\\ \rho\neq i}}^{r}\left\|\left\|f_\rho\right\|\right\|_{\infty,\prod_{\lambda=1}^{k}[a_\lambda,b_\lambda]}\right)\right]$$

$$\left[\sum_{\sum_{\lambda=1}^{k}\rho_\lambda=n+1}\frac{1}{\prod_{\lambda=1}^{k}\rho_\lambda!\prod_{\lambda=1}^{k}(\rho_\lambda+1)}\prod_{\lambda=1}^{k}\left((b_\lambda-x_{0\lambda})^{\rho_\lambda+1}+(x_{0\lambda}-a_\lambda)^{\rho_\lambda+1}\right)\right].$$

Ostrowski type inequalities have a lot of important applications in Numerical Analysis and Probability.

We are also inspired by [6].

In this chapter we establish multivariate Ostrowski type inequalities for several smooth functions from a compact convex subset of $\mathbb{R}^k$, $k \geq 2$, to von Neumann–Schatten class $\mathcal{B}_\gamma(H)$, $\gamma \geq 1$, which is a Banach algebra. These involve the norms $\|\cdot\|_p$, $1 \leq p \leq \infty$.

12.2 About Banach Algebras

All here come from [10].

We need

Definition 12.3 ([10], *p. 245*) A complex algebra is a vector space A over the complex field $\mathbb{C}$ in which a multiplication is defined that satisfies

$$x(yz) = (xy)z, \tag{12.7}$$

$$(x+y)z = xz + yz,\ x(y+z) = xy + xz, \tag{12.8}$$

and

$$\alpha(xy) = (\alpha x)y = x(\alpha y), \tag{12.9}$$

for all x, y and z in A and for all scalars α.

Additionally if A is a Banach space with respect to a norm that satisfies the multiplicative inequality

$$\|xy\| \leq \|x\|\|y\| \quad (x \in A,\ y \in A) \tag{12.10}$$

and if A contains a unit element e such that

$$xe = ex = x \ \ (x \in A) \tag{12.11}$$

and

$$\|e\| = 1, \tag{12.12}$$

then A is called a Banach algebra.

A is commutative iff $xy = yx$ for all $x, y \in A$.

We make

Remark 12.4 Commutativity of A will be explicated stated when needed.

There exists at most one $e \in A$ that satisfies (12.11).

Inequality (12.10) makes multiplication to be continuous, more precisely left and right continuous, see [10], p. 246.

Multiplication in A is not necessarily the numerical multiplication, it is something more general and it is defined abstractly, that is for $x, y \in A$ we have $xy \in A$, e.g. composition or convolution, etc.

For nice examples about Banach algebras see [10], p. 247–248, Sect. 10.3.

We also make

Remark 12.5 Next we mention about integration of A-valued functions, see [10], p. 259, Sect. 10.22:

If A is a Banach algebra and f is a continuous A-valued function on some compact Hausdorff space Q on which a complex Borel measure μ is defined, then $\int f d\mu$ exists and has all the properties that were discussed in Chap. 3 of [10], simply because A is a Banach space. However, an additional property can be added to these, namely: If $x \in A$, then

$$x \int_Q f \, d\mu = \int_Q xf(p) \, d\mu(p) \tag{12.13}$$

and

$$\left(\int_Q f \, d\mu \right) x = \int_Q f(p) \, x \, d\mu(p). \tag{12.14}$$

The vector integrals we will involve in our chapter follow (12.13) and (12.14).

12.3 p-Schatten Norms Background

In this advanced section all come from [6].

Let $(H, \langle \cdot, \cdot \rangle)$ be a complex Hilbert space and $\mathcal{B}(H)$ the Banach algebra of all bounded linear operators on H. If $\{e_i\}_{i \in I}$ an orthonormal basis of H, we say that $A \in \mathcal{B}(H)$ is of trace class if

$$\|A\|_1 := \sum_{i \in I} \langle |A| e_i, e_i \rangle < \infty. \tag{12.15}$$

The definition of $\|A\|_1$ does not depend on the choice of the orthornormal basis $\{e_i\}_{i \in I}$. We denote by $\mathcal{B}_1(H)$ the set of trace class operators in $\mathcal{B}(H)$.

We define the trace of a trace class operator $A \in \mathcal{B}_1(H)$ to be

$$tr(A) := \sum_{i \in I} \langle Ae_i, e_i \rangle, \tag{12.16}$$

where $\{e_i\}_{i \in I}$ an orthonormal basis of H. Note that this coincides with the usual definition of the trace if H is finite-dimensional. We observe that the series (12.16) converges absolutely and it is independent from the choice of basis.

The following result collects some properties of the trace:

Theorem 12.6 *We have:*

(i) If $A \in \mathcal{B}_1(H)$ then $A^* \in \mathcal{B}_1(H)$ and

$$tr(A^*) = \overline{tr(A)}; \tag{12.17}$$

(ii) If $A \in \mathcal{B}_1(H)$ and $T \in \mathcal{B}(H)$, then $AT, TA \in \mathcal{B}_1(H)$ and

$$tr(AT) = tr(TA) \text{ and } |tr(AT)| \le \|A\|_1 \|T\|; \tag{12.18}$$

(iii) $tr(\cdot)$ is a bounded linear functional on $\mathcal{B}_1(H)$ with $\|tr\| = 1$;

(iv) If $A, B \in \mathcal{B}_2(H)$ then $AB, BA \in \mathcal{B}_1(H)$ and $tr(AB) = tr(BA)$;

(v) $\mathcal{B}_{fin}(H)$ (finite rank operators) is a dense subspace of $\mathcal{B}_1(H)$.

An operator $A \in \mathcal{B}(H)$ is said to belong to the von Neumann–Schatten class $\mathcal{B}_p(H)$, $1 \le p < \infty$ if the p-Schatten norm is finite [14, pp. 60–64]

$$\|A\|_p := \left[tr\left(|A|^p\right)\right]^{\frac{1}{p}} < \infty,$$

$|A|^p$ is an operator notation and not a power.

For $1 < p < q < \infty$ we have that

$$\mathcal{B}_1(H) \subset \mathcal{B}_p(H) \subset \mathcal{B}_q(H) \subset \mathcal{B}(H) \tag{12.19}$$

and

$$\|A\|_1 \ge \|A\|_p \ge \|A\|_q \ge \|A\|. \tag{12.20}$$

For $p \ge 1$ the functional $\|\cdot\|_p$ is a norm on the $*$-ideal $\mathcal{B}_p(H)$, which is a Banach algebra, and $\left(\mathcal{B}_p(H), \|\cdot\|_p\right)$ is a Banach space.

Also, see for instance [14, pp. 60–64], for $p \ge 1$,

$$\|A\|_p = \left\|A^*\right\|_p, \; A \in \mathcal{B}_p(H) \tag{12.21}$$

$$\|AB\|_p \leq \|A\|_p \|B\|_p , A, B \in \mathcal{B}_p(H) \tag{12.22}$$

and

$$\|AB\|_p \leq \|A\|_p \|B\| , \|BA\|_p \leq \|B\| \|A\|_p , A \in \mathcal{B}_p(H) , B \in \mathcal{B}(H) . \tag{12.23}$$

This implies that

$$\|CAB\|_p \leq \|C\| \|A\|_p \|B\| , A \in \mathcal{B}_p(H) , B, C \in \mathcal{B}(H) . \tag{12.24}$$

In terms of p-Schatten norm we have the Hölder inequality for $p, q > 1$ with $\frac{1}{p} + \frac{1}{q} = 1$:

$$(|tr\,(AB)| \leq) \|AB\|_1 \leq \|A\|_p \|B\|_q , \quad A \in \mathcal{B}_p(H) , B \in \mathcal{B}_q(H) . \tag{12.25}$$

For the theory of trace functionals and their applications the interested reader is referred to [13, 14].

For some classical trace inequalities see [4, 5, 9], which are continuations of the work of Bellman [3].

12.4 Vector Analysis Background

(see [12], pp. 83–94)

Let $f(t)$ be a function defined on $[a, b] \subseteq \mathbb{R}$ taking values in a real or complex normed linear space $(X, \|\cdot\|)$, Then $f(t)$ is said to be differentiable at a point $t_0 \in [a, b]$ if the limit

$$f'(t_0) = \lim_{h \to 0} \frac{f(t_0 + h) - f(t_0)}{h} \tag{12.26}$$

exists in X, the convergence is in $\|\cdot\|$. This is called the derivative of $f(t)$ at $t = t_0$.

We call $f(t)$ differentiable on $[a, b]$, iff there exists $f'(t) \in X$ for all $t \in [a, b]$.

Similarly and inductively are defined higher order derivatives of f, denoted f'', $f^{(3)}$, ..., $f^{(k)}$, $k \in \mathbb{N}$, just as for numerical functions.

For all the properties of derivatives see [12], pp. 83–86.

Let now $(X, \|\cdot\|)$ be a Banach space, and $f : [a, b] \to X$.

We define the vector valued Riemann integral $\int_a^b f(t)\,dt \in X$ as the limit of the vector valued Riemann sums in X, convergence is in $\|\cdot\|$. The definition is as for the numerical valued functions.

If $\int_a^b f(t)\,dt \in X$ we call f integrable on $[a, b]$. If $f \in C([a, b], X)$, then f is integrable, [12], p. 87.

For all the properties of vector valued Riemann integrals see [12], pp. 86–91.

We define the space $C^n([a,b], X)$, $n \in \mathbb{N}$, of n-times continuously differentiable functions from $[a,b]$ into X; here continuity is with respect to $\|\cdot\|$ and defined in the usual way as for numerical functions·.

Let $(X, \|\cdot\|)$ be a Banach space and $f \in C^n([a,b], X)$, then we have the vector valued Taylor's formula, see [12], pp. 93–94, and also [11], (IV, 9; 47).

It holds

$$f(y) - f(x) - f'(x)(y-x) - \frac{1}{2} f''(x)(y-x)^2 - \ldots - \frac{1}{(n-1)!} f^{(n-1)}(x)(y-x)^{n-1} \tag{12.27}$$

$$= \frac{1}{(n-1)!} \int_x^y (y-t)^{n-1} f^{(n)}(t)\, dt, \ \forall x, y \in [a,b].$$

In particular (12.27) is true when $X = \mathbb{R}^m, \mathbb{C}^m$, $m \in \mathbb{N}$, etc.

A function $f(t)$ with values in a normed linear space X is said to be piecewise continuous (see [12], p. 85) on the interval $a \le t \le b$ if there exists a partition $a = t_0 < t_1 < t_2 < \ldots < t_n = b$ such that $f(t)$ is continuous on every open interval $t_k < t < t_{k+1}$ and has finite limits $f(t_0+0)$, $f(t_1-0)$, $f(t_1+0)$, $f(t_2-0)$, $f(t_2+0)$, ..., $f(t_n-0)$.

Here $f(t_k - 0) = \lim_{t \uparrow t_k} f(t)$, $f(t_k+0) = \lim_{t \downarrow t_k} f(t)$.

The values of $f(t)$ at the points t_k can be arbitrary or even undefined.

A function $f(t)$ with values in normed linear space X is said to be piecewise smooth on $[a,b]$, if it is continuous on $[a,b]$ and has a derivative $f'(t)$ at all but a finite number of points of $[a,b]$, and if $f'(t)$ is piecewise continuous on $[a,b]$ (see [12], p. 85).

Let $u(t)$ and $v(t)$ be two piecewise smooth functions on $[a,b]$, one a numerical function and the other a vector function with values in Banach space X. Then we have the following integration by parts formula

$$\int_a^b u(t)\, dv(t) = u(t)\, v(t) \,|_a^b - \int_a^b v(t)\, du(t), \tag{12.28}$$

see [12], p. 93.

We mention also the mean value theorem for Banach space valued functions.

Theorem 12.7 (see [8], p. 3) *Let $f \in C([a,b], X)$, where X is a Banach space. Assume f' exists on $[a,b]$ and $\|f'(t)\| \le K$, $a < t < b$, then*

$$\|f(b) - f(a)\| \le K(b-a). \tag{12.29}$$

Here the multiple Riemann integral of a function from a real box or a real compact and convex subset to a Banach space is defined similarly to numerical one however convergence is with respect to $\|\cdot\|$. Similarly are defined the vector valued partial derivatives as in the numerical case.

We mention the equality of vector valued mixed partials derivatives.

Proposition 12.8 (see Proposition 4.11 of [7], p. 90) *Let $Q = (a, b) \times (c, d) \subseteq \mathbb{R}^2$ and $f \in C(Q, X)$, where $(X, \|\cdot\|)$ is a Banach space. Assume that $\frac{\partial}{\partial t} f(s, t)$, $\frac{\partial}{\partial s} f(s, t)$ and $\frac{\partial^2}{\partial t \partial s} f(s, t)$ exist and are continuous for $(s, t) \in Q$, then $\frac{\partial^2}{\partial s \partial t} f(s, t)$ exists for $(s, t) \in Q$ and*

$$\frac{\partial^2}{\partial s \partial t} f(s, t) = \frac{\partial^2}{\partial t \partial s} f(s, t), \text{ for } (s, t) \in Q. \tag{12.30}$$

12.5 Main Results

We present a general Ostrowski type inequality result regarding several $\mathcal{B}_2(H)$-Banach algebra valued functions.

Theorem 12.9 *Let $p, q > 1 : \frac{1}{p} + \frac{1}{q} = 1$; $\mathcal{B}_2(H)$ is a $*$-ideal, which $(\mathcal{B}_2(H), \|\cdot\|_2)$ is a Banach algebra, and $f_i \in C^{n+1}(Q, \mathcal{B}_2(H))$, $i = 1, ..., r$; $r \in \mathbb{N}$, $n \in \mathbb{Z}_+$, and fixed $\overrightarrow{x_0} \in Q \subset \mathbb{R}^k$, $k \geq 2$, where Q is a compact and convex subset. Here all vector partial derivatives $f_{i\alpha} := \frac{\partial^\alpha f_i}{\partial z^\alpha}$, where $\alpha = (\alpha_1, ..., \alpha_k)$, $\alpha_\lambda \in \mathbb{Z}^+$, $\lambda = 1, ..., k$, $|\alpha| = \sum_{\lambda=1}^{k} \alpha_\lambda = j$, $j = 1, ..., n$, fulfill $f_{i\alpha}(\overrightarrow{x_0}) = 0$, $i = 1, ..., r$.*

Denote

$$D_{n+1}(f_i) := \max_{\alpha:|\alpha|=n+1} \left\| \|f_{i\alpha}\|_2 \right\|_{\infty, Q}, \tag{12.31}$$

$i = 1, ..., r$, and

$$\left\| \overrightarrow{z} - \overrightarrow{x_0} \right\|_{l_1} := \sum_{\lambda=1}^{k} |z_\lambda - x_{0\lambda}|. \tag{12.32}$$

Then

$$\left\| \sum_{i=1}^{r} \int_Q \left(\prod_{\substack{\rho=1 \\ \rho \neq i}}^{r} f_\rho(\overrightarrow{z}) \right) f_i(\overrightarrow{z}) \, d\overrightarrow{z} - \sum_{i=1}^{r} \left(\int_Q \left(\prod_{\substack{\rho=1 \\ \rho \neq i}}^{r} f_\rho(\overrightarrow{z}) \right) d\overrightarrow{z} \right) f_i(\overrightarrow{x_0}) \right\|_1 \leq \tag{12.33}$$

$$\frac{\max_{i \in \{1,...,r\}} D_{n+1}(f_i)}{(n+1)!} \sum_{i=1}^{r} \left(\int_Q \left(\prod_{\substack{\rho=1 \\ \rho \neq i}}^{r} \left\| f_\rho(\overrightarrow{z}) \right\|_2 \right) \left\| \overrightarrow{z} - \overrightarrow{x_0} \right\|_{l_1}^{n+1} d\overrightarrow{z} \right) \leq$$

$$\frac{\max_{i \in \{1,...,r\}} D_{n+1}(f_i)}{(n+1)!} \min \left\{ \left(\int_Q \left\| \overrightarrow{z} - \overrightarrow{x_0} \right\|_{l_1}^{n+1} d\overrightarrow{z} \right) \left[\sum_{i=1}^{r} \left(\prod_{\substack{\rho=1 \\ \rho \neq i}}^{r} \left\| \|f_\rho\|_2 \right\|_{\infty, Q} \right) \right], \right.$$

$$\left\| \left\| \cdot - \overrightarrow{x_0} \right\|_{l_1}^{n+1} \right\|_{\infty, Q} \left[\sum_{i=1}^{r} \left\| \left(\prod_{\substack{\rho=1 \\ \rho \neq i}}^{r} \left\| f_\rho \right\|_2 \right) \right\|_{L_1(Q)} \right],$$

$$\left\| \left\| \cdot - \overrightarrow{x_0} \right\|_{l_1}^{n+1} \right\|_{L_p(Q)} \left[\sum_{i=1}^{r} \left[\left\| \left(\prod_{\substack{\rho=1 \\ \rho \neq i}}^{r} \left\| f_\rho \right\|_2 \right) \right\|_{L_q(Q)} \right] \right] \Bigg\}. \tag{12.34}$$

Proof Take $g_{i\overrightarrow{z}}(t) := f_i\left(\overrightarrow{x_0} + t\left(\overrightarrow{z} - \overrightarrow{x_0}\right)\right)$, $0 \leq t \leq 1$; $i = 1, ..., r$. Notice that $g_{i\overrightarrow{z}}(0) = f_i\left(\overrightarrow{x_0}\right)$ and $g_{i\overrightarrow{z}}(1) = f_i\left(\overrightarrow{z}\right)$. The jth derivative of $g_{i\overrightarrow{z}}(t)$, based on Proposition 12.8, is given by

$$g_{i\overrightarrow{z}}^{(j)}(t) = \left[\left(\sum_{\lambda=1}^{k} (z_\lambda - x_{0\lambda}) \frac{\partial}{\partial z_\lambda} \right)^j f_i \right] (x_{01} + t(z_1 - x_{01}), ..., x_{0k} + t(z_k - x_{0k})) \tag{12.35}$$

and

$$g_{i\overrightarrow{z}}^{(j)}(0) = \left[\left(\sum_{\lambda=1}^{k} (z_\lambda - x_{0\lambda}) \frac{\partial}{\partial z_\lambda} \right)^j f_i \right] \left(\overrightarrow{x_0}\right), \tag{12.36}$$

for $j = 1, ..., n+1$; $i = 1, ..., r$.

Let $f_{i\alpha}$ be a partial derivative of $f_i \in C^{n+1}(Q, \mathcal{B}_2(H))$. Because by assumption of the theorem we have $f_{i\alpha}\left(\overrightarrow{x_0}\right) = 0$ for all $\alpha : |\alpha| = j$, $j = 1, ..., n$, we find that

$$g_{i\overrightarrow{z}}^{(j)}(0) = 0, \; j = 1, ..., n; \; i = 1, ..., r.$$

Hence by vector Taylor's theorem (12.27) we see that

$$f_i\left(\overrightarrow{z}\right) - f_i\left(\overrightarrow{x_0}\right) = \sum_{j=1}^{n} \frac{g_{i\overrightarrow{z}}^{(j)}(0)}{j!} + R_{in}\left(\overrightarrow{z}, 0\right) = R_{in}\left(\overrightarrow{z}, 0\right), \tag{12.37}$$

where

$$R_{in}\left(\overrightarrow{z}, 0\right) := \int_0^1 \left(\int_0^{t_1} ... \left(\int_0^{t_{n-1}} \left(g_{i\overrightarrow{z}}^{(n)}(t_n) - g_{i\overrightarrow{z}}^{(n)}(0) \right) dt_n \right) ... \right) dt_1, \tag{12.38}$$

$i = 1, ..., r$.

Therefore,

$$\left\| R_{in}\left(\overrightarrow{z},0\right)\right\|_2 \leq \int_0^1 \left(\int_0^{t_1} \cdots \left(\int_0^{t_{n-1}} \left\| \left\| g_{i\overrightarrow{z}}^{(n+1)}\left(\xi\left(t_n\right)\right)\right\|_2 \right\|_\infty t_n dt_n\right)\cdots\right) dt_1, \tag{12.39}$$

by the vector mean value Theorem 12.7 applied on $g_{i\overrightarrow{z}}^{(n)}$ over $(0, t_n)$. Moreover, we get

$$\left\| R_{in}\left(\overrightarrow{z},0\right)\right\|_2 \leq \left\| \left\| g_{i\overrightarrow{z}}^{(n+1)} \right\|_2 \right\|_{\infty,[0,1]} \int_0^1 \int_0^{t_1} \cdots \left(\int_0^{t_{n-1}} t_n dt_n\right) \ldots dt_1$$

$$= \frac{\left\| \left\| g_{i\overrightarrow{z}}^{(n+1)} \right\|_2 \right\|_{\infty,[0,1]}}{(n+1)!}. \tag{12.40}$$

However, there exists a $t_{i0} \in [0, 1]$ such that $\left\| \left\| g_{i\overrightarrow{z}}^{(n+1)} \right\|_2 \right\|_{\infty,[0,1]} = \left\| g_{i\overrightarrow{z}}^{(n+1)}\left(t_{i0}\right)\right\|_2$.

That is

$$\left\| \left\| g_{i\overrightarrow{z}}^{(n+1)} \right\|_2 \right\|_{\infty,[0,1]} = \left\| \left[\left(\sum_{\lambda=1}^k \left(z_\lambda - x_{0\lambda}\right)\frac{\partial}{\partial z_\lambda}\right)^{n+1} f_i\right]\left(\overrightarrow{x_0} + t_{i0}\left(\overrightarrow{z} - \overrightarrow{z_{0i}}\right)\right)\right\|_2$$

$$\leq \left[\left(\sum_{\lambda=1}^k \left|z_\lambda - x_{0\lambda}\right| \left\|\frac{\partial}{\partial z_\lambda}\right\|_2\right)^{n+1} f_i\right]\left(\overrightarrow{x_0} + t_{i0}\left(\overrightarrow{z} - \overrightarrow{z_{0i}}\right)\right).$$

I.e.,

$$\left\| \left\| g_{i\overrightarrow{z}}^{(n+1)} \right\|_2 \right\|_{\infty,[0,1]} \leq \left[\left(\sum_{\lambda=1}^k \left|z_\lambda - x_{0\lambda}\right| \left\|\left\|\frac{\partial}{\partial z_\lambda}\right\|_2\right\|_\infty\right)^{n+1} f_i\right], \tag{12.41}$$

$i = 1, \ldots, r$.

Hence by (12.41) we get

$$\left\| R_{in}\left(\overrightarrow{z},0\right)\right\|_2 \leq \frac{\left[\left(\sum_{\lambda=1}^k \left|z_\lambda - x_{0\lambda}\right| \left\|\left\|\frac{\partial}{\partial z_\lambda}\right\|_2\right\|_\infty\right)^{n+1} f_i\right]}{(n+1)!} \leq$$

$$\frac{D_{n+1}\left(f_i\right)}{(n+1)!}\left(\sum_{\lambda=1}^k \left|z_\lambda - x_{0\lambda}\right|\right)^{n+1} = \frac{D_{n+1}\left(f_i\right)}{(n+1)!}\left\|\overrightarrow{z} - \overrightarrow{x_0}\right\|_{l_1}^{n+1}, \tag{12.42}$$

$i = 1, \ldots, r$.

Therefore it holds

$$\left\| R_{in}\left(\overrightarrow{z}, 0\right)\right\|_2 \leq \frac{\max\limits_{i\in\{1,\dots,r\}} D_{n+1}\left(f_i\right)}{(n+1)!} \left\| \overrightarrow{z} - \overrightarrow{x_0} \right\|_{l_1}^{n+1}, \tag{12.43}$$

for $i = 1, \dots, r$.

By (12.37) we get that

$$\left(\prod_{\substack{\rho=1\\ \rho\neq i}}^{r} f_\rho\left(\overrightarrow{z}\right)\right) f_i\left(\overrightarrow{z}\right) - \left(\prod_{\substack{\rho=1\\ \rho\neq i}}^{r} f_\rho\left(\overrightarrow{z}\right)\right) f_i\left(\overrightarrow{x_0}\right) = \left(\prod_{\substack{\rho=1\\ \rho\neq i}}^{r} f_\rho\left(\overrightarrow{z}\right)\right) R_{in}\left(\overrightarrow{z}, 0\right), \tag{12.44}$$

for all $i = 1, \dots, r$.

Hence

$$\sum_{i=1}^{r}\left(\prod_{\substack{\rho=1\\ \rho\neq i}}^{r} f_\rho\left(\overrightarrow{z}\right)\right) f_i\left(\overrightarrow{z}\right) - \sum_{i=1}^{r}\left(\prod_{\substack{\rho=1\\ \rho\neq i}}^{r} f_\rho\left(\overrightarrow{z}\right)\right) f_i\left(\overrightarrow{x_0}\right)$$

$$= \sum_{i=1}^{r}\left(\prod_{\substack{\rho=1\\ \rho\neq i}}^{r} f_\rho\left(\overrightarrow{z}\right)\right) R_{in}\left(\overrightarrow{z}, 0\right). \tag{12.45}$$

Therefore we find

$$E\left(f_1, \dots, f_r\right)\left(x_0\right) :=$$

$$\sum_{i=1}^{r}\int_Q \left(\prod_{\substack{\rho=1\\ \rho\neq i}}^{r} f_\rho\left(\overrightarrow{z}\right)\right) f_i\left(\overrightarrow{z}\right) d\overrightarrow{z} - \sum_{i=1}^{r}\left(\int_Q \left(\prod_{\substack{\rho=1\\ \rho\neq i}}^{r} f_\rho\left(\overrightarrow{z}\right)\right) d\overrightarrow{z}\right) f_i\left(\overrightarrow{x_0}\right) =$$

$$\sum_{i=1}^{r}\int_Q \left(\prod_{\substack{\rho=1\\ \rho\neq i}}^{r} f_\rho\left(\overrightarrow{z}\right)\right) R_{in}\left(\overrightarrow{z}, 0\right) d\overrightarrow{z}. \tag{12.46}$$

Consequently, we have that

$$\left\| E\left(f_1, \dots, f_r\right)\left(x_0\right)\right\|_1 =$$

$$\left\| \sum_{i=1}^{r}\int_Q \left(\prod_{\substack{\rho=1\\ \rho\neq i}}^{r} f_\rho\left(\overrightarrow{z}\right)\right) f_i\left(\overrightarrow{z}\right) d\overrightarrow{z} - \sum_{i=1}^{r}\left(\int_Q \left(\prod_{\substack{\rho=1\\ \rho\neq i}}^{r} f_\rho\left(\overrightarrow{z}\right)\right) d\overrightarrow{z}\right) f_i\left(\overrightarrow{x_0}\right)\right\|_1 =$$

$$\left\| \sum_{i=1}^{r} \int_{Q} \left(\prod_{\substack{\rho=1 \\ \rho \neq i}}^{r} f_{\rho}\left(\overrightarrow{z}\right) \right) R_{in}\left(\overrightarrow{z}, 0\right) d\overrightarrow{z} \right\|_{1} \leq \tag{12.47}$$

$$\sum_{i=1}^{r} \left\| \int_{Q} \left(\prod_{\substack{\rho=1 \\ \rho \neq i}}^{r} f_{\rho}\left(\overrightarrow{z}\right) \right) R_{in}\left(\overrightarrow{z}, 0\right) d\overrightarrow{z} \right\|_{1} \leq$$

$$\sum_{i=1}^{r} \left(\int_{Q} \left\| \left(\prod_{\substack{\rho=1 \\ \rho \neq i}}^{r} f_{\rho}\left(\overrightarrow{z}\right) \right) R_{in}\left(\overrightarrow{z}, 0\right) \right\|_{1} d\overrightarrow{z} \right) \leq$$

(by (12.25), (12.22))

$$\sum_{i=1}^{r} \left(\int_{Q} \left(\prod_{\substack{\rho=1 \\ \rho \neq i}}^{r} \left\| f_{\rho}\left(\overrightarrow{z}\right) \right\|_{2} \right) \left\| R_{in}\left(\overrightarrow{z}, 0\right) \right\|_{2} d\overrightarrow{z} \right) \overset{(12.43)}{\leq} \tag{12.48}$$

$$\frac{\max\limits_{i \in \{1,\dots,r\}} D_{n+1}\left(f_{i}\right)}{(n+1)!} \sum_{i=1}^{r} \left(\int_{Q} \left(\prod_{\substack{\rho=1 \\ \rho \neq i}}^{r} \left\| f_{\rho}\left(\overrightarrow{z}\right) \right\|_{2} \right) \left\| \overrightarrow{z} - \overrightarrow{x_0} \right\|_{l_1}^{n+1} d\overrightarrow{z} \right).$$

So far we have proved

$$\left\| E\left(f_{1}, \dots, f_{r}\right)\left(x_{0}\right) \right\|_{1} \leq$$

$$\frac{\max\limits_{i \in \{1,\dots,r\}} D_{n+1}\left(f_{i}\right)}{(n+1)!} \sum_{i=1}^{r} \left(\int_{Q} \left(\prod_{\substack{\rho=1 \\ \rho \neq i}}^{r} \left\| f_{\rho}\left(\overrightarrow{z}\right) \right\|_{2} \right) \left\| \overrightarrow{z} - \overrightarrow{x_0} \right\|_{l_1}^{n+1} d\overrightarrow{z} \right) =: (\xi). \tag{12.49}$$

Furthermore it holds

$$(\xi) \leq \frac{\max\limits_{i \in \{1,\dots,r\}} D_{n+1}\left(f_{i}\right)}{(n+1)!} \left(\int_{Q} \left\| \overrightarrow{z} - \overrightarrow{x_0} \right\|_{l_1}^{n+1} d\overrightarrow{z} \right) \left[\sum_{i=1}^{r} \left(\prod_{\substack{\rho=1 \\ \rho \neq i}}^{r} \left\| \left\| f_{\rho} \right\|_{2} \right\|_{\infty, Q} \right) \right], \tag{12.50}$$

and

$$(\xi) \leq \frac{\max\limits_{i\in\{1,\dots,r\}} D_{n+1}(f_i)}{(n+1)!} \left\| \left\| \cdot - \overrightarrow{x_0} \right\|_{l_1}^{n+1} \right\|_{\infty,Q} \left[\sum_{i=1}^{r} \left\| \left(\prod_{\substack{\rho=1 \\ \rho\neq i}}^{r} \| f_\rho \|_2 \right) \right\|_{L_1(Q)} \right], \tag{12.51}$$

and finally

$$(\xi) \leq \frac{\max\limits_{i\in\{1,\dots,r\}} D_{n+1}(f_i)}{(n+1)!} \left[\sum_{i=1}^{r} \left[\left\| \left(\prod_{\substack{\rho=1 \\ \rho\neq i}}^{r} \| f_\rho \|_2 \right) \right\|_{L_q(Q)} \right] \right] \left\| \left\| \cdot - \overrightarrow{x_0} \right\|_{l_1}^{n+1} \right\|_{L_p(Q)}, \tag{12.52}$$

proving (12.33), (12.34). ■

We give

Corollary 12.10 (to Theorem 12.9) *All as in Theorem 12.9, with $f_1 = \dots = f_r = f$, $r \in \mathbb{N}$. Then*

$$\left\| \int_Q f^r(\overrightarrow{z})\, d\overrightarrow{z} - \left(\int_Q f^{r-1}(\overrightarrow{z})\, d\overrightarrow{z} \right) f(\overrightarrow{x_0}) \right\|_1 \leq$$

$$\frac{D_{n+1}(f)}{(n+1)!} \left(\int_Q \| f(\overrightarrow{z}) \|_2^{r-1} \| \overrightarrow{z} - \overrightarrow{x_0} \|_{l_1}^{n+1} d\overrightarrow{z} \right) \leq \tag{12.53}$$

$$\frac{D_{n+1}(f)}{(n+1)!} \min \left\{ \left(\int_Q \| \overrightarrow{z} - \overrightarrow{x_0} \|_{l_1}^{n+1} d\overrightarrow{z} \right) \left(\left\| \| f \|_2 \right\|_{\infty,Q} \right)^{r-1}, \right.$$

$$\left. \left\| \| \cdot - \overrightarrow{x_0} \|_{l_1}^{n+1} \right\|_{\infty,Q} \left\| \| f \|_2^{r-1} \right\|_{L_1(Q)}, \left\| \| \cdot - \overrightarrow{x_0} \|_{l_1}^{n+1} \right\|_{L_p(Q)} \left\| \| f \|_2^{r-1} \right\|_{L_q(Q)} \right\}. \tag{12.54}$$

We make

Remark 12.11 Of great interest are applications of Theorem 12.9 when $Q = \prod_{\lambda=1}^{k} [a_\lambda, b_\lambda]$, where $[a_\lambda, b_\lambda] \subset \mathbb{R}$, $\lambda = 1, \dots, k$.

We observe that by the multinomial theorem we get:

$$\int_{\prod_{\lambda=1}^{k} [a_\lambda, b_\lambda]} \left(\sum_{\lambda=1}^{k} |z_\lambda - x_{0\lambda}| \right)^{n+1} dz_1 \dots dz_k = \sum_{\rho_1+\rho_2+\dots\rho_k=n+1} \frac{(n+1)!}{\rho_1!\rho_2!\dots\rho_k!}$$

$$\int_{\prod_{\lambda=1}^{k} [a_\lambda, b_\lambda]} |z_1 - x_{01}|^{\rho_1} |z_2 - x_{02}|^{\rho_2} \dots |z_k - x_{0k}|^{\rho_k} dz_1 \dots dz_k = \tag{12.55}$$

$$\sum_{\rho_1+\rho_2+\ldots\rho_k=n+1} \frac{(n+1)!}{\rho_1!\rho_2!\ldots\rho_k!} \prod_{\lambda=1}^{k} \left(\int_{a_\lambda}^{b_\lambda} |z_\lambda - x_{0\lambda}|^{\rho_\lambda} dz_\lambda \right) =$$

$$\sum_{\sum_{\lambda=1}^{k} \rho_\lambda=n+1} \frac{(n+1)!}{\prod_{\lambda=1}^{k} \rho_\lambda!} \prod_{\lambda=1}^{k} \left(\int_{a_\lambda}^{x_{0\lambda}} (x_{0\lambda} - z_\lambda)^{\rho_\lambda} dz_\lambda + \int_{x_{0\lambda}}^{b_\lambda} (z_\lambda - x_{0\lambda})^{\rho_\lambda} dz_\lambda \right) =$$

$$\sum_{\sum_{\lambda=1}^{k} \rho_\lambda=n+1} \frac{(n+1)!}{\prod_{\lambda=1}^{k} \rho_\lambda!} \prod_{\lambda=1}^{k} \left(\frac{(x_{0\lambda} - a_\lambda)^{\rho_\lambda+1} + (b_\lambda - x_{0\lambda})^{\rho_\lambda+1}}{\rho_\lambda + 1} \right). \tag{12.56}$$

We have found that

$$\int_{\prod_{\lambda=1}^{k} [a_\lambda, b_\lambda]} \left\| \overrightarrow{z} - \overrightarrow{x_0} \right\|_{l_1}^{n+1} d\overrightarrow{z} = \tag{12.57}$$

$$\sum_{\sum_{\lambda=1}^{k} \rho_\lambda=n+1} \frac{(n+1)!}{\prod_{\lambda=1}^{k} \rho_\lambda!} \prod_{\lambda=1}^{k} \left(\frac{(b_\lambda - x_{0\lambda})^{\rho_\lambda+1} + (x_{0\lambda} - a_\lambda)^{\rho_\lambda+1}}{\rho_\lambda + 1} \right).$$

Based on (12.33), (12.34) and (12.57) we conclude:

Theorem 12.12 *Let* $(\mathcal{B}_2(H), \|\cdot\|_2)$ *the* $*$*-ideal and* $f_i \in C^{n+1}\left(\prod_{\lambda=1}^{k} [a_\lambda, b_\lambda], \mathcal{B}_2(H) \right)$, $i = 1, ..., r$; $r \in \mathbb{N}$, $n \in \mathbb{Z}_+$, *and fixed* $\overrightarrow{x_0} \in \prod_{\lambda=1}^{k} [a_\lambda, b_\lambda] \subset \mathbb{R}^k$, $k \geq 2$. *Here all vector partial derivatives* $f_{i\alpha} := \frac{\partial^\alpha f_i}{\partial z^\alpha}$, *where* $\alpha = (\alpha_1, ..., \alpha_k)$, $\alpha_\lambda \in \mathbb{Z}^+$, $\lambda = 1, ..., k$, $|\alpha| = \sum_{\lambda=1}^{k} \alpha_\lambda = j$, $j = 1, ..., n$, *fulfill* $f_{i\alpha}(\overrightarrow{x_0}) = 0$, $i = 1, ..., r$.

Denote

$$D_{n+1}(f_i) := \max_{\alpha:|\alpha|=n+1} \left\| \|f_{i\alpha}\|_2 \right\|_{\infty, \prod_{\lambda=1}^{k} [a_\lambda, b_\lambda]}, \tag{12.58}$$

$i = 1, ..., r$.

Then

$$\left\| \sum_{i=1}^{r} \int_{\prod_{\lambda=1}^{k} [a_\lambda, b_\lambda]} \left(\prod_{\substack{\rho=1 \\ \rho \neq i}}^{r} f_\rho(\overrightarrow{z}) \right) f_i(\overrightarrow{z}) \, d\overrightarrow{z} - \right.$$

$$\sum_{i=1}^{r}\left(\int_{\prod_{\lambda=1}^{k}[a_\lambda,b_\lambda]}\left(\prod_{\substack{\rho=1\\ \rho\neq i}}^{r} f_\rho\left(\overrightarrow{z}\right)\right)d\overrightarrow{z}\right)f_i\left(\overrightarrow{x_0}\right)\Bigg\|_1 \leq \tag{12.59}$$

$$\left(\max_{i\in\{1,\dots,r\}} D_{n+1}(f_i)\right)\left[\sum_{i=1}^{r}\left(\prod_{\substack{\rho=1\\ \rho\neq i}}^{r}\left\|\left\|f_\rho\right\|_2\right\|_{\infty,\prod_{\lambda=1}^{k}[a_\lambda,b_\lambda]}\right)\right]$$

$$\left[\sum_{\sum_{\lambda=1}^{k}\rho_\lambda=n+1}\frac{1}{\prod_{\lambda=1}^{k}\rho_\lambda!\prod_{\lambda=1}^{k}(\rho_\lambda+1)}\prod_{\lambda=1}^{k}\left((b_\lambda-x_{0\lambda})^{\rho_\lambda+1}+(x_{0\lambda}-a_\lambda)^{\rho_\lambda+1}\right)\right].$$

We continue with a general Ostrowski type inequality result regarding several $\mathcal{B}_\gamma(H)$, $\gamma \geq 1$, Banach algebra valued functions.

Theorem 12.13 *Let* $\gamma \geq 1$, $p, q > 1 : \frac{1}{p}+\frac{1}{q}=1$; $\mathcal{B}_\gamma(H)$ *is a* $*$*-ideal, which* $\left(\mathcal{B}_\gamma(H), \|\cdot\|_\gamma\right)$ *is a Banach algebra, and* $f_i \in C^{n+1}\left(Q, \mathcal{B}_\gamma(H)\right)$, $i = 1, \dots, r$; $r \in \mathbb{N}$, $n \in \mathbb{Z}_+$, *and fixed* $\overrightarrow{x_0} \in Q \subset \mathbb{R}^k$, $k \geq 2$, *where* Q *is a compact and convex subset. Here all vector partial derivatives* $f_{i\alpha} := \frac{\partial^\alpha f_i}{\partial z^\alpha}$, *where* $\alpha = (\alpha_1, \dots, \alpha_k)$, $\alpha_\lambda \in \mathbb{Z}^+$, $\lambda = 1, \dots, k$, $|\alpha| = \sum_{\lambda=1}^{k}\alpha_\lambda = j$, $j = 1, \dots, n$, *fulfill* $f_{i\alpha}\left(\overrightarrow{x_0}\right) = 0$, $i = 1, \dots, r$.

Denote

$$D_{n+1}(f_i) := \max_{\alpha:|\alpha|=n+1}\left\|\left\|f_{i\alpha}\right\|_\gamma\right\|_{\infty,Q}, \tag{12.60}$$

$i = 1, \dots, r$, *and*

$$\left\|\overrightarrow{z}-\overrightarrow{x_0}\right\|_{l_1} := \sum_{\lambda=1}^{k}|z_\lambda - x_{0\lambda}|. \tag{12.61}$$

Then

$$\left\|\sum_{i=1}^{r}\int_Q\left(\prod_{\substack{\rho=1\\ \rho\neq i}}^{r} f_\rho\left(\overrightarrow{z}\right)\right)f_i\left(\overrightarrow{z}\right)d\overrightarrow{z}-\sum_{i=1}^{r}\left(\int_Q\left(\prod_{\substack{\rho=1\\ \rho\neq i}}^{r} f_\rho\left(\overrightarrow{z}\right)\right)d\overrightarrow{z}\right)f_i\left(\overrightarrow{x_0}\right)\right\|_\gamma \leq \tag{12.62}$$

$$\frac{\max_{i\in\{1,\dots,r\}} D_{n+1}(f_i)}{(n+1)!}\sum_{i=1}^{r}\left(\int_Q\left(\prod_{\substack{\rho=1\\ \rho\neq i}}^{r}\left\|f_\rho\left(\overrightarrow{z}\right)\right\|_\gamma\right)\left\|\overrightarrow{z}-\overrightarrow{x_0}\right\|_{l_1}^{n+1}d\overrightarrow{z}\right) \leq$$

$$\frac{\max\limits_{i\in\{1,\dots,r\}} D_{n+1}(f_i)}{(n+1)!} \min \left\{ \left(\int_Q \left\| \overrightarrow{z} - \overrightarrow{x_0} \right\|_{l_1}^{n+1} d\overrightarrow{z} \right) \left[\sum_{i=1}^{r} \left(\prod_{\substack{\rho=1 \\ \rho\neq i}}^{r} \left\| \left\| f_\rho \right\|_\gamma \right\|_{\infty,Q} \right) \right], \right.$$

$$\left\| \left\| \cdot - \overrightarrow{x_0} \right\|_{l_1}^{n+1} \right\|_{\infty,Q} \left[\sum_{i=1}^{r} \left\| \left(\prod_{\substack{\rho=1 \\ \rho\neq i}}^{r} \left\| f_\rho \right\|_\gamma \right) \right\|_{L_1(Q)} \right],$$

$$\left. \left\| \left\| \cdot - \overrightarrow{x_0} \right\|_{l_1}^{n+1} \right\|_{L_p(Q)} \left[\sum_{i=1}^{r} \left[\left\| \left(\prod_{\substack{\rho=1 \\ \rho\neq i}}^{r} \left\| f_\rho \right\|_\gamma \right) \right\|_{L_q(Q)} \right] \right] \right\}. \tag{12.63}$$

Proof As similar to Theorem 12.9 is omitted. Use of (12.22). ■

We also give

Corollary 12.14 (to Theorem 12.13) *All as in Theorem 12.13, with $f_1 = ... = f_r = f$, $r \in \mathbb{N}$. Then*

$$\left\| \int_Q f^r\left(\overrightarrow{z}\right) d\overrightarrow{z} - \left(\int_Q f^{r-1}\left(\overrightarrow{z}\right) d\overrightarrow{z} \right) f\left(\overrightarrow{x_0}\right) \right\|_\gamma \leq$$

$$\frac{D_{n+1}(f)}{(n+1)!} \left(\int_Q \left\| f\left(\overrightarrow{z}\right) \right\|_\gamma^{r-1} \left\| \overrightarrow{z} - \overrightarrow{x_0} \right\|_{l_1}^{n+1} d\overrightarrow{z} \right) \leq \tag{12.64}$$

$$\frac{D_{n+1}(f)}{(n+1)!} \min \left\{ \left(\int_Q \left\| \overrightarrow{z} - \overrightarrow{x_0} \right\|_{l_1}^{n+1} d\overrightarrow{z} \right) \left(\left\| \left\| f \right\|_\gamma \right\|_{\infty,Q} \right)^{r-1}, \right.$$

$$\left. \left\| \left\| \cdot - \overrightarrow{x_0} \right\|_{l_1}^{n+1} \right\|_{\infty,Q} \left\| \left\| f \right\|_\gamma^{r-1} \right\|_{L_1(Q)}, \left\| \left\| \cdot - \overrightarrow{x_0} \right\|_{l_1}^{n+1} \right\|_{L_p(Q)} \left\| \left\| f \right\|_\gamma^{r-1} \right\|_{L_q(Q)} \right\}. \tag{12.65}$$

In particular we obtain:

Theorem 12.15 *Let $\left(\mathcal{B}_\gamma(H), \|\cdot\|_\gamma\right)$, $\gamma \geq 1$, the $*$-ideal and $f_i \in C^{n+1}\left(\prod_{\lambda=1}^{k}[a_\lambda, b_\lambda], \mathcal{B}_\gamma(H)\right)$, $i = 1, ..., r$; $r \in \mathbb{N}$, $n \in \mathbb{Z}_+$, and fixed $\overrightarrow{x_0} \in \prod_{\lambda=1}^{k}[a_\lambda, b_\lambda] \subset \mathbb{R}^k$, $k \geq 2$. Here all vector partial derivatives $f_{i\alpha} := \frac{\partial^\alpha f_i}{\partial z^\alpha}$, where $\alpha = (\alpha_1, ..., \alpha_k)$, $\alpha_\lambda \in \mathbb{Z}^+$, $\lambda = 1, ..., k$, $|\alpha| = \sum_{\lambda=1}^{k} \alpha_\lambda = j$, $j = 1, ..., n$, fulfill $f_{i\alpha}\left(\overrightarrow{x_0}\right) = 0$, $i = 1, ..., r$.*

Denote

$$D_{n+1}(f_i) := \max_{\alpha:|\alpha|=n+1} \left\| \|f_{i\alpha}\|_\gamma \right\|_{\infty, \prod_{\lambda=1}^{k}[a_\lambda, b_\lambda]}, \tag{12.66}$$

$i = 1, ..., r$.

Then

$$\left\| \sum_{i=1}^{r} \int_{\prod_{\lambda=1}^{k}[a_\lambda,b_\lambda]} \left(\prod_{\substack{\rho=1 \\ \rho\neq i}}^{r} f_\rho(\vec{z}) \right) f_i(\vec{z})\, d\vec{z} - \right.$$

$$\left. \sum_{i=1}^{r} \left(\int_{\prod_{\lambda=1}^{k}[a_\lambda,b_\lambda]} \left(\prod_{\substack{\rho=1 \\ \rho\neq i}}^{r} f_\rho(\vec{z}) \right) d\vec{z} \right) f_i(\vec{x_0}) \right\|_\gamma \leq \tag{12.67}$$

$$\left(\max_{i\in\{1,...,r\}} D_{n+1}(f_i) \right) \left[\sum_{i=1}^{r} \left(\prod_{\substack{\rho=1 \\ \rho\neq i}}^{r} \left\| \|f_\rho\|_\gamma \right\|_{\infty, \prod_{\lambda=1}^{k}[a_\lambda,b_\lambda]} \right) \right]$$

$$\left[\sum_{\sum_{\lambda=1}^{k}\rho_\lambda = n+1} \frac{1}{\prod_{\lambda=1}^{k}\rho_\lambda! \prod_{\lambda=1}^{k}(\rho_\lambda+1)} \prod_{\lambda=1}^{k} \left((b_\lambda - x_{0\lambda})^{\rho_\lambda+1} + (x_{0\lambda} - a_\lambda)^{\rho_\lambda+1} \right) \right].$$

Proof As similar to Theorem 12.12 is omitted. ∎

When $r = 2$ we derive the following p-Schatten norm operator related Ostrowski type inequalities.

Theorem 12.16 *Let* $p, q > 1 : \frac{1}{p} + \frac{1}{q} = 1$; *and let the* $*$*-ideals* $\mathcal{B}_p(H)$, $\mathcal{B}_q(H)$, *for which* $(\mathcal{B}_p(H), \|\cdot\|_p)$, $(\mathcal{B}_q(H), \|\cdot\|_q)$ *are Banach algebras,* $n \in \mathbb{Z}_+$; $\vec{x_0} \in Q \subset \mathbb{R}^k$, $k \geq 2$, *where* Q *is a compact and convex subset, and* $f_1 \in C^{n+1}(Q, \mathcal{B}_p(H))$, $f_2 \in C^{n+1}(Q, \mathcal{B}_q(H))$. *Here all vector partial derivatives* $f_{i\alpha} := \frac{\partial^\alpha f_i}{\partial z^\alpha}$, *where* $\alpha = (\alpha_1, ..., \alpha_k)$, $\alpha_\lambda \in \mathbb{Z}^+$, $\lambda = 1, ..., k$, $|\alpha| = \sum_{\lambda=1}^{k} \alpha_\lambda = j$, $j = 1, ..., n$, *fulfill* $f_{i\alpha}(\vec{x_0}) = 0$, $i = 1, 2$.

Denote by

$$D_{n+1}(f_1) := \max_{\alpha:|\alpha|=n+1} \left\| \|f_{1\alpha}\|_p \right\|_{\infty,Q}, \tag{12.68}$$

and

$$D_{n+1}(f_2) := \max_{\alpha:|\alpha|=n+1} \left\| \|f_{2\alpha}\|_q \right\|_{\infty,Q}, \tag{12.69}$$

and

$$\left\| \overrightarrow{z} - \overrightarrow{x_0} \right\|_{l_1} := \sum_{\lambda=1}^{k} |z_\lambda - x_{0\lambda}| ; \ \gamma, \delta > 1 : \frac{1}{\gamma} + \frac{1}{\delta} = 1. \tag{12.70}$$

Then

$$\left\| \int_Q f_2 \left(\overrightarrow{z} \right) f_1 \left(\overrightarrow{z} \right) d\overrightarrow{z} + \int_Q f_1 \left(\overrightarrow{z} \right) f_2 \left(\overrightarrow{z} \right) d\overrightarrow{z} - \right.$$

$$\left. \left(\int_Q f_2 \left(\overrightarrow{z} \right) d\overrightarrow{z} \right) f_1 \left(\overrightarrow{x_0} \right) - \left(\int_Q f_1 \left(\overrightarrow{z} \right) d\overrightarrow{z} \right) f_2 \left(\overrightarrow{x_0} \right) \right\|_1 \leq$$

$$\frac{\max \left(D_{n+1} \left(f_1 \right), D_{n+1} \left(f_2 \right) \right)}{(n+1)!}$$

$$\left\{ \int_Q \left\| f_2 \left(\overrightarrow{z} \right) \right\|_q \left\| \overrightarrow{z} - \overrightarrow{x_0} \right\|_{l_1}^{n+1} d\overrightarrow{z} + \int_Q \left\| f_1 \left(\overrightarrow{z} \right) \right\|_p \left\| \overrightarrow{z} - \overrightarrow{x_0} \right\|_{l_1}^{n+1} d\overrightarrow{z} \right\} \leq \tag{12.71}$$

$$\frac{\max \left(D_{n+1} \left(f_1 \right), D_{n+1} \left(f_2 \right) \right)}{(n+1)!}$$

$$\min \left\{ \left(\int_Q \left\| \overrightarrow{z} - \overrightarrow{x_0} \right\|_{l_1}^{n+1} d\overrightarrow{z} \right) \left[\left\| \left\| f_2 \right\|_q \right\|_{\infty, Q} + \left\| \left\| f_1 \right\|_p \right\|_{\infty, Q} \right], \right.$$

$$\left\| \left\| \cdot - \overrightarrow{x_0} \right\|_{l_1}^{n+1} \right\|_{\infty, Q} \left[\left\| \left\| f_2 \right\|_q \right\|_{L_1(Q)} + \left\| \left\| f_1 \right\|_p \right\|_{L_1(Q)} \right],$$

$$\left. \left\| \left\| \cdot - \overrightarrow{x_0} \right\|_{l_1}^{n+1} \right\|_{L_\gamma(Q)} \left[\left\| \left\| f_2 \right\|_q \right\|_{L_\delta(Q)} + \left\| \left\| f_1 \right\|_p \right\|_{L_\delta(Q)} \right] \right\}. \tag{12.72}$$

Proof Similar to Theorem 12.9. Use of (12.25). ■

We finish with

Corollary 12.17 (to Theorem 12.16) *All as in Theorem 12.16, with* $Q = \prod_{\lambda=1}^{k} [a_\lambda, b_\lambda]$. *Then*

$$\left\| \int_{\prod_{\lambda=1}^{k} [a_\lambda, b_\lambda]} f_2 \left(\overrightarrow{z} \right) f_1 \left(\overrightarrow{z} \right) d\overrightarrow{z} + \int_{\prod_{\lambda=1}^{k} [a_\lambda, b_\lambda]} f_1 \left(\overrightarrow{z} \right) f_2 \left(\overrightarrow{z} \right) d\overrightarrow{z} - \right.$$

$$\left. \left(\int_{\prod_{\lambda=1}^{k} [a_\lambda, b_\lambda]} f_2 \left(\overrightarrow{z} \right) d\overrightarrow{z} \right) f_1 \left(\overrightarrow{x_0} \right) - \left(\int_{\prod_{\lambda=1}^{k} [a_\lambda, b_\lambda]} f_1 \left(\overrightarrow{z} \right) d\overrightarrow{z} \right) f_2 \left(\overrightarrow{x_0} \right) \right\|_1 \leq$$

$$\max\left(D_{n+1}\left(f_1\right), D_{n+1}\left(f_2\right)\right)\left[\left\|\left\|f_1\right\|_p\right\|_{\infty, \prod\limits_{\lambda=1}^{k}\left[a_\lambda, b_\lambda\right]} + \left\|\left\|f_2\right\|_q\right\|_{\infty, \prod\limits_{\lambda=1}^{k}\left[a_\lambda, b_\lambda\right]}\right] \quad (12.73)$$

$$\left[\sum_{\sum\limits_{\lambda=1}^{k} \rho_k = n+1} \frac{1}{\prod\limits_{\lambda=1}^{k} \rho_k! \prod\limits_{\lambda=1}^{k} \left(\rho_\lambda + 1\right)} \prod_{\lambda=1}^{k} \left(\left(b_\lambda - x_{0\lambda}\right)^{\rho_\lambda+1} + \left(x_{0\lambda} - a_\lambda\right)^{\rho_\lambda+1}\right)\right].$$

Proof Similar to Theorem 12.12. ■

References

1. Anastassiou, G.A.: Multivariate Ostrowski type inequalities for several Banach algebra valued functions. J. Comput. Anal. Appl. **30**(1), 95–106 (2022)
2. Anastassiou, G.A.: γ-Schatten norm Multivariate Ostrowski type inequalities for several Neumann-Schatten class $B_\gamma(H)$ valued functions. Adv. Nonlinear Var. Inequal. Accepted (2022)
3. Bellman, R.: Some inequalities for positive definite matrices. In: Beckenbach, E.F. (ed.) General Inequalities 2, Proceedings of the 2nd International Conference on General Inequalities, pp. 89–90. Birkhauser, Basel (1980)
4. Chang, D.: A matrix trace inequality for products of Hermitian matrices. J. Math. Anal. Appl. **237**, 721–725 (1999)
5. Coop, I.D.: On matrix trace inequalities and related topics for products of Hermitian matrix. J. Math. Anal. Appl. **188**, 999–1001 (1994)
6. Dragomir, S.S.: p-Schatten norm inequalities of Ostrowski's type. RGMIA Res. Rep. Coll. **24**, Art. 108, 19 (2021)
7. Driver, B.: Analysis Tools with Applications. Springer, New York (2003)
8. Ladas, G., Laksmikantham, V.: Differential Equations in Abstract Spaces. Academic, New York(1972)
9. Neudecker, H.: A matrix trace inequality. J. Math. Anal. Appl. **166**, 302–303 (1992)
10. Rudin, W.: Functional Analysis, 2nd edn. McGraw-Hill Inc, New York (1991)
11. Schwartz, L.: Analyse Mathematique. Hermann, Paris (1967)
12. Shilov, G.E.: Elementary Functional Analysis. Dover Publications Inc., New York (1996)
13. Simon, B.: Trace Ideals and Their Applications. Cambridge University Press, Cambridge (1979)
14. Zagrebvov, V.A.: Gibbs Semigroups. Operator Theory: Advances and Applications, vol. 273, Birkhauser (2019)

Chapter 13
Conclusion

During the last 50 years fractional calculus due to its wide applications to many applied sciences has become a main trend in mathematics. By using generalized Caputo and Canavati fractional left and right vectorial Taylor formulae we prove a great variety of generalized fractional inequalities. More precisely, we derive generalized Caputo Fractional Ostrowski and Grüss type inequalities involving several Banach algebra valued functions. Also we establish generalized Canavati fractional Ostrowski, Opial, Grüss and Hilbert–Pachpatte type inequalities for multiple Banach algebra valued functions. By applying the p-Schatten norms over the von Neumann–Schatten classes we produce the analogous refined and elegant inequalities. Many applications follow. This monograph's results are expected to find applications in many areas of pure and applied mathematics, especially in fractional inequalities and fractional differential equations. Other interesting applications can be in applied sciences like geophysics, physics, chemistry, economics and engineering, etc. The central conclusion we derive from this monograph is that generalized fractional calculus penetrates almost any branch of mathematics with great applications, especially in natural phenomena.

G. A. Anastassiou, *Fractional Inequalities In Banach Algebras*,
Studies in Systems, Decision and Control 441,
https://doi.org/10.1007/978-3-031-05148-7_13